John J. Ratey, M.D., is Associate Clinical Professor of Psychiatry at Harvard Medical School. He is the co-author of *Driven to Distraction*, *Answers to Distraction*, and *Shadow Syndromes* and lives in Wellesley, Massachusetts.

D1346698

ALSO BY JOHN J. RATEY, M.D.

Mental Retardation: Developing Pharmacotherapies

The Neuropsychiatry of Personality Disorders

Driven to Distraction (with Edward M. Hallowell, M.D.)

Answers to Distraction (with Edward M. Hallowell, M.D.)

Shadow Syndromes (with Catherine Johnson, Ph.D.)

Delivered from Distraction (with Edward M. Hallowell, M.D.)

Spark (with Eric Hagerman)

A USER'S GUIDE TO THE BRAIN

PERCEPTION, ATTENTION, AND THE FOUR THEATERS OF THE BRAIN

JOHN J. RATEY, M.D.

ABACUS

First published in the United States of America in 2001 by
Pantheon Books, a division of Random House, Inc., New York
First published in Great Britain by Little, Brown and Company in 2001
This edition published by Abacus in 2003
Reprinted 2005, 2007, 2009 (twice)

Copyright © 2001 by John J. Ratey, M.D.

The moral right of the author has been asserted.

Grateful acknowledgement is made to Allyn & Bacon for permission to reprint figures 8.8 and 8.9 from *Physiology of Behavior*, 5th edition, by Neil R. Carlson
Copyright © 1994 by Allyn & Bacon.
Reprinted by permission of Allyn & Bacon.

All rights reserved.
No part of this publication may be reproduced, stored in a retrieval system, or transmitted, in any form or by any means, without the prior permission in writing of the publisher, nor be otherwise circulated in any form of binding or cover other than that in which it is published and without a similar condition including this condition being imposed on the subsequent purchaser.

A CIP catalogue record for this book
is available from the British Library.

ISBN 978-0-349-11296-1

Book design by Jo Anne Metsch
Illustrations by Hilda R. Muinos
Printed and bound in Great Britain by Clays Ltd, St Ives plc

Papers used by Abacus are natural, renewable and recyclable products sourced from well-managed forests and certified in accordance with the rules of the Forest Stewardship Council.

Abacus
An imprint of
Little, Brown Book Group
100 Victoria Embankment
London EC4Y 0DY

An Hachette UK Company
www.hachette.co.uk

www.littlebrown.co.uk

FOR TEMPLE GRANDIN AND MY WIFE, NANCY.
TWO BRAINS WHO HAVE TAUGHT ME MUCH.

CONTENTS

INTRODUCTION 3

1. DEVELOPMENT 14
2. PERCEPTION 48
3. ATTENTION AND CONSCIOUSNESS 110
4. MOVEMENT 147
5. MEMORY 182
6. EMOTION 222
7. LANGUAGE 252
8. THE SOCIAL BRAIN 290
9. THE FOUR THEATERS 336
10. CARE AND FEEDING 356

ACKNOWLEDGMENTS 379

SUGGESTED READING 381

INDEX 385

A USER'S GUIDE TO THE BRAIN

INTRODUCTION

FOR THE FIRST TIME EVER, discoveries in the neurosciences are beginning to overlap with fields as different as anthropology, philosophy, linguistics, and psychology. The media seem obsessed with one particular aspect of these discoveries, namely, that there is a biological basis for our personalities, behaviors, and mental disorders. This is exciting stuff, of course, but it misses the real thrill of what we are beginning to uncover.

Along with better drugs, our new discoveries are beginning to generate speculative theories about how the brain itself works. If several of these theories are even remotely close to the truth, they will change the way we think about ourselves forever. An observer of the developing scene must feel as Balboa must have when he first saw the Pacific Ocean: we don't yet know the full meaning of what we are seeing in the neurosciences, but we do know it means the beginning of a new age. To see yourself through the eyes of the modern neuroscientist, it turns out, not only provides a means of self-understanding that is new and inspiring but also points to a unification, rather than a contradiction, of what psychologists, anthropologists, linguists, and philosophers have been saying all along. Suddenly, for us, the world has gone from flat to round! Because the brain is ultimately responsible for personality, culture, language, and reason, this emerging unity is hardly surprising, but it is breathtakingly new, for only recently has research allowed us to speculate on how the brain actually works.

How, then, are we to set about translating such a complex topic into a useful primer for everyone? At first glance, the neuroscientific mapping of the brain is a confusing hodgepodge of different classification systems. Because individual brains vary so significantly, even professional neurogeographers often disagree as to where the lines should be drawn, and function is an entirely different matter. Imagine trying to read a road map of Pennsylvania that had, in tiny print, all of the combined data from voting belts, income concentrations, population density, ethnic composition, and geologic surveys, and all of it recording historical changes over the last century. It would be so much easier, you sigh, to call the tourist office and get specific directions, but then you would miss out on the rich culture the state has to offer. Brain science is its own priestly sect, a mystical order quite closed off to the uninitiated. I will be working my hardest throughout this book to give to you, the reader, a chance to learn about the brain's complexity without jargon. I hope you begin to get excited as you realize that what we now confront in the neurosciences is more enthralling than the computer or cyberspace in all its glory. Discoveries in the next thirty years will transform not merely our world but our very selves. How and when these changes in our world will occur is, of course, unpredictable, but we all need to be ready and very much a part of what is coming.

The confusing terminology that neuroscience applies to the brain and its functions will itself eventually need to change—and it will as our understanding of the brain deepens. Scientists looking at pathology are still caught up in the unitary hunt for the broken neural component they imagine to be at fault, and are doing their best to match up specific brain functions with specific neurogeographical locations. The sooner we replace our mechanistic model of the brain with an ecologically centered, systems-based view, the better off we will be, for such a model better accounts for much of human experience. But changing the terminology is not the purpose of the book; that really is work for neuroscientists, and work they must, for their own prejudices and categories are a big barrier to progress in the field. Terms like "memory" or "happiness" or "plausibility" are not brain functions, they are semantic categories generated by brain functions for which we have no names and of which, as yet, we have a poor understanding. Fortunately, most scientists in the field are aware of how urgent it is that we revise our models of how the brain works and find a corresponding new language to express those ideas. In these pages I will have to translate the jargon into some better form, and so I will, in a sense, be finding a new way to talk about the brain.

Much of the language used in discussing the brain, particularly in the cognitive sciences, comes from computation, and it is inconsistent with what we know about the brain. The brain is nothing like the personal computers it has designed, for it does not process information and construct images by manipulating strings of digits such as ones and zeros. Instead, the brain is largely composed of maps, arrays of neurons that apparently represent entire objects of perception or cognition, or at least entire sensory or cognitive qualities of those objects, such as color, texture, credibility, or speed. Most cognitive functions involve the interaction of maps from many different part of the brain at once; it is the bane of cognitive scientists that bananas are not located in a single structure of the brain. The brain assembles perceptions by the simultaneous interaction of whole concepts, whole images. Rather than using the predicative logic of a microchip, the brain is an analog processor, meaning, essentially, that it works by analogy and metaphor. It relates whole concepts to one another and looks for similarities, differences, or relationships between them. It does not assemble thoughts and feelings from bits of data.

Consequently, I have decided that I will have to replace much of the technical language about the brain with a language more akin to what the brain itself uses. Throughout this book I will be making constant use of metaphors and analogies, as well as anecdotes from my life and from those of patients. I do not include a glossary on purpose, as I try to use clarity and repetition to solidify in the reader's memory the names, functions, and approximate locations of the many suborgans or parts of the brain. Although metaphor and analogy are unconventional in scientific circles, I am firmly convinced that a more nonlinear kind of thought will eventually supplant much of the logical reasoning we use today. Chris Langton, one of the primary researchers in the field of complexity theory, has speculated that in the future science will become more poetic. Our troubled world, too, is becoming too complex for logical argumentation, and may have to change its thinking: real trust, when emotions are running high, is based on analogy, not calculation. Meanwhile, we must concentrate our attention on learning all we can about the brain, as a way to get the jump on where the field, and our world, are headed.

Ever since Freud invented the technique of psychoanalysis, the human psyche has been regarded as an object of such complexity that only those few individuals trained to interpret the hieroglyphic communications of dreams have been deemed fit to delve into its depths. The mental health profession has always been shrouded in mystery, as if its members belonged to a secret priestly sect. These days, of course, science is beginning to replace several aspects of the Freudian model with biological explanations. While psychotherapy is still an essential part of treating mental disorders such as depression and anxiety, we know much more than we used to about how the brain may assist or fail us. Many aspects of the way we are, formerly blamed on the environment, on bad parenting, or on early childhood trauma, are now more correctly recognized as deficits in the brain. Autism, once attributed to a child's being raised by an emotionally cold mother, is now known to be an extreme case of a developmental pattern whose causes have little to do with the environment. Bedwetting was once blamed on parents' lack of proper love and discipline; recently, a Dutch research group has found a genetic marker for the disorder. What we are seeing is a gradual replacement of the traditional psychodynamic-centered approach to healing by a biology-centered one.

For those with disabilities once considered to be their own fault, this is good news. Our new science has shown that as an organ, a part of the body, the brain is subject to the same kinds of influences and dysfunctions as other organs. Like a set of muscles, it responds to use and disuse by either growing and remaining vital or decaying, and thus, for the first time, we are learning to see mental weaknesses as physical systems in need of training and practice. The brain is a dynamic, highly sensitive yet robust system that may adapt, for better or worse, to almost any element of its environment. If we are going to set about training our brains to succeed in the world, we certainly need to learn about the various factors that can influence brain functions.

Most people already have some sense of the kinds of strategies that exist to change how the brain works, for Prozac, melatonin, and biofeedback machines are advertised everywhere. In fact, almost anything we do, eat, or drink can affect the brain. What is not so widely understood, however, is the brain itself. Before people can really begin to understand why they think, speak, love, laugh, cry, or see the world as they do, they must first come to terms with who, and what, they really are. Unfortunately, it is here that the limitations of this biological model really show up. If before we were searching for a hidden trauma through the mists of dreams, today we are hunting for the single gene, the defective piece of brain tissue, or the unbalanced neurotransmitter that we suppose to be behind our misery. This perspective makes it look as if our mental life is wholly determined by the genetic hand fate deals us. If biology is behind who we are, how can we feel that we have any free will or hope for a different life? A better understanding of how the brain works will give us all a better way to get a handle on who we are and how we can take an active hand in shaping our lives, without having to place all our hopes on a single, often imaginary, miracle cure. Every brain is different, and no brain is perfect; it is our responsibility to learn about ourselves and about what gives us each a unique way of seeing the world.

The transition from trauma to biology has unfortunately failed to wean clinicians from affect-centered diagnosis. If you are unhappy and decide to seek help, the main thrust of the diagnostic process begins with an inquiry into how you feel. From this initial information, diagnosis and treatment proceed, as a rule, by either sifting through your psyche for sources of guilt, anger, or unfulfilled longing, or by attempting to modify the affective symptoms pharmacologically, or both. The entire approach here is, in my opinion, quite misguided. Many brain disorders, particularly those that produce deficits in perception and cognition, can turn the lives of patients into abject misery. Historically, virtually all mental disorders were associated with character flaws. While clinicians, thank goodness, no longer frame psychopathology in such cruel terms, in the popular mind there is still enough residual confusion about the psyche to make it all too easy for us to feel ashamed of our physiological shortcomings. If your hearing is crippled by an undiagnosed auditory problem, the natural conclusion is that you are either stupid, indifferent, or both. When such patients seek help, they are indeed in search of a cure for their unhappiness, but to begin with the unhappiness itself can lead nowhere. The world today is filled with dyslexic or otherwise learning-impaired people who wonder why years of antidepressants and analysis have failed to improve their lives.

An alternative is for the clinician to begin therapy by looking for how the patient experiences the world, asking not "How do you feel?"

but "How do you know the world?" If a clinician first attempts to determine the functioning of the apparatus by which a person experiences life, he or she is, in a sense, beginning at the beginning. A problem with perception may cause a cognitive deficit, which may then lead to social impairments, loss of self-esteem, and a failed life. By identifying where in the brain problems arise, the therapist may devise a treatment plan that targets the true source of the patient's unhappiness. Once a source of guilt and self-blame is recognized as a developmental deficit, the shame is lifted away, often curing the patient of many affective troubles on the spot.

In order to bring about this kind of shift in how we see psychopathology, however, we all need to learn how to study our behavior in terms of the organ behind it. Therapists, of course, need to know how to recognize the signature of an organic, perceptual, or cognitive deficit, no matter how slight, and what questions are most likely to reveal the nature of the problem. Yet patients need to know a fair amount themselves, for they are the only witnesses to their own subjective experience. Patients are the only ones who can describe how they see, hear, think, and feel in a given situation. The accepted wisdom has always been that the key to success in life is to know yourself, and this still remains one of the most basic truths of philosophy, psychology, and religion. To know ourselves, we must become good self-observers, and it is for this reason, more than any other, that we must learn about the object that drives our logic, imagination, and passion. Biological determinism, in recent years, has begun to erode our confidence in our knowledge of what is and is not an issue of morality. It is a paradoxical age in which we live, for there seems to be a mounting tension between advances in human biology, with their power to heal, and the principles of responsibility that hold our society together. Whatever the future may hold for us, we must all, at the very least, be prepared to make informed decisions.

The other main problem with our new science of the brain, as I mentioned earlier, is that neuroscientists have, in a sense, simply taken over the elite, almost clerical office once held by analysts. The language used to describe the brain is, if anything, more opaque than any of the old psychoanalytic terminology, which was itself so obscure that only trained professionals could wade through the literature. Most people never even bother to learn such terminology, deeming that, like the language of the computer scientists of the early 1970s, it is better left to the nerds. If anyone should doubt it, a brief glance into a modern textbook on neurophysiology is all that is needed to make one want to run and hide. Although most disciplines, including the sciences, tend to replace older descriptive language with new terminology that better conveys the current understanding of the field, the neurosciences have simply added layer upon successive layer of jargon in the hundred or so years that I consider modern.

When early explorers peered inside the human skull, they had no idea as to the actual function of the regions and organs they saw: the specifics of function would come much later. Consequently, they simply gave names to parts of the brain based on their shape, much as we name things that we see in the clouds. Meanwhile, comparative anatomists were busy discovering that in structural terms the human brain contains the evolutionary history of all brains. Its core is, in appearance, quite similar to the entire brains of modern reptiles. As the fetal brain develops, each stage of brain evolution is reenacted in miniature, as successively more modern layers are added to the more primitive layers beneath. On top of our fanciful anatomical language, therefore, we find the language of evolution. And on top of this layer of description we have the last fifty years of genuinely functional brain-mapping, the neurophysiological terminology that describes what parts of the brain actually do. Like the brain itself, the language of the neurosciences represents a history of building upon and enriching, rather than replacing, previous structures.

It's not surprising that language about the brain is complex, for the brain is the most complex object in the universe. There are a hundred billion neurons in a single human brain, and roughly ten times as many other cells that have noncomputational roles. Each of these neurons is connected to others by branching treelike projections known as axons and dendrites, most of which terminate in tiny structures called synapses. Synapses are the subject of much current brain research, for it is believed that most learning and development occurs in the brain through the process of strengthening or weakening these connections. Each one of our hundred billion neurons may have anywhere from 1 to 10,000 synaptic connections to other neurons. This means that the theoretical number of different patterns of connections possible in a single brain is approximately 40,000,000,000,000,000—forty quadrillion. It is in the tiny synaptic gaps, where an electrical signal is briefly transformed into a chemical one and back again, that our psychoactive

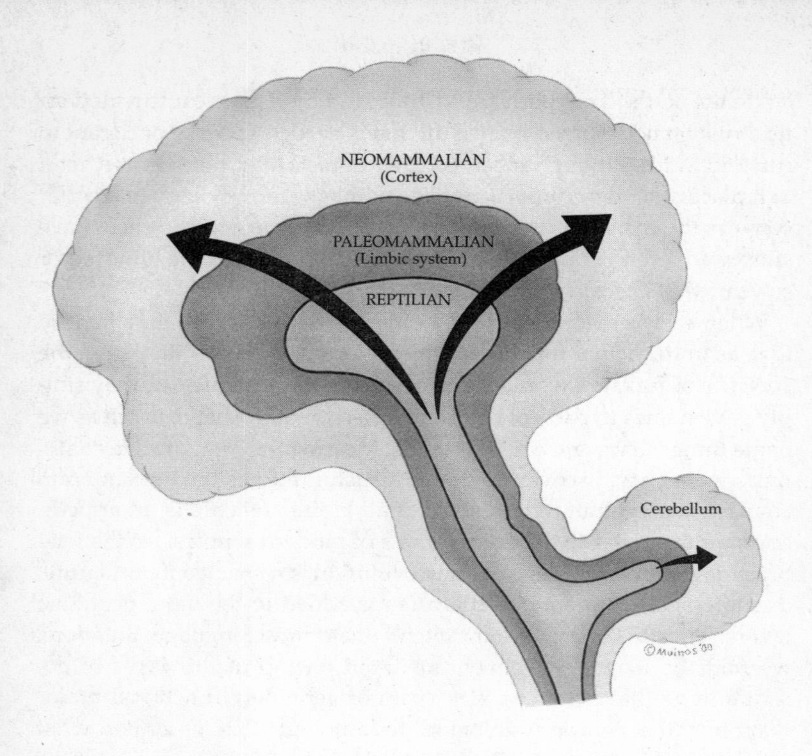

THE TRAJECTORY OF EVOLUTION The brain evolved from the bottom up, as illustrated here. This model of the so-called triune brain proposed by Paul MacLean in 1967 suggests that our brain developed by keeping those areas of our predecessors' brains that had proved useful and building new structures that helped the species to dominate in the evolutionary struggle. Through the process of chance mutation and survival of the fittest, evolution tinkered with what came before to arrive at the most adaptive mechanism in the universe.

The base of the brain, called the reptilian brain, is where the necessary command centers for living are located. These control sleep and waking, respiration, temperature regulation, and basic automatic movements and are way stations for sensory input. Next, the paleomammalian brain (including the limbic system) promotes survival and refines, amends, and coordinates movements. Here we also see development of the apparatuses for memory and emotions, which further enhance internal regulation of the body while beginning to deal with the social world. Finally the neomammalian brain, or cortex, developed. This area is responsible for the fine-tuning of our lower functions and for our associations, abstract thinking, and planning abilities, and allows us to respond to new challenges.

The cerebellum has evolved too, reflecting the fact that it has a role in thinking, speaking, memory, and our emotional life.

drugs, from aspirin to Prozac, work their magic. If changes in synaptic strengths (not merely the different arrangements of synapses) are the primary mechanism behind the brain's ability to represent the world, and if each synapse has, say, ten different strengths, then the different electrochemical configurations in a single brain come to a staggering number: ten to the trillionth power. This is an unimaginably large number: most astrophysicists calculate the volume of the known universe, in cubic meters, to be roughly ten to the eighty-seventh power. An added dimension of this complexity is that the figure of ten different synaptic strengths is merely a convenience; there are, as of this year, fifty-three different known neurotransmitters, the substances that carry information across the synaptic clefts. The very term "number of strengths" is, quite probably, an inaccurate description of what is going on. And finally, the brain is changing its connective patterns every second of our lives in response to everything we perceive, think, or do!

Happily, this dynamic complexity is actually the solution to many people's fears that our nature is genetically "hard-wired." The brain is so complex, and so plastic, that it is virtually impossible, except in the broadest fashion, to predict how a given factor will influence its state. Genes do contain directions for much of the brain's initial development, yet they have no absolute power to determine how the brain will respond. The brain is, according to current theories, more like an ecosystem than a machine, and many of its systems are in constant competition with one another all of our lives, in a process that the Nobel laureate Gerald Edelman has called "neural Darwinism." These networks of synapses, Edelman argues, are more than a vast communicative infrastructure; each network in the brain is striving against the others for feedback from the outside world. Thus, scientists can no more predict how a given brain will express a gene than predict what a tropical jungle will look like in thirty years. Even if you know the exact number of tigers, beetles, parrots, monkeys, and banana trees, you have no hope of knowing which species will fare best in the long run. Every single event has the potential to upset the balance of power and thus to change every subsequent event. In any such complex system, it really is up to the monkeys and tigers to see who gets the upper hand. In a system as complex as our brain, it really is up to us, and this is why it is so crucial that we learn about our brains. We do have free will, in a sense, for everything we do affects everything that follows, and the brain develops in a largely unpredictable way. Genetics are important but not determinative, and the kinds of exercise, sleep, diet, friends, and activities we choose, as well as the goals we set for ourselves, have perhaps equal power to change our lives. And neurological self-awareness is the most important first step we can take.

Unfortunately, the second most complex object in the universe is the body of language we use to talk about the brain, which is one reason why social workers, psychologists, or other mental health professionals shy away from having to consider it. Many wish that they could practice without having to take the brain into account at all, and a pervasive attitude in the field holds that the brain should be treated like plumbing: forget about it unless it backs up. This wish-it-away thinking is analogous to a business executive ignoring the Internet as a fad that will soon go away. Ignoring the brain actually cripples any psychological theory.

For holdouts, there is a consolation. It is possible to see psychodynamics, not as an alternative to the genetic model of human behavior, but merely as an extension of that model. Most genes behind mental disorders, for example, vary tremendously in the way and degree to which they are expressed in individuals, and recent research indicates that childhood stress may trigger the expression of genes that might otherwise have lain dormant throughout development. To take a hypothetical example, what if the "cold parent" supposedly responsible for autism, a disorder we now know is developmental and almost certainly genetic in origin, were in fact mildly autistic herself, and noticeably unemotional by nature? While she might not actually worsen her child's condition, a mildly autistic parent might be particularly ill-equipped to handle the extreme emotional commitment required by the most effective (at best, partially effective) therapies available to parents. In any case, a neo-Freudian observer would not be wrong to speculate on a connection between the mother's coldness and the child's autism; it's just that the connection might be genetic rather than causal. A sound knowledge of the biological foundations of our experience won't ruin our theory; it may only enrich it.

Debates as to which genetic, cultural, or environmental factor is the true cause of phenomenon X are often a waste of time; the brain is the binding principle behind it all. In a developing brain, each component initially develops on its own. After a certain internal degree of complexity is reached, however, and after the environment itself becomes more challenging, the separate parts of the brain begin to form more extensive connections with one another. Perhaps, ironically, the human brain itself may become the catalyst for a multidisciplinary maturation process in the fields of psychology, anthropology, linguistics, and philosophy.

To begin, the least we can do is learn what is now known about how our brains do what they do. Contrary to popular belief, this is a moral as well as a scientific pursuit, for self-knowledge makes it all the more our responsibility to live lives that maximize our strengths and minimize our weaknesses. There is no blame involved. Reformed alcoholics don't berate themselves for not being able to hang out in bars with their drinking friends; they know themselves too well for that. Likewise, blaming yourself for the physiological shortcomings of your brain, whatever they may be, is misdirected energy, energy better spent in changing your habits and lifestyle to live the most productive life you can.

1

DEVELOPMENT

SHE WAS DOING IT AGAIN. That young woman who periodically showed up dressed in a Western shirt and kerchief was standing in front of the automatic sliding doors at the Safeway supermarket. She'd look intently straight ahead, take five abrupt steps toward the doors, and try to restrain herself from walking through until they had fully opened. Sometimes she couldn't stop herself and nearly slammed right into the glass. Other times she'd wait long enough and then lunge through. Regardless, she'd back out and do it again. And again. Regular shoppers at the Phoenix, Arizona, store would hesitate beside her, then scurry past, eyeing her while trying not to stare. Once inside they'd shake their heads and make the usual comments: "Must be insane." They didn't know that Temple Grandin would go on to earn a doctorate in animal sciences and become an internationally recognized expert in animal handling. Or that she was autistic.

Temple had a normal birth, but by the time she was six months old she'd stiffen at her mother's touch and claw to free herself from her mother's hug. Soon she could not stand the feeling of other skin touching hers. A ringing telephone and a car driving by her house while a conversation was going on inside caused such severe confusion and hurt in the toddler's ears that she would tantrum, hitting whoever was within reach.

When she was three the doctors said that Temple had "brain damage." Her parents hired a stern governess, who structured the child's day around physical exercise and repetitive play such as "marching band." Occasionally the routine allowed Temple to focus on what she was doing, even speak. She taught herself to escape the stimuli around her, which caused pain in her overly sensitive nervous system, by daydreaming in pictures of places far away. By the time she reached high school she had made great progress. She could handle some of the academic subjects, and sometimes she could control her hypersensitive reactions to the chaos around her, primarily by shutting down to reduce the constant anxiety and fear. This made the other kids regard her as cold and aloof. She grew agonizingly lonely and would often tantrum or engage in pranks to combat her feelings of rejection. The school expelled her.

When she was sixteen Temple's parents sent her to an aunt's cattle ranch in Arizona. The rigid daily schedule of physical work helped her focus. She became fixated on the cattle chute, a large machine with two big metal plates that would squeeze a cow's sides. The high pressure apparently relaxed the animals, calming them enough for a vet to examine them. She visualized a squeeze machine for herself to give her the tactile stimulation she craved but couldn't get from human contact because the stimulation from physical closeness to another person was too intense, like a tidal wave engulfing her.

By this time Temple and her doctors had realized that she had a photographic memory. She was an autistic savant. When she returned to a special school for gifted children with emotional difficulties—the only school option left—her advisors allowed her to build a human squeeze machine. The project got her hooked on learning mechanical engineering and mathematics and on problem-solving, and she excelled at them all. She built a prototype, and would climb into it and use a lever to control the degree and duration of the pressure on her body. Afterward, she would feel relieved, more empathic, and more in touch with feelings of love and caring, even more tolerant of human touch. She started controlled experiments with the device and became skilled in research and lab techniques—which provided the impetus to apply to college.

Temple's state of hyperarousal and her inability to manage environmental stimuli impaired her ability to cope with the normal surroundings of her family or peers. The repetitive exercises as a child, the squeeze machine, and her academic successes gradually gave her the ability to control her offending behavior. Yet by her late twenties she still had had no success in creating social relationships. She was in a constant state of stage fright. She would get so anxious about approaching someone that she would literally race up and knock the person over, unable to restrain her muscles as she got emotionally energized. If she did manage to stop in time she would stand collar to collar, talking three inches from the person's face, an instant turnoff.

Then Temple put it all together. Walking up to someone in a socially acceptable way was the same as approaching the automatic doors at the supermarket. Both had to be done at the same relaxed pace. So she started showing up at the Safeway. She practiced approaching the doors for hours on end, until the process became automatic. The exercise helped. She found that she could approach people properly if she visualized herself approaching the doors. The doors were like a physical map; they provided a concrete visual picture of an abstract idea about approaching social interactions carefully.

Temple used another rehearsal technique to learn how to negotiate with people, a stressful interaction that usually sent her reeling. She read the *New York Times*'s accounts of the Camp David peace talks conducted by President Jimmy Carter with Egypt's Anwar Sadat and Israel's Menachem Begin. She read every word and memorized them at once; being a savant came in handy. She played the conversations over and over again in her brain, like watching an internal videotape, and used them to guide her own conduct when negotiating with real people.

Today Temple Grandin, at fifty-one, leads a fulfilling professional and social life. It is twenty-five years since her training in front of the Safeway doors, and she has learned to pay attention to certain stimuli while ignoring others so that she does not become overly aroused. She also takes low doses of an antidepressant drug that alleviates her pent-up discomfort even better than the squeeze machine.

Temple resorted to a host of unusual practices to rewire her faulty brain circuits in order to control her conduct. She developed the circuits that enabled her to approach the supermarket doors, and then used these newly trained circuits to help position herself with relation to other people. She mastered each technique with practice, made it automatic, and then applied the newly imprinted pattern to other cognitive skills. Temple developed in adulthood the brain circuits her physical childhood development did not provide.

THE BRAIN IS not a computer that simply executes genetically predetermined programs. Nor is it a passive gray cabbage, victim to the environmental influences that bear upon it. Genes and environment interact to continually change the brain, from the time we are conceived until the moment we die. And we, the owners—to the extent that our genes allow it—can actively shape the way our brains develop throughout the course of our lives. There is a great ongoing debate between different schools of neuroscientists as to whether the brain is merely a "ready-to-respond-to-environment" machine, an idea championed by a group who identify themselves as "connectionists," and those who would say that the brain is genetically made up of "ready-to-access" modules that the environment merely stimulates. However, the majority of neuroscientists see a hybrid, where the broad outlines of the brain's development are under genetic control, while the fine-tuning is up to the interaction of brain and environment.

Certainly, much of the course of our brains' development is determined while we are fetuses and young children. But as we will see, there are many other factors that can alter the process—in pregnancy, childhood, adulthood, and old age. A father's smile, exercise before the workday, a game of chess in the retirement home—everything affects development, and development is a lifelong process.

We are not prisoners of our genes or our environment. Poverty, alienation, drugs, hormonal imbalances, and depression don't dictate failure. Wealth, acceptance, vegetables, and exercise don't guarantee success. Our own free will may be the strongest force directing the development of our brains, and therefore our lives. As Temple's experience shows, the adult brain is both plastic and resilient, and always eager to learn. Experiences, thoughts, actions, and emotions actually change the structure of our brains. By viewing the brain as a muscle that can be weakened or strengthened, we can exercise our ability to determine who we become. Indeed, once we understand how the brain develops, we can train our brains for health, vibrancy, and longevity. Barring a physical illness, there's no reason that we can't stay actively engaged into our nineties.

Research on the brain's development has been fast and furious in this decade. The subject has become so popular that in the last few years it has rated a cover story in *Time* (three times), *Newsweek* (twice), and *Life,* as well as in other major magazines. New imaging technologies and scores of studies are providing enormous insight into ways to help the brain develop in babies, children, and adults, even in fetuses in the womb. Of course, there is also the chance here for misdirection.

The research has even prompted political action at the highest levels of government. In April 1997, Hillary Clinton hosted an all-day White House scientific conference, an unusual event, on new findings indicating that a child's acquiring language, thinking, and emotional skills is an active process that may be largely finished before age three. This premise is in stark contrast to the common wisdom of only a few years ago: that infants are largely passive beings who are somewhat unaware of their surroundings or who simply record everything in the environment without editing it. If infants are in fact editing and processing environmental stimuli, it behooves us to make these stimuli such good ones that they can move through them quickly and on to other learning.

The problem here is that such notoriety can cause sweeping action that may run ahead of sound clinical trials and testing of new hypotheses. Based on research that is not fully confirmed, panelists at the White House conference urged the adoption of federal programs to increase wages and training for day-care workers, improve parenting education, broaden training of pediatricians, and expand prenatal health-care coverage.

The best example of running ahead of research involves the "proof" that exposing infants to classical music enhances their brain development. Several recent studies indicate that this is so, yet others do not, and replication of the positive studies is not yet conclusive. Nonetheless, Georgia governor Zell Miller added $105,000 to his 1998 state budget proposal so that a cassette or compact disk of classical music could be included with the bag of free goodies that hospitals send home with each of the 100,000 babies born in the state each year. Miller's proposal, and his press conference about it, made national headlines. "No one questions that listening to music at a very early age affects the spatial, temporal reasoning that underlies math and engineering, and even chess," he said. "Having that infant listen to soothing music helps those trillions of brain connections to develop."

While the governor's awareness of brain research was commendable, his action might have been premature. The worry is not that it may waste the state's money. As Sandra Trehaub, a professor of psychology at the University of Toronto who studies infants' perception of music, said in response, "If we really think you can swallow a pill, buy a record or book, or have any one experience that will be the thing that gets you into Harvard or Princeton, then that's an illusion." John Breuer, president of the McDonnell Foundation, a funding organization for biomedical and behavioral research as it affects education, warns that though there may be great advantages to early education programs, neuroscience does not provide reasons for it yet. The link is just beginning to become clear. And as Michael Gazzaniga, a noted neuroscientist at Dartmouth, cautions, we are in danger of being overdone with "politically correct pseudoscience babble" when we allow our enthusiasm to outstrip facts.

Hillary Clinton realized herself that the White House conference could lead to premature and irresponsible decision-making and that the enthusiasm it catalyzed had to be tempered. Appearing on ABC's *Good Morning America* a week later, she admitted that the hyperfocus on properly stimulating babies "does ratchet up the guilt" about what parents ought and ought not to do.

That is why we will take a careful look at research findings throughout this book, and particularly in this chapter. There is much we can learn about how to improve the development of our brains and those of our children. But we have to keep a trained eye out to distinguish research that can be applied to our daily lives from that which is, for now, simply interesting.

A JUNGLE OF NEURONS

THE HUMAN BRAIN is responsible for the painting of Van Gogh, the creation of democracy, the design of the atomic bomb, psychosis, and the memory of one's first vacation and of the way that hot dog tasted. How does this organ encompass such diversity?

The brain is not a neatly organized system. It is often compared to an overgrown jungle of 100 billion nerve cells, or neurons, which begin as round cell bodies that grow processes called axons and dendrites. Each nerve cell has one axon and as many as 100,000 dendrites. Dendrites are the main way by which neurons get information (learn); and axons are the main way by which neurons pass on information to (teach) other neurons. The neuron and its thousands of neighbors send out roots and branches—the axons and dendrites—in all directions, which intertwine to form an interconnected tangle with 100 trillion constantly changing connections. There are more possible ways to connect the brain's neurons than there are atoms in the universe. The connections guide our bodies and behaviors, even as every thought and action we take physically modifies their patterns.

This description of the developing brain was heresy until recently. For decades scientists maintained that once its physical connections were completed during childhood, the brain was hard-wired. The tiny neurons and their interconnections were fixed; any neuron or link could die, but none could grow stronger, reorganize, or regenerate. Today, these axioms have been amended and enhanced. Thanks to sharp imaging technology and brilliant clinical research, we now have proof that development is a continuous, unending process. Axons and dendrites, and their connections, can be modified up to a point, strengthened, and perhaps even regrown.

Temple Grandin's achievement demonstrates that the brain has great plasticity. But what was actually going on, physically, inside her head? We get a strong clue from Michael Merzenich at the University of California, San Francisco.

Merzenich implanted electrodes in the brains of six adult squirrel monkeys, in the region that coordinates the movement of their fingers. Using computer imaging, he created a map of the neurons that fired when the monkeys manipulated objects with their hands. He then placed four food cups of decreasing diameter outside each of their cages. He put a single banana-flavored food pellet in the widest cup. The monkeys would reach through the bars and work their fingers into the cups until each was able to grasp its pellet and eat it. They practiced dozens of times for several days.

Once they had mastered the widest cup, Merzenich put the food pellets in the next smaller cup. After several days of repetition, the pellets were moved to the third cup, then the fourth. By the end of the experiment the monkeys were extremely skilled with their fingers.

After only one day, the computer images showed that the area of the brain that became active when the monkeys moved their fingers had increased in size. As the animals conquered successively smaller cups, the area got bigger; the number of cells that participated in the task increased. But after the neurons in the cortex mastered the fourth cup, the area shrank again; as the skill became more automatic, it was delegated to other parts of the brain lower down in the chain of command. The expanded portion of the executive part of the brain, the cerebral cortex, was no longer needed to carry out the skill and guide the hand. This commanding part of the brain, the control center, reverted back to its original size, freeing up neurons to learn other things.

There is evidence that the same thing happens in humans as in Merzenich's monkeys. Alvaro Pascual-Leone of Harvard Medical School, Boston, and Avi Karni of Hadassah University Hospital, Israel, have independently shown, using mapping techniques—for example, magnetic resonance imaging (MRI) and transcranial magnetic stimulation—in living human subjects, that skill acquisition recruits more cortical neurons to master the skill, and that as the skill becomes more automatic, less of the recruited cortex is used.

Thus, the brain has a tremendous ability to compensate and rewire with practice. Temple trained at the Safeway doors for hours each day for several months until the skill became automatic. At first it was incredibly difficult; by the end she did it with little concentration. Once she had mastered the initially higher-order activity, it was probably pushed down into lower regions of the brain, freeing her cortex to learn a new skill. The same would seem to be true about her rehearsing the Sadat-Begin tapes and applying them to her own conversations. Practice counts.

What we learn from Temple's story and Merzenich's monkeys is that our brains are wonderfully plastic throughout adulthood. Brain structure is not predetermined and fixed. We can alter the ongoing development of our brains and thus our capabilities. This is not always beneficial, however, as sometimes in the brain's attempt to adapt, the rewiring can make things worse.

MASSIVE CELL DEATH

THE HUMAN BRAIN HAS EVOLVED, thanks to natural selection, always in the direction of pushing our genes forward. Different sections of the brain expanded and specialized from the less complex swelling at the end of a nerve cord in primitive vertebrates in order to adapt to different environments across evolutionary history. In fish and amphibians, the visual perception of motion was important to track prey or escape predators, so the parts of the brain responsible for this sense expanded over time in these animals. In monkeys and early humans, color perception was needed in order to tell which fruits were ripe and which were not, and perception of form was needed even in the absence of movement. Thus, a large expansion of cortex evolved to handle these complex visual challenges. Similarly, the need to manipulate these objects in trees and to get from one branch to another led to specialized motor systems not useful in the aquatic environment.

Despite specializations typical only of our species, our brains retain the three basic components found in the simplest vertebrates: the hindbrain at the top of our spinal cord, which controls sensation and movement of the muscles of our face and throat; the midbrain, farther into the center of the head, which deals with some movements of the eyes and some rudimentary hearing and vision; and the forebrain, which achieves its most glorious development in human beings and which contains the cerebral cortex, the white-matter fibers connecting neurons of the cortex with each other and with other neurons, as well as those areas deep in the center of the brain that coordinate automatic sensory and motor functions. The cortex is the layers of neurons lying immediately below the bones of the skull, arching from just behind the forehead and over the top and sides, back to where the back of the head meets the neck. The cortex has evolved and expanded, adding many new functional areas, which participate in activities from playing basketball to designing software. Yet we retain our ancestral past; the seasonal depression many people experience in the dreary darkness of January may stem from animals that survived cold, foodless winters by slowing down their metabolism and hibernating. This ancient pathway remains in our brains despite electric heat and convenience stores.

The human brain has the same organization, the same types of neurons, and the same set of neurotransmitters—the chemical messengers between neurons—as other mammalian brains, which is why rats and monkeys are so widely used to test theories about human brain function. In fact, the basic control mechanisms for developing the brain are shared among all species. Thus we can study worms, fish, and even flies to help us uncover the genetic and chemical processes that guide the development of the human brain.

However, the cortex, which is dwarfed in most species by other brain areas, makes up a whopping 80 percent of the human brain. Compared with other animals, our huge cortex also has many more regions specialized for particular functions, such as associating words with objects or forming relationships and reflecting on them. The cortex is what makes us human.

Human brain development starts soon after the sperm penetrates the egg. The zygote begins to divide—two, four, eight, sixteen—until there are hundreds of cells. By the fourteenth day, the tiny ball of multiplying cells begins to fold in on itself. The process resembles a finger being pressed into the center of a soft balloon; cells from the outer surface begin to move inside the sphere. This movement activates the genes in cells that will form the nervous system. The compressed balloon lengthens and continues folding in on itself to form a tube. One end of the tube will become the spinal cord, and the other will become the brain. Cell division continues, and by the eighth week the brain has developed its three parts. The first weeks and months are a time of furious cell production and overproduction, with 250,000 neuroblasts, or primitive nerve cells, being created every minute.

During and after this period, neurons differentiate to perform distinct functions, first by traveling to a specific site, and then by extending an open hand to neighboring neurons. From the beginning of its being built, the brain is a social brain, the neurons making connections with their neighbors or dying for lack of contact. Little colonies begin developing on their own, and then reach out to other migratory communities. Continually dividing cells on the inside of the neural tube produce incredible numbers of neurons, which migrate out to the various regions of the brain. Most neurons migrate straight out until they reach the developing cortex. However, some go sideways a fair distance away from the original community, or clone, of neurons. Presumably these migrants will set up house in other communities and open the way for communication between the two sites, like an ambassador.

The migration can mean the difference between normal and crippled function. As recently as the early 1980s scientists thought that each cell in the fetal brain had a predetermined function and location in the adult brain. Today we know that the migration itself affects how neurons gain their identity and organize the brain's architecture. For instance, visual neurons become visual neurons not entirely because they are born visual neurons but because they migrate to a part of the brain where visual information arrives. Proper migration of neurons, therefore, is important for the development of normal brain function. There is a lengthening list of disorders, including autism, dyslexia, epilepsy, and schizophrenia, that may be caused in part by a migration problem. Plenty can go wrong during the journey, as a neuron becomes functional. Other cells it comes into contact with along the way and the specific genes within them that are turned on and off in response to the fetal environment all contribute to the form and function neurons take. Thus hormones, growth factors, cell adhesion molecules that cause neurons to stick together, other signals between cells not as yet well understood, and substances in the mother's blood all have an effect on determining where the neurons will end up and how they will perform. The inner environment guides the genes to make the brain.

During their journey, the neurons are fed and guided by caretaker glial cells, which form a scaffolding along which the neurons migrate—a lattice of support, guidance, protection, and nourishment. After the neurons reach their final places, the glial cells remain, although they change their shape and molecular properties in order to perform different functions. Two types of glia appear: one type controls the metabolism and function of the neurons; the other, which coats the axons with a fatty substance called myelin, controls how fast axons conduct information. The two main types of cells, neurons and glia, make up the brain, which is pretty much complete by the eighth month of pregnancy. At this point there are twice as many neurons as in the adult brain. As the brain ages, neurons that are weak or unused or simply don't fit the job that needs to be done are pruned away, to leave more efficient connections for those that are performing brain work. The principle of "Use it or lose it" begins, with nonworking, "couch-potato" cells dying off while those that are exercised get stronger and develop more connections.

Millions of the neurons travel amazing distances, the equivalent of a hike from New York City to San Francisco. Where they settle helps determine our individual temperament, talents, foibles, and quirks as well as the quality of our thinking processes. If neurons lose their way on their long journeys, developmental disorders may result, which is why it is so important that a pregnant woman not ingest harmful substances; a particular chemical in the brain at a critical moment will send neurons down the wrong fork in the road, or simply stop the process and cause havoc. Alcohol, nicotine, drugs and toxins, infections such as German measles, and lack of certain nutrients such as folic acid can interrupt the migration.

Once neurons have settled in at their final home—why they stop where they do is still a mystery—they grow dendrites and axons to communicate with other dendrites and axons. The tentacles reach for each other but don't quite touch. Like the outstretched fingers of God and Adam on the ceiling of the Sistine Chapel, they remain separated by a small gap, called a synaptic cleft. The axons and dendrites communicate by sending chemical messengers—neurotransmitters—back and forth across the synapse. A single neuron may be communicating across 100,000 synapses.

Chemical signals, called trophic factors, tell the axons where and how to connect. Whether or not electrical stimulation becomes sustained determines whether a connection between neurons survives or even whether a given neuron lives or dies. Because of the huge overproduction of neurons, there is not enough biochemical juice to support all of the axons searching for connections. Axons battle for limited sites, and those that lose the competition perish. Others that try to connect with the wrong kind of neuron are cut off from nourishment. However, there is no mindless competition of neurons for survival. Instead, forces external to each element in question (receptor, synapse, etc.) determine its degree of use and hence its survival. At first the activity that determines survival is random and spontaneous, but it becomes more organized as the fetus, and then the baby, receives input from its environment.

Two sequential pruning processes then fine-tune the initial neuronal networks that are formed. One causes the loss of entire neurons and the other the loss of branches and synapses. Both seem to involve competition for limited amounts of specific chemical signals released by the target cells. In the first process, neurons that fail to get enough signals from their target cells undergo cell death. This eliminates neurons that have made inappropriate connections and helps match the number of neurons to the number of target cells. In the second process, the connections between surviving neurons are refined by the removal of some dendrites and their synapses and the stabilization of others in a process that depends on electrical activity along the axons and competition among neighboring target cells. As the brain matures, the synaptic connections are further modified by use.

A period of cell death during the later stages of pregnancy wipes out almost half the neurons in the brain, which are probably phagocytized, or eaten up, by the support cells of the brain and the molecules recycled locally. There is a drop from about 200 billion neurons to 100 billion. This widespread cell death is normal, for it eliminates the wrong and weak connections that could inhibit efficient and proper brain function. This is a classic example of the incredible resourcefulness of evolution, which makes us highly adaptable creatures. It also points to the fact that even at the very beginning of development the brain is a social organ: where there is no connection, there is no life.

When a baby is born, it has millions of good connections waiting for a specific assignment. As the world makes demands, many of the connections are enlisted for specific jobs: seeing, babbling, remembering, throwing a ball. Connections that aren't used are eventually pruned. In the absence of the proper stimulation, a brain cell will die, but offer it a diet of enriched experiences and its neural synapses sprout new branches and connections.

Neurons that survive communicate rapid-fire across the synapses. The more firing that occurs across a specific connection, the stronger that pathway becomes. Billions of these exchanges take place continuously throughout the brain. Some connections transmit and receive signals often, others only occasionally, and the messages change constantly. The exact web of connections among neurons at a particular moment is determined by a combination of genetic makeup, environment, the sum of experiences we've imposed on our brains, and the activity we are bombarding it with now and each second into the future. What we do moment to moment greatly influences how the web continually reweaves itself.

DRUGS, MALNUTRITION, AND STRESS

THOUGH WE SHOULD all heed the lesson to follow, expectant mothers should take it very seriously. The developing brain in the fetus is extremely sensitive to its environment. Most pregnant women are aware of the dangers they can pose to their unborn babies. But they may not realize just how potent their actions can be. Let's consider a few of the most striking cases of environmental influence.

Smoking

CIGARETTES ARE PROBABLY the "drug" most commonly used during pregnancy. Despite warnings, 20 to 25 percent of pregnant women still smoke. Nicotine can reduce blood flow in the uterus and placenta by causing constriction of the blood vessels. It decreases the fetus's heart rate and breathing movements and exposes it to carbon monoxide.

Smoking raises considerably the risk that a baby will be born premature and underweight. The risk of spontaneous abortion is 1.7 times higher for smoking than for nonsmoking mothers. The risk of congenital abnormalities is 2.3 times higher. Research also shows that there is a 50 percent greater incidence of mental retardation among the children of mothers who smoked during pregnancy and that the more a woman smoked when she was expecting, the greater the chance of retardation. Importantly, the children of mothers who smoke show a threefold increase in attention deficit disorder (ADD), and a well-known reduction in birth weight, which is thought to have a great effect on the development of the brain. The incidence of sudden infant death syndrome is also higher among babies whose mothers smoked during pregnancy. Prenatal use of marijuana has similar effects. A mother's smoking affects her unborn baby because certain substances in her blood are passed to the fetus across the placenta. Research indicates that nicotine actually concentrates in the fetus, exposing it to an even higher level of the drug than the mother experiences.

The leading theory as to how nicotine affects the fetus's brain development is that the drug interferes with the natural migration of neurons, their connections, and their proper pruning during fetal development, although a direct link has not yet been proven. There is also evidence that nicotine can deregulate the dopamine system, undermining the modulating effect dopamine has on the brain's development.

Alcohol

ALCOHOL CONSUMPTION DURING pregnancy can have devastating consequences. Microscopic studies of fetal brains show that alcohol causes faulty cell migration. Once they begin to travel, the neurons do not know when to stop, miss their proper destinations, and often die. As a result, the brains of babies whose mothers drink regularly are frequently small, shrunken, and malformed, with a lower density of neurons. These fetal alcohol syndrome (FAS) babies have low IQ scores in childhood and severe reading and math disabilities by the time they reach high school and adulthood, as well as maladaptive behavior, hyperactivity, and depression.

The really unfortunate news is that, as with every other developmental toxin, the most significant effects of alcohol come early in pregnancy: the first six weeks are the most crucial. If a woman is drinking during this period, by the time that she becomes aware that she is pregnant the damage may have been done. Given this, there may be hundreds of thousands of people who have some degree of mental or physical impairment owing to *in utero* exposure to alcohol.

Research also shows that the effects associated with FAS continue and indeed increase as children become adults. There is also a more subtle version of fetal damage known as fetal alcohol effects (FAE). A recent study of 253 people diagnosed with FAS and FAE found that 90 percent had mental health problems; 60 percent experienced disrupted educations; 60 percent had trouble with the law; 50 percent had been accused of inappropriate sexual behavior. This points to a theme that will be repeated time and again in this book: Some types of antisocial and even criminal behavior could be linked to, if not caused by, physical problems in the brain.

Meanwhile, as treatments develop, society can do a great deal to prevent FAS and FAE in the first place, by educating all its citizens about the risks of drinking during pregnancy.

Cocaine

ONLY A FEW studies of cocaine use during pregnancy have been completed. More are needed. But the early results indicate effects similar to those of alcohol. Cocaine interferes with the transfer of nutrients and can shrink the amount of oxygen that travels from the placenta to the fetus, causing impaired growth of the fetus's body and brain. However, most recent studies show that many of the effects of cocaine disappear as the infant matures.

Malnutrition

DURING PREGNANCY, the fetus is more readily harmed by foreign substances than by poor nutrition. Nonetheless, a shortage of certain nutrients in the mother's diet, such as iron, vitamin B_{12}, folic acid, and essential fatty acids, can retard the brain's development. For example, research has shown a clear correlation between an insufficient intake of folic acid and a high incidence of spina bifida. If essential nutrients are not available, neurons stop forming, resulting in smaller brains, less overproduction of neurons in the fetus and subsequently less pruning or fine-tuning, and less cognitive development. Once born, these babies have lower birth weights, slower rates of growth, less coordination, a higher incidence of poor sight, and more learning difficulties. Malnutrition in young children also slows the brain's development, impairing cognition.

However, anxious pregnant mothers must also be careful not to overindulge. The vitamin craze that continues to sweep Western cultures makes it too easy, and seemingly imperative, to take massive doses of vitamins. Consumption of excessive amounts of some vitamins, particularly A and D, can lead to toxicity, which interferes with the brain's neurochemistry. A recent study shows that the consumption of too much vitamin A by pregnant women may cause birth defects. Large quantities of retinol and retinyl esters, the forms of vitamin A commonly used in dietary supplements, cause birth defects in tests on many animals. Pregnant women, with their doctors, should make sure they are getting enough vitamins, but not too many.

Toxins

MOST OF US REMEMBER the frightening warnings we received as children about not eating paint. Toxins such as lead can severely disrupt brain chemistry. During pregnancy, lead, pesticides, anesthesia gases, antibiotics, over-the-counter and prescription medications, and even acne medicine containing large amounts of vitamin A can all act as toxins on the fetus's brain. Ionizing radiation, such as X-rays and drugs used in treating cancer, have the same effect.

Some chemicals may not interfere with the brain's development. A small 1996 study at the University of Toronto of three widely used antidepressants—paroxatine, sertraline, and fluvoxamine—showed that these drugs do not appear to cause birth defects, which agrees with research in animals and previous studies of the antidepressant Prozac among pregnant women. However, because the study looked at only 267 expectant mothers, it was far too limited to establish that these drugs are safe during pregnancy. The study also did not explore behavioral differences in the babies born.

A previous study of Prozac found no effects on IQ, language, or behavior among babies exposed to the drug as fetuses. However, other research indicates an increased rate of "minor anomalies" at birth, such as abnormal creases in the palm of the hand. Until more research is done, women who are taking antidepressants or any medication and are considering pregnancy should consult with their doctors about the risks of continuing or stopping medications on the health of mother and fetus.

For their part, would-be fathers would also be wise to avoid exposure to smoking, alcohol, drugs, and toxins for at least three months prior to conception—the life span of the sperm.

NEURAL DARWINISM

IN THE EARLY STAGES of development, neurons travel freely in the brain, though guided in general pathways by genetic instruction. As they float around, some divide into more neurons, some die, and others settle down at permanent sites and make connections with neighbors, building the brain's complex circuitry. Genes provide the basic guidelines that control how the neurons form functioning networks. But the precise chemical environment influences which neurons connect with which.

All of our brains have the same general features that make us human, but each neural connection is unique, reflecting a person's special genetic endowment and life experience. Circuit connections are made stronger or weaker throughout a lifetime according to use. Neurologist and Nobel laureate Gerald Edelman, head of the Neurosciences Institute at the Scripps Clinic in La Jolla, California, calls the process neural Darwinism. Connections that cope well with the sensory inputs they receive, which they can convert into effective actions, stay intact and become strong. Those that do not, die off in a process that resembles natural selection. Neurons and the circuits they form part of compete with other neurons for survival, and those that are best adapted to the environment survive. The environment around us—what we ingest and inhale, the amount and type of light and sound—actually changes the physical interconnection of synapses within the brain, providing us with more efficient circuitry, and allowing each of us to develop an exclusive brain suited to our particular needs.

Neural Darwinism is the theory that explains why the brain needs to be plastic, that is, able to change as our environment and experiences change. That is why we can learn in the first place, and unlearn too, and why people with brain injuries can recover lost functions. The concept also underlies two of the mantras of this book. "Neurons that fire together wire together" means that the more we repeat the same actions and thoughts—from practicing a tennis serve to memorizing multiplication tables—the more we encourage the formation of certain connections and the more fixed the neural circuits in the brain for that activity become. "Use it or lose it" is the corollary: if you don't exercise brain circuits, the connections will not be adaptive and will slowly weaken and could be lost.

NATURE OR NURTURE

BEING ALTOGETHER HUMAN, which means in part understanding who and what we are, we are curious about the answer to the question of which force plays a stronger developmental role: genes or environment. The debate over "nature or nurture" has raged for two thousand years. At opposite extremes are euthenists, who cite bad parenting and the evils of society as the cause of all mental problems, and the proponents of eugenics, who blame faulty genes for all of society's ills and want to prevent all "bad" people from reproducing.

In reality there is no debate. Most of who we are is a result of the interaction of our genes and our experiences. In some cases, the genes are more important, while in others the environment is more crucial. We tend to oversimplify because we want to identify a single cause of a particular problem, so we can pour our efforts into one "cure." Some people hope that programs such as Head Start, which are designed to change a child's environment, will improve intellectual development. Others hope that one drug or one gene alteration will cure all aggressive behavior. Such simplistic approaches will rarely work. The real question is how genes and the environment influence each other, brain structure, and behavior. Untangling each factor's contribution is difficult because we can never fully understand an individual's genes outside an environment, and we can never study the effects of environment on a person while "isolating out" his genes.

Since the 1990s the pendulum has swung toward nature. It seems as though we hear almost daily of a new discovery; genes are now linked to Alzheimer's disease, bedwetting, obesity, and even to overall happiness. Many aspects of development that were previously attributed to learning, bad habits, or environment are now thought to be determined by genes. Many of us are fascinated by the international Genome Project, which is mapping the function of the 100,000 genes in the human genome, some 30,000 to 50,000 of which are designated for the brain. But we must remember that genetics is not destiny. A mere 50,000 genes for the brain are not nearly enough to account for the 100 trillion synaptic connections that are made there. Genes set boundaries for human behavior, but within these boundaries there is immense room for variation determined by experience, personal choice, and even chance.

The point to remember is that genes can be active or inactive and that everything we do affects the activity of our genes. We tend to think of genes as tiny entities that are isolated from the rest of the body, but they reside in every cell and are affected by anything that affects that cell, whether the cell is in the thigh or the cortex. For example, genes activate the exploratory network in a child's brain, and the more enriched the child's environment, the more these genes turn on, and the more the child explores. Adults experience many similar effects: learning increases the activation of genes that turn on the production of proteins in the brain needed to solidify memory.

In a few cases, one gene has complete control over whether you will develop a particular trait; if a man has the gene for color blindness or Huntington's disease, he will suffer these ills. Otherwise, it's rare that a single gene controls anything. In a few other cases, such as heart disease, genes predispose you to possible trouble, but your lifestyle can be the more important determining factor.

Most of our traits are caused by the interaction of many genes as influenced by the environment. That is why it is highly unlikely that specific patterns of behavior, such as stealing or brilliance in math, can be wholly inherited. If sons act like fathers in these cases it is primarily because the son is raised in a criminal environment or is praised for solving math problems and encouraged to play games such as chess that promote spatial thinking. Environment can even negate strong genetic predispositions. For example, Type II diabetes is highly genetic, but if the susceptible person can avoid becoming overweight in midlife there is a good chance that the genes for the disease will not be activated. In another example, it has been found that even though twins carry an identical set of genes, it is not uncommon for one in a pair to exhibit severe Tourette syndrome—a neurological disorder characterized by the presence of tics and streams of nasty language—while the other twin's case is hardly observable. Interaction with the environment accounts for the great difference in severity.

Studies of identical twins separated at birth are often used to test the debate between nature and nurture. While these can be valuable, this test is hamstrung from the start for several reasons, one of which is that differences in the position of each twin in relation to the placenta may bring with them differences in blood supply, hormonal levels, and other factors that are not intrinsic to the genes of the twins. Whether a twin is a "front child" or a "back child," a "spleen child" or a "liver child," makes a difference.

We have much more to learn before we will be able to draw conclusions about which one, nature or nurture, is more important, and for what areas of development. If environment is all-important, it is hard to explain child prodigies—children who seem simply to sit down and play the piano or chess at a very early age, and are able to learn very fast, very well, with little or no instruction. It seems that prodigies must possess an inborn "talent," which means a gene or genes for the intellectual and physical capabilities needed to play the piano or chess.

The remarkable twinning effect directly contradicts the notion that environment is more important than genes. In these cases, twins who are raised apart (with no contact) and are reunited years later find that their lives are very similar. This was the case for a pair of twin brothers separated five weeks after birth and raised eighty miles apart in Ohio. When Jim Lewis and Jim Springer were reunited at the age of thirty-nine, they found they had both married women named Linda, divorced, and remarried women named Betty. Both chain-smoked Salem cigarettes, drank Miller Lite, loved stock-car racing, hated baseball, and vacationed on the same stretch of beach in Florida. Studies of 7,000 sets of twins by the Minnesota Center for Twin and Adoption Research show that a number of traits may be driven by genes, including alienation, leadership, vulnerability to stress, and even religious conviction and career choice.

But then it turns out that for some twins separated at birth, one twin ends up as a schizophrenic adult and the other does not. How is this possible if they have the same genes? Environment may be the answer. Other twin studies show that environment can mitigate or exaggerate the effect of genes; twin halves raised by parents living in a tough inner city demonstrate more aggressive and violent behavior than the other twin halves raised in suburbs.

The point to remember is that the issue is not nature *versus* nurture. It is the balance between nature *and* nurture. Genes do not make a man gay, or violent, or fat, or a leader. Genes merely make proteins. The chemical effect of these proteins may make the man's brain and body more receptive to certain environmental influences. But the extent of those influences will have as much to do with the outcome as the genes themselves. Furthermore, we humans are not prisoners of our genes or our environment. We have free will. Genes are overruled every time an angry man restrains his temper, a fat man diets, and an alcoholic refuses to take a drink. On the other hand, the environment is overruled every time a genetic effect wins out, as when Lou Gehrig's athletic ability was overruled by his ALS. Genes and the environment work together to shape our brains, and we can manage them both if we want to. It may be harder for people with certain genes or surroundings, but "harder" is a long way from predetermination.

LEARNING TO CHANGE

THE NEURAL PATHWAYS that control the basic functions we need to survive—heartbeat, temperature control, breathing—are already connected at birth, but many more pathways are determined by the greatest environmental factor in our lives: learning. Although the brain's flexibility may decrease with age, it remains plastic throughout life, restructuring itself according to what it learns.

The brains of children three to ten years old consume twice as much of the blood nutrient glucose as those of adults, in part because their brains are less efficient and are in the business of forming a vast number of connections. Studies also show that children who exercise regularly do better in school. New research indicates that adult exercise juices the brain with more glucose too, which may promote an increase in neural connections.

Because the young brain prunes weak connections, environmental input in a child's early years can have amazing or devastating effects on the brain's wiring, and thus on future behavior. Geraldine Dawson at the University of Washington followed 160 children from infancy until age six. She found that infants raised by depressed mothers—and so not exposed to many smiles or sounds of excitement in response to their actions—showed reduced activity in the left frontal region of the brain, the area responsible for the expression of positive emotions. At three and a half years of age, the children were more likely to exhibit behavioral problems. In cases such as these, intervention by positive fathers or other caregivers and having the mothers undergo treatment could help to strengthen the neural connections before they are eliminated permanently.

Neurons are constantly competing to make connections. Many maps have been drawn that match each region of the brain to the function it controls: one area for speech, another for spatial skills, and so on. However, changes in environmental input continually move the boundaries. An accurate map of the brain would be different for each of us, and would shift over time. Connections that receive input from frequently used body parts will expand and take up more area than those that receive input from infrequently used parts. Magnetic resonance imaging (MRI) shows that the brains of violin players devote much more area to pathways representing the thumb and fifth finger of the left hand—the fingering digits—which are used extensively in hours of training. The younger a child begins practicing, the more area her cortex devotes to these fingers.

The competition to gain more representation in the brain explains why babies born with cataracts that cloud their vision must have them removed by six months or never gain sight. The brain must learn to see, making connections and stimulating them with inputs from the retina. If these pathways aren't stimulated, they will be eliminated as not useful. Many of us who need glasses have a different prescription for each eye to make the eyes comparably strong. Otherwise, neurons serving the stronger eye will branch out their connections, beating out the neurons serving the weaker eye and making the latter permanently weak. This condition is known as amblyopia. To stimulate the neurons of a weaker eye and prevent its becoming amblyopic, eye doctors will patch the stronger eye. If one eye of a newborn kitten is sewn closed, the eye's neural connections will wither and disappear from lack of use. If the eye is later opened it will never gain sight, because the stronger eye has permanently taken over the available synapses, and, more important, the weaker eye has permanently lost its ability to make connections.

Changing your pattern of thinking also changes the brain's structure. Jeffrey Schwartz at the University of California at Los Angeles School of Medicine found that obsessive-compulsive patients who changed their problematic behavior by repeatedly not giving in to an urge, and deliberately engaging in another activity instead, showed a decrease in brain activity associated with the original, troublesome impulse. It is theorized that neurons contain tiny electromagnetic fields that become misaligned, or "locked," for the duration of a disease or disorder. The neurons get stuck in a rut of abnormal patterns of activity, becoming underactive or overactive or just nonperforming, it being either too easy or too hard for them to fire. A person who forcibly changes his behavior can break the deadlock by requiring neurons to change connections to enact the new behavior. Changing the brain's firing patterns through repeated thought and action is also what is responsible for the initiation of self-choice, freedom, will, and discipline. The drug Prozac can be helpful in breaking these kinds of deadlocks.

We always have the ability to remodel our brains. To change the wiring in one skill, you must engage in some activity that is unfamiliar, novel to you but related to that skill, because simply repeating the same activity only maintains already established connections. To bolster his creative circuitry, Albert Einstein played the violin. Winston Churchill painted landscapes. You can try puzzles to strengthen connections involved with spatial skills, writing to boost the language area, or debating to help your reasoning networks. Interacting with other intelligent and interesting people is one of the best ways to keep expanding your networks—in the brain and in society.

Some of these activities help owing to a neurological phenomenon called cross-modal influences—cross-training in the sports world. For certain sets of skills, training one part of the brain also benefits another. As we will see in the discussion of language in Chapter 7, dyslexic children who repeatedly listen to elongated sounds generated by a computer can improve their ability to spell and read. A study performed by a team at the University of California showed that college students who listened to 10 minutes of Mozart's piano sonatas just prior to taking spatial reasoning tests scored higher than students who listened to relaxation tapes or the more hypnotic music of Philip Glass. This "Mozart effect" lasted only 15 minutes, though, and other studies have shown weaker improvement or none at all. More research is needed before we all start strapping Sony Walkmans to our heads or sending Mozart tapes home with newborn babies.

There is stronger evidence, however, that children who listen to and play music at ages younger than eight do better on spatial reasoning tests. For example, the California team studied a class of three-year-olds. Half the class attended piano or singing lessons for eight months. Their scores on puzzles, tests of spatial reasoning, and drawing of geometric figures shot up to 80 percent higher than those of their classmates who did not attend music lessons. The musical children gradually became faster and more accurate at spatial reasoning over the school year and boosted their spatial intelligence. The theory is that as music is structured in space and time, practicing it will strengthen circuits that help the brain think and reason in space and time, important for math. If the effect of sustained practice during childhood is permanent, the improved ability will help children in complex math and engineering problems when they grow up. It is theorized that the music triggers neural firing patterns over large regions of the cortex that are also used for spatial reasoning.

Activities that challenge your brain actually expand the number and strength of neural connections devoted to the skill. But as Merzenich's monkeys showed, when complex motor tasks become routine they are pushed down to the subcortical areas, where they reside as more automatic programs. Once a procedure is stored in this lower memory it becomes hard-wired. That's why we can get on the proverbial bike and pedal away after a decade of not riding. If these skills had stayed in the higher cortex and been unused, the connections would have withered and been lost. Adults who gave up their rock-'n'-roll bands in high school find that when they pick up a guitar years later they can still play, and when their children bring home their first algebra problems, they can still set up an algebraic equation.

The more that higher skills such as bike-riding and cognition are practiced, the more automatic they become. When first established, these routines require mental strain and stretching—the formation of new and different synapses and connections to neural assemblies. But once the routine is mastered, the mental processing becomes easier. Neurons initially recruited for the learning process are freed to go to other assignments. This is the fundamental nature of learning in the brain.

The brain's ability to rewire means that in principle it can recover from damage. Young children who have had an entire brain hemisphere removed because of severe epilepsy manage to compensate with only slight mental or physical disabilities. Intense physical and mental rehabilitation allows circuits in the remaining hemisphere to gradually rewire, taking over many of the functions that the lost hemisphere used to perform. Things won't ever be "normal," as the original well-trained and appropriately placed circuits have been lost, but even complex functions such as language and reasoning are relatively spared after this sudden, massive loss of neurons. We will encounter several examples of such dramatic rewiring throughout this book.

The brain is amazingly plastic. In the past it was commonly accepted that any brain damage was permanent; once a brain region died, the function it controlled was gone forever. More than 500,000 Americans have strokes each year, killing many neurons and cutting many connections, yet in many of them undamaged neurons take over, changing the number, variety, and strength of the messages they send, rerouting traffic around the accident site. Rewiring is possible throughout life.

New connections take time to form and strengthen. They gradually learn what is most useful and adapt. Many stroke victims lose language abilities, but neighboring circuits or neurons in the nondamaged hemisphere try to take over and compensate for the lost function. Of course, these patients are relying on different neural connections that are probably less efficient for language, so their speech may never be as natural or easy. In cases where brain damage occurs slowly, such as Alzheimer's disease, the brain has more time to compensate, and many deleterious effects can be postponed, though the progressive march of this devastating disease cannot yet be stopped.

The brain reacts differently to injury during different periods of development. Prenatal or early childhood brain damage is often less problematic since many neural circuits are not yet committed to specific skills, knowledge, or memories. The brain can readily rewire on a widespread scale. Although the damage may result in a smaller adult brain or one of lesser overall intellectual abilities, it will seldom cause specific deficits. Most lasting problems are actually due to misconnections from neurons that try to branch out and fill new roles. In later childhood, major damage will be more permanent, although many skills can still be recovered. Plasticity at multiple levels is more active in early life, so that damage at one site produces changes at many other sites, thus changing the brain and its functioning in a more widespread manner. In later life, with less capacity for remodeling at multiple levels, effects at a distance from the site of damage are less likely and specific deficits are more common. From mid-adolescence on, there is less rapid growth of new synapses that allow for flexibility, and by then neurons are completely myelinated, or sheathed. Damage will cause deficits in specific skills with varying degrees of recovery. The more we learn about how the brain restructures itself, the better we will be able to direct other brain areas to take over faulty functions, resulting in greater recovery from trauma and disease.

LIMITS TO PLASTICITY

DESPITE THE BRAIN'S amazing ability to adapt, there are limits to its flexibility. Age does make it harder to reroute and establish new circuits. Music teachers, chess champions, and star athletes all advise parents to start their disciples young. We've all seen how much easier it is for young children to pick up second languages. In an American family that is transferred to Tokyo for a year, the preschool child will learn to converse in Japanese while her mother is still struggling with basic communication.

Children who are exposed to two languages from birth learn to speak both fluently. Linguistic researcher Patricia Kuhl, at the University of Washington in Seattle, likes to say that all babies are born "citizens of the world," meaning that they can learn any language perfectly. She has tested newborn infants with sounds unique to African languages, English, and Japanese. No matter where a baby is being reared, he or she can distinguish the fine auditory cues typical of any non-native language and is presumably ready to learn any language heard.

From about six months on, however, if babies have not heard a particular speech sound, they can no longer distinguish it. Infants whose parents speak English have formed different linguistic connections than infants whose parents speak Japanese, based on the phonemes they hear: the long "oooo" and abrupt "ba" of English, the clipped "toh" and slurred "rr/ll" of Japanese. By its first birthday an infant can no longer process phonemes it hasn't heard; it is functionally deaf to foreign sounds, having learned to ignore sound distinctions not necessary for its native language. In fact its babbling, though not yet words, is confined to sounds that the infant has already heard in its own language. To learn Japanese after childhood, we conjugate long lists of verbs and endlessly repeat dialogue from language tapes, but we can never speak like a native, because our language circuits are unable to form new basic connections.

Brain development in the fetus and baby occurs through a series of critical periods, "windows of opportunity," when the connections for a function are extremely receptive to input. Once the window closes, neural connections are pruned down to the most efficient, according to how much they are used. Then the battle is over: the closed eye and the deciphering of foreign phonemes will never regain space in the brain. It is clear that it is possible for adults to learn to speak a new language with little or no accent, but it is also clear that they do not do this the way a baby does, and instead use altogether different systems to learn. The adult systems are not nearly as good as the baby ones.

These precious windows of opportunity are also times of great vulnerability to irreversible damage. "Closet kids" found by police provide the strongest evidence. These children have been locked in closets or basements for years by psychotic or brutal parents. They grow up without hearing human conversation and are never able to master the sounds and grammatical rules necessary for smooth speech. Through long instruction after they are found, other pathways compensate to some extent, but tragically, owing to the extreme deprivation, the critical period for natural speech development has been missed.

One girl, Genie, was discovered in 1970 in Los Angeles at age thirteen. She had spent her entire life, from babyhood on, in one room, often chained for hours to a potty chair and beaten if she made a noise. Imprisoned and isolated by her psychotic father, she had effectively grown up without human contact. All she was able to hear was blurred conversation through the walls of her room. After four years of subsequent experiments and training she had learned a vocabulary and sign language, but her syntax remained disrupted. She could produce pidgin-like sentences such as "Applesauce buy store," but was permanently incapable of mastering grammar. She had already passed the limited window of opportunity for language acquisition. (Unfortunately, Genie did not come out of this story well. Funding ran out and Genie regressed after passing through a string of foster homes, where she was beaten and abused.)

In contrast is Isabelle, who was six years old when she and her mute, brain-damaged mother escaped the silent imprisonment of her grandfather's house. With training, a year and a half later she had a 1,500-word vocabulary and could form complex sentences like "What did Miss Mason say when you told her I cleaned my classroom?" She had not yet passed through the window of opportunity for attaining syntax.

University of Chicago psychologist Janellen Huttenlocher has found that the frequency with which normal parents speak to and around their child during the child's second year significantly affects the size of the child's vocabulary for the rest of his or her life. The more words a child hears during this sensitive period, whether it's "cat" or "existentialism," the stronger the basic language connections.

Constraints on plasticity for many sensory and motor functions also depend on critical time periods. Most humans move all their body parts during the first two years of life. By age two the motor circuits become hard-wired. If for some reason a child never moved his arms these circuits would be lost and he would never be able to move his arms in a natural way. Regions for basic vision are complete by six months. Acquisition of other functions, however, such as academic learning, takes place over a lifetime, unconstrained by windows of development.

Understanding when the brain's circuits are most receptive to learning a particular skill can help create the optimum environment for a child's development. Psychiatrist Dan Stern at the University of Geneva believes that the critical period for developing emotions occurs from ten to eighteen months. Stern was one of the first and most long-term baby watchers, and has looked for evidence of emotional and social critical periods. His work indicates that if parents regularly respond to their baby with delight, the child's circuits for positive emotions are reinforced. If parents repeatedly respond with horror, the child will shut down those circuits and instead reinforce the fear circuits; the research shows that early fright conditions the baby's brain for more fright. Prolonged depression in the mother conditions the baby for depression, too. The key words here are "repeatedly" and "prolonged." An occasional scowl won't set the child up for a miserable life.

Stress management is also learned during a critical period early in life, according to research on newborn rats, which have neurons very similar to humans. The studies reveal that the more the rats are gently handled, the more they produce serotonin, a brain chemical that controls aggressive behavior. As adults, the rats who had received gentle handling were better able to cope with stress, had stronger immune systems, and actually lived longer than rats who had not been treated gently.

Many cognitive functions share pathways in our brain's complex tangle of neural connections. The development of one skill can therefore profoundly influence another that is seemingly unrelated. As the Mozart effect shows, music and spatial reasoning appear to be linked. Listening to words and reading share some of the same circuits, too.

THE NUNS OF MANKATO

THE BRAIN'S PLASTICITY not only helps with recovery but may actually play a role in preventing brain disease. For evidence, just visit the School Sisters of Notre Dame nunnery in remote Mankato, Minnesota. Many are older than ninety, and a surprising number reach one hundred; on average they live much longer than the general public. They also suffer far fewer, and milder, cases of dementia, Alzheimer's, and other brain diseases. David Snowdon, the University of Kentucky professor who has been studying them for years, thinks he knows why.

Spurred on by their belief that "an idle mind is the devil's plaything," the nuns doggedly challenge themselves with vocabulary quizzes, puzzles, and debates about health care. They hold current-events seminars every week, and write often in their journals. Sister Marcella Zachman, featured in *Life* magazine in 1994, didn't stop teaching at the nunnery until she was ninety-seven. Sister Mary Esther Boor, also pictured in *Life,* still worked the front desk at ninety-nine. Snowdon, who has examined more than 100 brains donated at death by nuns in Mankato and other School Sisters locations across the nation, maintains that the axons and dendrites that usually shrink with age branch out and make new connections if there is enough intellectual stimulation, providing a bigger backup system if some pathways fail.

Snowdon has found that the nuns who earned college degrees, taught school, and constantly challenged their minds into old age lived longer and resisted Alzheimer's disease better than the nuns who had lower levels of formal education and spent most of their time cleaning rooms and preparing food. Snowdon's conclusion, and that of other scientists who have studied aging and the brain, is that any intellectually challenging activity stimulates dendritic growth, which adds to the neural connections in the brain. The more mentally challenged sisters have more neural connections, which allows them to reroute messages when the brain is damaged by stroke or disease, counteracting the debilitating effects on the brain and thus keeping them healthier and more active for more years. Given that the sisters have led otherwise similar lives in the same environment for decades minimizes the influence of any other factor.

The hypothesis that more academic challenge leads to a more flexible brain in old age is supported by gerontologist Denis Evans, who studied elderly residents in the working-class community of East Boston, Massachusetts. He gave them a series of memory and mental status tests, and repeated the testing three years later. Residents who had fewer years of formal education consistently showed a greater decline in test scores, independent of age, birthplace, occupation, income, or native language.

REGENERATION

HOPE IS NOW GROWING that brain damage can be treated by inducing neurons to regenerate; recent discoveries indicate that regeneration might be possible. If so, brains damaged by stroke, Parkinson's disease, and Alzheimer's disease could actually produce new brain cells to fill the roles of the cells that have died. It is true that most neurons can't be regrown when they die; our brains would have a hard time holding on to memories and skills if cells were easy to replace. However, in a few specific regions, such as the hippocampus, the birth and differentiation of neurons continues through old age.

In studies with adult rats, Alan Lewis of Signal Pharmaceuticals found that new, undifferentiated neurons taken from brain areas that are still growing and moved to another part of the brain can learn to perform the local function there. Lewis grafted immature nerve cells from memory areas into smell areas; the cells used cues from the local environment to develop into mature olfactory neurons. Since neurons aren't totally preprogrammed to perform specific functions, neurons moved from one area may be able to take over functions lost to brain damage in another.

Genetic manipulation may also help. Evan Snyder at Harvard Medical School removed newly formed brain cells from newborn mice. He injected them with a gene that would cause them to divide and then inserted them into different areas of brains of adult mice in which stroke had been induced. The implanted neurons migrated to the damaged areas, divided, and took on local specialized functions. One theory is that the cells responded to chemical signals released by dying neurons or by ordinary brain cells that no other cells were responding to. Because actively dividing cells have not yet specialized their function, if they are directed to the right location they may be able to fill in for neurons lost to stroke, disease, or accidents.

Implanting dividing cells that could renew brain function could be a key to fighting Parkinson's, the disease that has ravaged boxer Muhammad Ali and that affects 500,000 to 1 million Americans a year, most over age fifty-five. Parkinson's results from cell death in the substantia nigra, which produces dopamine and sends it to a second brain structure called the striatum, which coordinates movement. No dopamine means no muscular coordination.

One way of restoring the dopamine supply would be to replace the faltering substantia nigra neurons with ones that work. The question is where to get the immature neurons. Until recently the only source was brain tissue from aborted human fetuses. The first fetal transplants a decade ago were promising but prompted isolated instances of public outcry, reported in the media with persistent Frankenstein clichés. President Bush promised to ban the technique, just as President Clinton would later vow to do for human cloning. However, research continued in Europe and in privately funded ventures in America. In 1993 the U.S. presidential disapproval was lifted. No conclusive clinical trial of the process has yet been completed. Only 200 transplants have been carried out at the universities of Lund, in Sweden, and Colorado, at Denver, so the real value of the treatment is only just being discerned.

The anecdotal evidence is somewhat encouraging. Curt Freed at the University of Colorado says that roughly two-thirds of the patients have improved. Half can abandon their medication altogether while keeping up normal appearance. A study by Olle Lindvall at Lund suggests that remission can last for up to six years.

However, the results remain open to debate, since techniques for doing the transplants have not been standardized. Researchers do not agree on how much fetal tissue to implant, and they do not know how best to scatter the cells in the patient's brain. None of the researchers has much idea what is happening in the third of the patients who do not improve.

Another recent breakthrough has been the parallel work of teams at Johns Hopkins and at the University of Wisconsin who have successfully cultured embryonic stem cells in the laboratory. These cells may someday be able to be harvested for successful transplantation into the brain. We are coming into an age when mature nerve cells can be changed into undifferentiated young neurons in a Petri dish by manipulating the function of their genes, which will allow more rapid progress of this line of research and will get around the ethical issues that have marred its progress thus far.

Meanwhile, researchers are trying the same approach in patients with Huntington's disease, the hereditary disorder that killed Woody

Guthrie, which also involves the striatum. Unlike the symptoms of Parkinson's, which can often be checked with a drug called L-dopa (making transplants a treatment of last resort), the symptoms of Huntington's disease cannot be counteracted, and the disease affects intellectual function as well as movement, leading to severe dementia and death.

More positive results are needed before transplantation of fetal nerve cells can be said to be effective. The ethical objections to the use of cells from aborted fetuses persist, too; up to eight fetuses are used for some Parkinson's patients. Legislators have tried to pass laws to ensure that a woman's decision to have an abortion is not influenced by the idea that her fetus's tissue may help a Parkinson's victim, and many states still prohibit fetal-tissue transplants. Clearly, we need new ways for getting immature brain cells that do not involve fetal tissue. Another way around the problem would be to find an animal source of nerve cells.

Fetal pigs are one possibility. One of the problems is that the brain may reject any cells that are not its own, more so if they are not even human. Like pig insulin, which is rejected by diabetic humans much less than beef insulin, pig's brain cells are less likely to be rejected. When transplanted into rats with Parkinson-like traits, pig donor cells can accurately rewire damaged portions of the striatum.

Undifferentiated fetal pig cells have already been implanted in the brains of some Parkinson's patients at the Lahey Hitchcock Medical Center in Burlington, Massachusetts, in hopes that the cells will take on local functions, such as the production of dopamine. The first person to receive such cells was Tony Johnson, a fifty-seven-year-old civil engineer who had endured Parkinson's for twenty-seven years. Six months after a tiny drop of cells was injected into his brain, Johnson's wife told the *Boston Globe* that her husband's speech and walking were better, though he still needed drugs to control his symptoms.

Researchers at Harvard Medical School have shown that nerve cells from fetal pigs can mature in the human brain, but so far the number of trials has been small. About a dozen Parkinson's and a dozen Huntington's patients have had fetal-pig-cell transplants. Their recovery rate is comparable to that of patients receiving human fetal tissue; over half have regained some of their motor control six months after surgery.

While using pig cells overcomes the ethical questions raised in connection with the use of human fetal tissue, it brings problems of its own, as it carries the risk of transplanting pig diseases. Therefore, a third possibility is being pursued: regenerating replacement cells taken from a patient's own brain.

Recent research has overturned the old neurological dogma that adult brains cannot renew themselves. It used to be thought that neural stem cells (neuroblasts)—which divide to produce nerve cells in the fetal and child brain—shut down in adulthood. But Brent Reynolds and Sam Weiss at Neurospheres, a Canadian biotechnology company, have shown that stem cells may be inactive in adults but are still alive, and might be prompted to create new neurons. They have coerced stem cells in a test tube to churn out new cells by adding "growth factors," molecules that stimulate tissue growth. If this were sustained, cells already in the patient's brain could be triggered, or moved and triggered, to create new neurons to replace the lost brain function.

Scientists at Ontogeny, a company in Cambridge, Massachusetts, are trying to leapfrog this procedure by working with a potent growth-factor protein that, in a Petri dish, can transform stem cells into mature dopamine producers. They have named the protein "sonic hedgehog," after a fast-moving children's video-game character. They are currently implanting it into the brains of mice to see what happens. One of the oldest biotech companies, Amgen, is experimenting with growth factors derived from glial cells, which has been effective in slowing the onslaught of Parkinson-like symptoms in monkeys. Amgen's and Ontogeny's approaches would require regular injections of growth factors into the brain, which means a patient would have to have a hole drilled in his skull and be fitted with a catheter.

It will be some time before efforts to regenerate brain cells become part of established medicine. Meanwhile, for the vast majority of us, who are not debilitated but are coping with everyday problems and with aging, the lesson about brain development is that we have the power to influence our brain's ability to renew itself. The human brain's amazing plasticity enables it to continually rewire and learn—not just through academic study, but through experience, thought, action, and emotion. As with our muscles, we can strengthen our neural pathways with brain exercise. Or we can let them wither. The principle is the same: Use it or lose it!

2

PERCEPTION

RICKIE'S FATHER STILL REMEMBERS an odd incident that took place when his forty-five-year-old daughter was only three. They were standing together in front of a giant picture window, looking out into a forest. Suddenly, Rickie began to tremble. She grabbed her father's hands and stood virtually paralyzed, deathly afraid. Her father hurriedly asked her what was wrong. "The trees are coming into the house!" she yelled. "They're all coming in here!"

Her father was taken aback by Rickie's strange behavior, but then again, she was only three, and young children certainly have their moments. He dismissed it. But it was a defining moment that would cast Rickie into a tortured life, though at the time neither she nor her father knew it.

The pain of the next twenty years of Rickie's life is chronicled in the 1990 book *Rickie,* by Frederic Flach, a psychiatric researcher at Cornell University. As a child Rickie was lively and fun-loving. She was a bright student when she first entered elementary school in the early 1960s. But by the time she had reached third grade she heard the teachers talking behind her back: "She's stupid. We're going to have to put her with the 'special' kids." Rickie was terrified. She loved schoolwork, even though it took her a long time to read, write, and do math.

Rickie had one great friend, but that was it. At home she was vibrant and full of life. But she never played with more than one child at a time, and avoided any kind of group activity. She heard the labels again: "She's a loner."

Rickie's parents had her tested. Her eyesight and hearing were fine. Her cognitive abilities were well within the norms. She was a mystery to her teachers, but one that they did not explore further. She was placed with other children who had learning disabilities because she just couldn't keep up with the regular class.

As Rickie got older she became mystified herself. At times everybody and everything seemed far away. She felt as though she was sitting in a closed box and seeing things "through a pinhole." But she didn't tell anyone; everyone already thought she was strange, and she certainly didn't want to say anything that made her seem even more like a "weirdo."

Then her grandfather, the only adult who didn't judge her, died, pushing her into a descent from sadness to despair. She was admitted to a psychiatric hospital when she was thirteen. She was first diagnosed as mildly autistic and then as borderline schizophrenic. Her fate was sealed, for in psychiatry in those days there was an awful truism: A diagnosis is destiny. It was impossible to shake, and it directed all subsequent treatment.

For ten years Rickie was in and out of mental hospitals. She became severely frustrated because no one could tell her what was wrong. The insults, the scolding, the drugs they gave her, and the sense of futility made her more and more depressed. She got so bad at one point that she was given a series of eighteen shock treatments, which jolted her out of clinical depression, but only long enough for her to try to kill herself. One psychiatrist even suggested a lobotomy.

The damnedest thing was that often, when she was talking with a doctor, she was lucid and calm. The doctors were impressed with how clearheaded she seemed. But then the frustration would get the better of her and she would scream and carry on and cut her arms with knives, just like the other "crazies" in the hospital. She knew it would get her attention. The doctors figured her lucid spells were deliberate cover-ups.

When Rickie was twenty-three she was referred to Melvin Kaplan, a developmental optometrist—one who specializes not in how the eyes work alone but how they work together with the brain. Although standard eye exams had shown that Rickie had 20/30 vision in each eye—

nearly perfect—she was prone to accidents and falling because she walked into or tripped on objects that someone with good vision would easily have avoided. Dr. Kaplan started with the usual vision tests. They were fine. He proceeded to the test everyone hates—when the doctor shines the bright light into your eye and tells you to stare at it. Your eyes water and your brow wrinkles as you feel the pain of the searing beam, but you manage to endure, perhaps swearing under your breath, until it's over. You wipe away the tears, blink a few times, and the room looks normal again.

This brief pain comes from the retina, and the eye reacts to save itself by trying to move laterally to get away from the intense light. When Kaplan shined the light in Rickie's eye, her retina went into spasm. When the light was turned off, she could hardly see; her vision dropped to 20/200 and stayed that way for minutes, sometimes even an hour.

In his book, Flach describes what happened during a subsequent test. Kaplan asked Rickie to focus on an object. She stared for a minute or so and then looked away. Kaplan asked her, "When you look at something, how long does the image stay?"

Rickie seemed puzzled.

"Does it stay or does it disappear, vanish?" Kaplan asked.

"It stays. I mean, I can make it stay."

"What happens when you look at me?"

"Well, if I look at you for a minute or so, you start to disappear. But if I get my will power going, I can keep you in sight for a long time."

"And what happens to the rest of the things in the room?"

"At first I see them, and you. Then, as I concentrate harder on seeing you, they get dimmer and dimmer, until I can't see them at all."

Kaplan expressed his surprise.

"Isn't that the way everybody sees?" Rickie asked.

"My God, Rickie," said her father, who was in the room with them. "You mean you thought that was normal?"

"Isn't it?" she asked sheepishly.

After further tests, Kaplan concluded that Rickie could not sustain a visual image for more than a minute without beginning to shut down everything else. She had to muster all of her brain power to keep seeing. No routine eye exam would have revealed this problem.

Together, Kaplan and psychologists began the arduous task of recreating Rickie's experiences from childhood on. What they concluded was a shock to all: Rickie had been struggling with a serious visual perception problem that probably became evident when she was about three years old—around the time of that fateful afternoon when she became frightened that the trees were "coming into the house," an early sign of the eventual collapse of her depth perception.

For twenty years—her entire childhood, adolescence, and young adulthood—Rickie endured an ever-worsening psychological fate at the hands of teachers, doctors, and even her parents, who had been too quick to interpret her early difficulty as psychological, not physical. Sadly, the early misdiagnoses snowballed, and Rickie, battered and isolated by it all, sank into real depression and paranoia.

Rickie may well have had mood problems, but the terrible sequence of events that almost led to her suicide could have been avoided with the proper detection and understanding of what was initially a pure perception problem.

When she was three, Rickie saw the trees "coming at her," but of course had no idea that she wasn't seeing correctly. As an older child, she didn't think her visual experiences were abnormal. When she was a teenager, she knew the world sometimes didn't look right, but didn't want to say anything about it because she was afraid of the consequences. On one occasion she had complained to her psychiatrist that when she tried to read, the letters would suddenly crumble. He told her she was phobic—that she was a poor reader and was afraid to read because she knew she would fail. Rickie knew that if she objected more vehemently she would be rehospitalized, so she simply shut up.

It is difficult to convey in words what things looked like to Rickie. Looking through the peephole of an apartment door is a fair approximation. When she focused on an object in the distance, everything in her peripheral field of vision became empty. She got tunnel vision. And if she focused on a near object, the image only lasted for a short time, then fell apart.

That's why Rickie had such trouble reading. She would look at the words in a book, and soon they would crumble—just fold and collapse on the page, melting into a black wash. If she tried to focus on the blackboard, the room would soon get dim; the teacher would start to appear farther away. She wasn't stupid or lazy—she literally couldn't see the words, or the board. Of course it took her forever to do her work, if she could do it at all. She wasn't a loner, either. She loved people and friends and yet because she could only focus on one object at a time, dealing with more than one person in front of her was confusing at best and frightening enough that she never got involved in any group activity.

When she was young—before becoming afraid of reprisal—wasn't Rickie smart enough to say, "Hey, I'm looking at the page and the letters are melting"? No, because she thought it was normal, just like kids who are dyslexic and see letters reversed; they think it's normal because they've always perceived things that way. Her brain got "tired," just like all our brains do. She was doing a lot of mental work to hold it all together, and in times of stress she lost it and the world flew apart for her. Her vision gave up. She thought it was normal to struggle to see and hold her visual field together, because that is what she had always known. She thought it was that way for everyone.

Once she was properly diagnosed, Rickie started the long road to recovery. Dr. Kaplan was aware of the many studies showing that subjects who were fitted with a pair of glasses that turned everything in their visual field upside down, and then discarded the glasses after wearing them for several hours, would continue to see things upside down for another few hours before normal vision returned. After only several hours of distortion, their brains were already adapting to the new reality, and they held on to it for a while before reverting to what they had learned from the years before. Recent research shows that the visual brain can reorganize in small ways in as little as thirty minutes.

Kaplan fitted Rickie with a special pair of glasses that enabled her to focus on an object for longer and longer periods of time, without the image breaking up or dragging her into tunnel vision. Her eyes, it must be remembered, were fine; the problem was her brain's interpretation of the images it was receiving. She had to train her brain with the glasses for months, tracing shapes and tracking moving objects, more and more and for longer and longer periods, until it became used to the new imagery and could process her vision properly. After six months Rickie's brain corrected itself, and she didn't need the glasses. She had trained her brain to see.

However, the truly more difficult part of Rickie's recuperation was correcting the psychological problems she had developed owing to her failures and the stigma-driven abuse she had suffered. After two decades of learning that people were cruel and could not be trusted, it took her years to reverse her depression and improve her social skills permanently. Not until she was forty was Rickie able to feel normal.

But she succeeded. She went to work as a rehabilitation counselor for people recovering from mental illness. She got married, had a baby, and then twins. Her comeback is a brilliant success.

At one point, however, Rickie nearly lost it all. When she began a job and moved out on her own, her vision suddenly collapsed. She panicked, horrified that all her problems would come flooding back. She was readmitted to the hospital, calmed down, trained with her glasses again, recovered quickly, and never again returned to the psychiatric halls.

The episode taught Rickie and her doctors a vital lesson: stress can shut down her visual system. "When I am under a lot of stress," Rickie says in the Flach book, "whether because of car payments or a nasty virus, my vision can go on me just like that." Since that first anxious experience, however, she has learned how to cope. "Now I don't panic. I shut my eyes, go to sleep for a while, or put on my special glasses, and it passes. I know that it will go away."

Rickie has recovered not only visually, but psychologically as well. She knows now that life is not so uncertain and that she can indeed count on people. Unfortunately, there are other Rickies in the world who have perception problems that have not been diagnosed. There are still others who have actually discovered their peculiar perception problem and had it corrected, but have not been able to rebound psychologically or socially after years of insult from the world around them.

Rickie's odyssey shows that perception is much more than simply sensing stimuli from the outside world. It is a huge factor in personality development. Even the smallest perception problem can lead to a cascade of changes in a person's psychological life. Abnormal perception can corrupt a person's experience. If perception distorts our picture of the world, everything that lies downstream from the senses can cause eventual brain dysfunction. Sometimes no amount of effort can help a child or an adult with a undiagnosed perceptual problem to learn what he or she needs to do to keep up with peers, and the person's vulnerable sense of self is damaged forever.

Rickie's rehabilitation with the special glasses shows that ongoing perception affects the very fabric of the brain. The brain is shaped by the perceptions it experiences, so we may be able to get our brains into better shape by becoming aware of what and how we perceive. The even bigger lesson is that all of us, as people, should not be so quick to label others as psychologically troubled, because their difficulties could be the result of physical problems.

Every Psychology 101 course begins with a chapter on perception, which usually mystifies freshmen. "Why study how information enters the mind," they wonder, "when we are really interested in what happens to it afterward?" The young, introspective, answer-seeking students in search of self-discovery get pretty turned off by lectures about pigeons' and rats' vision or hearing. Instructors even apologize for having to teach perception in the first place, and ask students to be patient while they rush through it so they can get to "the good stuff" of psychology, like Freud and sex, drugs, and rock-'n'-roll.

It is unfortunate that so many instructors fail to tell students why they need to know about perception—why it is essential for a deeper understanding of ourselves. The plain fact is that the way information enters the brain affects its final state as much as any other step in cognition.

CHANGING THE BRAIN

AS WE HAVE already learned, the brain is a dynamic ecosystem. The various neurons and networks are engaged in fierce competition for incoming stimuli. Networks that succeed in processing new experiences or behaviors end up as strong, permanent members of the neuronal neighborhood, while unused networks, cut off from the ebb and flow of information, wither away and die. In effect, the brain's structure becomes the information that it receives, and so how it perceives that information determines its future state. The adage we introduced in Chapter 1 applies equally well to perception: Use it or lose it. We must use the senses and their neurons or lose them forever to premature death or to be recruited for another function.

It is easy to understand how a disturbance in hearing, vision, or touch might impair normal brain development. The brain is constantly receiving information about its current state, both from the senses, concerning events in the environment, and from internal messages about the position of the body, its level of arousal, the activities of the various organs, and the chemical and nutritive state of the blood. Because the brain seeks to maintain a condition of internal constancy (homeostasis) in the face of a changing world, it is constantly interpreting all these incoming stimuli as instructions to modify the levels of neurotransmitters and hormones, the rates of electrical firing, and the chemical excitability of its own neural networks.

The development of personality itself is firmly rooted in the sensory apparatus. Even people with extraordinary perceptual abilities often exist in a love/hate relationship with their gifts, because it can be alienating to "see" the world differently than most people. Yet that different view is the defining characteristic of any great artist—indeed, the characteristic that makes each of us unique. The great American architect Buckminster Fuller, perhaps best known for his creation of the geodesic dome, often felt tremendously overloaded by visual stimulation. He would routinely wear glasses that allowed only part of the visual spectrum to enter his eyes, and when he put them on he found it easier to think. When out on a building site or out for a stroll in the city, he would wear earplugs because the noise left him unable to deal with the world.

Understanding how we see, hear, touch, smell, and even taste the world can tell us a lot about how we function in it. The sensations—or qualia—that come in from the environment are fitted into the categories or constructs that we have learned. We are constantly priming our perceptions, matching the world to what we expect to sense and thus making it what we perceive it to be. The first unsuspecting bite of a chili pepper causes our mouths to burn. The next time we are dared to taste a pepper we begin sweating, and we proclaim it to be just as hot as the first one—even if a tricky lab technician in a taste laboratory has presented us with a milder one. In this booming era of discovery, we have learned that the brain's neural networks respond in a pattern that is established by past experience: the more often a specific pattern is fired in response to a stimulus, the more firm the nerve assembly becomes. Hence the axiom: Neurons that fire together wire together. Input shapes the way we experience the next input. It is not an exaggeration to state that after you have an experience, you are not the same person you were before the experience. Experience colors perception.

What happens to our brains during new experiences? First, we must reject the idea that our brains are static storage depots of information. Rather, the nerves are constantly making new connections that will serve us better in the things we do frequently. The brain can be shaped by experiences, just as particular muscles respond to particular exercises. As we rehearse lines in a play or memorize multiplication tables, we build nerve assemblies, just as we do when we repeat a dance step or karate move. As our brains train, the tasks become easier and more automatic.

This training is accomplished in a fascinating and intricate way. The brain's nerve cells self-organize when they have been trained enough by repeated contact with a particular stimulus. A baby learns to distinguish its mother's voice from those of others; a second violinist learns to pick out his part when listening to a symphony. Test subjects learn to see the world upside down for a day. The neurons become "primed," prejudiced to expect the same old song. When they meet unfamiliar stimuli, they perceive the input as new and disturbing. Thankfully. For the disturbance leads to reorganization, which is why Rickie was able to recover and why we can delight in a new pattern of sunset on a new day.

Evidence of the brain's ability to adjust perception was one of the early influences that led me to a career in neuropsychiatry. In 1958 my brother applied for admission into the Naval Academy but was sent back because his vision was not perfect. However, they said he could reapply if his vision improved, and they gave him eye exercises to develop more visual acuity. A month later he passed the entrance eye exam. The exercises had trained his eye muscles and his brain's visual circuits. This was in 1959, long before most of us knew much about the plasticity of the brain.

An act of perception is a lot more than capturing an incoming stimulus. It requires a form of expectation, of knowing what is about to confront us and preparing for it. Without expectations, or constructs through which we perceive our world, our surroundings would be what William James called a "booming, buzzing confusion," and each experience truly would be a new one, rapidly overwhelming us. We automatically and unconsciously fit our sensations into categories that we have learned, often distorting them in the process.

For example, our "coherent" vision of the world actually comes from millions of bits of fragmented visual information. Even as you stare at this sentence, your eyes are constantly darting a bit in various directions, rarely focusing on any word or letter for more than a split second. Furthermore, the rest of your peripheral vision is fuzzy. That's because only one tiny pinhole region in the center of the eye, the fovea, can see with absolute clarity. In the fovea, the photoreceptors known as cones are jammed together, thus making their message loud and clear. The photoreceptors in the periphery, on the other hand, are more dispersed, so the messages they send are less clear. If you concentrate on your peripheral vision, you'll notice that there is hardly any detail. Yet we perceive our visual environment as a seamless, detailed reality all around us. How? The retina splits incoming information into specialized systems that carry only specific types of details, like a special highway up to the brain for motion, color, form, and so forth. Why? Because our brains would overload with visual clutter without the retina acting as a triage of sorts.

The fovea sees with absolute clarity only a thumbnail-size portion of a scene. It sees only bits of shapes, portions of curves, sections of edges, and parts of colors. It does not see whole shapes or colors. The brain predicts final shapes from the fragmented parts that the fovea sees. Nerve impulses that reflect fragments of images, movements, and wavelengths are sent to visual memory centers in the brain, which contain permanently stored image patterns. If the fovea's fragmented image can be matched to a pattern stored in the memory center, *voilà!*—the object is recognized.

The brain's need to predict, in order to fill the gaps between the fragments of images we see, is also the very reason we are prone to visual illusions. We think we see something that isn't there because the cues trigger our prediction models to tell us that it is. This is what magicians and sleight-of-hand experts depend on. We fill in visual information all the time. Each of us has a blind spot in our visual field that occurs because the place where the optic nerve comes into the retina has no rods or cones. We do not see in a big area of our field of vision in either eye. Having two eyes, and thus binocular vision, makes up for this. However, if you cover one eye, you do not see the scene in front of you with a hole in the middle, because our brain fills in this blind spot and does it well. We also fill in details and patterns all the time; when we spot our dog through a lattice fence, we do not see just parts of the dog but experience its whole visual image.

Our perceptive apparatus also filters signals from internal "noise," just like the tuning knob on a radio picks out a station from the static. Background noise is everywhere in a population of neurons. Neurons are generally firing all the time, but in a random manner; stimuli merely cause them to fire faster and to do it in an organized, synchronized way. The neurons are like members of an orchestra, warming up and tuning in chaotic fashion until the conductor suddenly signals the first downbeat, at which point they immediately sound the harmonious opening note.

The ongoing noise and the fact that neural networks become primed to expect certain sensations help explain why amputees suffer from phantom limb syndrome, a feeling that the missing body part is still there. These patients often experience intermittent pain or tingling sensations in the missing limb. This is because the neural networks that allowed that limb to feel sensations before the amputation are still intact and primed to respond; if a random signal comes in that is close to the signal that used to indicate cause for pain, the network fires and interprets the stimulus as real pain. Unfortunately for many amputees, abnormal pain and prickly sensations can become more pronounced as time goes on, because as the false pain network fires more and more, it is more ready to respond inappropriately to a random stimulus as pain. Neurons that fire together wire together.

It is interesting to note, however, that this background noise is also essential to our ability to perceive stimuli we might not otherwise pick up on. A team led by James Collins at the NeuroMuscular Research Center and the Department of Biomedical Engineering at Boston University demonstrated that the presence of a certain level of background noise, or stochastic resonance, can enhance an individual's ability to feel a subthreshold stimulus that he or she ordinarily would not feel. In essence, if neurons are already busily firing, they're more able to pick up on weak stimuli because some of them are already firing along that pathway. Collins purposely tickled the ends of subjects' fingers and added slight pressure while this was occurring; they were better able to sense the pressure than when their fingertips were not being tickled.

The ability to enhance detection of a stimulus that is normally below an ordinary threshold of detection by introducing a specific level of noise or stimulation that helps ready the receptors to be more acute only works to a degree, because if the noise level continues to increase it masks the subject's ability to feel the test stimulus. Collins hopes that his findings will lead to the development of noise-based techniques to improve the tactile perception of people who have lost sensation owing to injury or aging.

Collins collaborated in another study of stochastic resonance with Paul Cordo of the R. S. Dow Neurological Sciences Institute in Portland, Oregon. That study showed that for some amputees, the phantom sensations may be triggered by cortical maps that control muscle reactions to touch stimuli, which border the maps of the former limb. One patient whose left arm was missing from just above the elbow could feel a light touch on his phantom hand every time his cheek was stroked. The layout of cortical maps is rather different from the way our body parts are actually arranged, so although it may seem odd that the cheek and hand could be so closely connected, these cortical maps indeed lie right next to each other.

The pain and tingling sensations are likely the result of nearby neural networks that are trying to recruit the neurons of the missing limb area, since they have become inactive and therefore available for a different use. The occurrence of this invasion has been confirmed by EEG and PET-imaging studies, which show that after several months of disuse maps of amputated areas begin to show activation when neighboring areas are stimulated. This is another illustration of the "use it or lose it" principle: if neurons are not used for their original purpose, they may be recruited for other brain processes. The brain, being a marvelously thrifty manager, retrains the idle networks to perform a new job. During this co-opting, however, other parts of the old networks—in this case the thalamus—remain loyal to their historical, well-learned pattern of firing and misinterpret stimuli as instructions to fire, sometimes in patterns that would have represented discomfort or pain.

Other evidence of co-opting has been found in blind people. Studies in 1996 showed that when blind people read the raised dots of Braille, it turned on not just the usual brain areas for touch, but also an expansive area in the back of the brain devoted to vision in seeing persons: the visual cortex. The neurons there, originally wired for sight, were recruited for touch. A further study in 1997 showed that the visual cortex in blind people actually enhances their sense of touch; when researchers temporarily blocked the visual cortex's ability to function, the subjects had great difficulty reading Braille, and even indicated they felt dots on the page that researchers could see were not actually there.

On a related note, neurological evidence is mounting to prove the widely held belief that the blind hear better than the sighted. The notion makes sense: the area for hearing in blind people may develop more, since they depend on that sense much more than sighted people.

This idea has been proven in other instances. For example, research has shown that the area of the brain devoted to controlling the movements of the fingers on the left hand is much larger for violinists than for other people, because of the fingers' excessive use. The brain's natural adaptability aids the blind by enabling them to distinguish auditory cues with a much greater degree of discrimination. They can learn how to map the layout of a room based on echoes from a tapping cane, which sighted people can't do—and don't need to. Thus, a corollary to our rule of use it or lose it: Extra use means extra cortex. The lesson, again, is that ongoing perception reshapes the ongoing brain. Practice makes new brain.

SIGNAL AND NOISE

AS IS ALREADY CLEAR, the brain's ability to distinguish signal from noise is crucial to proper perception. A neuron only starts a firing process across its synapse when it is "ignited" by a signal from an adjoining neuron. For example, when a bird suddenly flies into your field of view, the light rays bouncing off the animal enter your eye and stimulate the receptor cells in the retina. The energy in the light waves triggers the release of a neurotransmitter from one neuron to the next. The receiving neuron sends neurotransmitters across the next synapse, and so on down the line, in a chain reaction that continues along the optic nerve to the visual cortex.

However, a chain reaction can get started without any trigger from external stimuli. In fact, it happens so often that most "idle" neurons are actually sending out neurotransmitters on their own many times a second. The reason is that neurons are "turned on" by any electrical activity around them—even the firing of a nearby but unconnected neuron. Each of the thousands of channels in each of our billions of neurons may close or open in response to electrical activity in the general area. Consequently, most neurons are firing more or less randomly, in a sort of electrochemical free-for-all. This makes it easier for the nerve to be trained, as it is easier to add more "oomph" to a firing nerve than to start from ground zero, or a nerve that is shut down.

Participating in this humdrum noise is the normal state of a neuron, until some form of more specific information passes its way. When light waves from a bird or a signal from an idea, or any other internally generated representation, comes in, the neurons suddenly begin to fire much more rapidly and, more important, in synchronized and ordered ways. It is as if an entire chorus was made up of neurons, and each member has been faintly humming his or her favorite song. Suddenly someone claps his hands—the stimulus—and the entire group bursts into a coherent song, in three-part harmony. In this way, we smell something good as we pass a local restaurant and rapidly identify it as pizza. This ability to synchronize electrical activity is the very basis of cognition. Our brains are always filled with noise. Whether we perceive something through the din depends on whether the pattern stands out strongly enough from the background.

In my psychiatric practice, I have long considered the concept of noise to be a valuable means of understanding both everyday experience and psychopathology. Different factors—genetic predispositions, childhood events, environmental factors, or altered levels of neurotransmitters or drugs—can change the intensity and tone of neural activity, including the spontaneous firing that goes on much of the time. When the signal-to-noise ratio on your favorite evening TV show gets too low, perhaps during a passing thunderstorm, so that you can no longer easily distinguish the images and voices from the snow, you get annoyed and change the channel. If there is too much background noise in your neurons you can also get annoyed, but you can't just switch to another station.

An excess of mental noise in the brain can make it difficult to perceive what's going on, overloading other circuits of attention, memory, learning, cognition, emotional stability, or any other brain function. The system goes into information overload. If random neuronal firings are too fast and furious, incoming stimuli might fail to activate and assemble the neurons into properly synchronized behavior. This, in turn, could result in the incorrect processing of a stimulus, and neurons would misfire accordingly. This is what can happen when highly anxious people take tests. The heightened anxiety drives up the mental noise, so much so that such people may literally see less of their environment, as though the brain space usually open for perception is busy with the internal noise. They will look at a test question and literally not see certain words, which causes them to misinterpret it and give the wrong answer. They may even miss seeing entire questions on the page. Their brains are so busy dealing with the noise that the visual channels in the brain aren't open to perceive accurately. Our brains are not infinite. They run out of space, run out of gas, as it were. If the brain is busy trying to filter uncomfortable and frustrating noise, worries, or other concerns, there is less "brain stuff" available for perceiving.

SMELL

OF ALL THE WAYS of getting sensory information to the brain, the olfactory system is the most ancient and perhaps the least understood. If there's any doubt about the contribution of smell to perception, consider people who suffer from anosmia, the complete loss of the ability to smell. They can't even taste their food. One patient with anosmia declared that the brief return of her sense of smell was like "the moment in *The Wizard of Oz* when the world is transformed from black-and-white to Technicolor."

Smells can have powerful effects. They can frighten us, intrigue us, or comfort us. Because the olfactory system in the brain has a short and direct connection to the memory centers, smells can take us right back to a vivid scene from the past. Different people can detect the same odor and come away with vastly divergent experiences. For a man who was a former boy scout and spent wonderful weekends in the forest, the smell of wood smoke coming from his neighbor's chimney can evoke feelings of pleasure and bring back fond memories. For a man passing by who was caught in a house fire as a boy, the smell of wood smoke can arouse intense anxiety. Each person's relational experiences are different. One woman in a research study was pleased with an odor that all other subjects found repugnant—a mix of garlic, natural gas, and motor oil. The woman had spent a summer vacation in Alaska near an oil refinery and had adapted while there; the smell reminded her of the wonderful time she had had.

Not all of us can detect all smells. If we are not exposed to certain scents during our early development, we may permanently lose our ability to recognize them. In the same way as with our other senses, we can train ourselves to smell better. For instance, perfumers have had years of experience at recognizing novel and interesting scents.

They have trained noses and make a living detecting just the right blend.

Smell also accounts for much of what we taste. Most of what our gustatory apparatus tells us about our favorite curry dinner is actually picked up by the olfactory system. Receptors in the nose are specialized for detecting chemical information in the air we inhale, as well as what is shunted upward as we chew our food. The taste buds on our tongues merely add a measurement of the presence of sugars, salts, acids, and bases (sweet, salty, sour, and bitter, respectively). The nose does the rest.

The neuroanatomy of the olfactory tract is unique among the senses. Hearing, a small part of vision, touch, and taste all enter the brain through the brainstem and are passed up to the thalamus. From this central way station, millions of neural networks transfer signals to regions of the cortex specialized for each sense. The signals are bounced around, then sent on for further processing to the limbic system, which is central to emotions, memory, pleasure, and learning. The limbic system often adds an emotional tag: joy to a former golden retriever owner who meets an unknown retriever on the street or fear to a person who was once bitten by an unknown dog. It calls up memories and may initiate a bodily response: a smile for the former dog owner or an accelerating heartbeat for the bite victim. As an emotional response is evoked, a person may begin planning a proper course of action: kneeling to pet the dog or walking away quickly to avoid it.

Given the complexity of visual and auditory information, and the corresponding potential for misinterpretation of ambiguous situations, the brain tries to make sense of fine details before making a judgment call. In contrast, olfactory nerves project directly into the amygdala and olfactory cortex, parts of the limbic system, without any mediation through the thalamus. The olfactory nerves have a hotline to the emotional brain, and only then is the information sent to the orbitofrontal cortex for more associating, inhibiting, and further processing. The "smell" connection is much faster and more decisive than the systems for the other senses and not much filtering goes on before action is called for by emotional memory. Indeed, the nostrils are positioned directly above the mouth because they serve as a last-resort alarm system. If you are about to eat something that is disgusting and would make you sick, the olfactory system must be able to detect the telltale odor, match it to a memory encoded in the limbic system, and alter your behavior, all in the fraction of a second that it takes for a morsel of food to pass beneath the nose to the lips.

The direct route traveled by olfactory information is a holdover from early evolution, when quick, emotional responses to odors played a crucial role in survival. Smell is also different from the other senses in that its machinery (the olfactory network) remains uncrossed. All the other senses send most of their information through the thalamus to the opposite hemisphere of the brain for processing.

Smell was a major catalyst in the evolution of the primitive brain (nicknamed the "smell brain" since it basically consisted of olfactory tissue on top of a nerve cord) into the modern, more complex brain. The entire limbic system is thought to have evolved solely from its original function of interpreting odors and emitting pheromones—chemical scents that send social or sexual messages to other members of the same species. As other brain connections evolved, our abilities to process sensory information grew more refined. The olfactory cortex has not shrunk, but the rest of the brain has expanded.

The actual sensing of molecules in the air begins high within each nostril at a patch of yellowish tissue called the olfactory epithelium, which contains olfactory receptors. Each receptor is covered with roughly twenty cilia, or microscopic hairy tentacles, that are constantly in motion, waving about randomly in a bath of moist secretions that help to dissolve the substances we inhale. When a substance diffuses through the mucus layer and binds to an olfactory receptor, it alters the firing pattern of neurons leading from there to the primary olfactory cortex, which lies at the heart of the limbic system.

Our olfactory system recognizes certain smells from birth, notably ones that signal danger, such as those of rotting foods. But the system is also trained by experience, as witness a perfumer or a competent wine steward. Since humans are capable of recognizing and differentiating as many as 10,000 odors, there cannot possibly be enough built-in receptors in the nostrils to be specialized for individual smells. The olfactory system appears to be similar to the immune system in that it is capable of acquiring recognition of a virtually unlimited range of molecular signals. How it accomplishes this feat is not yet clear. But Walter Freeman, who heads the neurophysiology lab at the University of California at Berkeley, and Luca Turin, a biophysicist at University College, London, performed experiments that yielded tantalizing clues. They found that after exposure to a smell, the neuronal firing across the surface of the olfactory receptor begins to settle into a characteristic pattern. However, if a second smell is introduced, the pattern is interrupted, and if the original smell is reintroduced later, a new firing pattern appears for it.

This discovery is intriguing because it implies two things. First, each perception influences all subsequent perceptions and therefore what the brain is ready to perceive. Second, the same stimulus may be represented entirely differently from one moment to the next, which suggests that we are still far from a complete understanding of how perception works.

However information is being represented in the olfactory receptors, its power to affect the brain emotionally is well known. The limbic system contains the brain's pleasure centers, many of which can be activated by the scents of food and sex. Therefore, the olfactory system can establish a link between intended behaviors and rewards, causing a person to pursue a mate or a good dinner. The reward center is central to learning and provides the motivation for doing something or the sense of feeling satisfied. Because the olfactory apparatus is wired directly to this system that determines pleasure and disgust, it is a powerful trigger that can motivate us very quickly and directly, without the associations or abstract thinking necessary for us to respond to vision or hearing. It is simple, direct, and powerful.

Many species use pheromone signals to direct essential behaviors such as mating, feeding, flight, combat, and nurturing the young. Pheromones can be detected in extremely small quantities over long distances. For example, prior to mating, the female silkworm releases 0.01 millionth of a gram of the chemical bombykol, which is enough to attract one billion males up to three kilometers away! Other pheromones control aggressive behavior—they are what compel salmon to attempt the amazing feat of swimming back upstream to the place in the river where they were born.

Humans emit pheromones in all body fluids. Our ability to communicate by pheromones is intriguing, even mysterious. One phenomenon is menstrual synchrony, in which the timing of women's menstrual cycles is affected by the presence of other women. This was discovered in 1971 in research that indicated that the menstrual cycles of women who spent a lot of time together, such as roommates or close friends, tended to start within a day or two of each other. Further studies found that when sweat from some women was rubbed onto other women's upper lips, these women's cycles became synchronized.

Recent experiments have also shown that an area in the nose can detect pheromones that carry no consciously perceptible odor; the subjects of these experiments reported that they smelled nothing at all, and yet electrodes picked up measurable changes in their autonomic nervous system, and they did note mild impressions of contentment or uneasiness, depending on the pheromone. Obviously, while our higher rational abilities do allow us to control our behaviors, we still have much to learn about what is involved when two people meet on the street!

Aromatherapy, an alternative healing method, is based on another phenomenon that has been widely investigated in humans. The theory is that because olfactory nerves transmit signals directly to the limbic system, eliciting an immediate emotional response, certain smells might calm us, stimulate us, help us sleep, or influence our eating habits. One study of children who went to schools in areas with persistent air pollution even showed that the scents increase aggression. Primarily, aromatherapy aims to find pleasant scents that can relieve stress. Using plant extracts, odors are targeted to elicit specific emotional responses. For example, the scents from certain extracts cause the brain to release enkephalins, naturally occurring fragments of morphine-like proteins; their release reduces pain and creates a feeling of well-being. The scents of other extracts cause the release of endorphins, which are naturally occurring neuropeptides that can heighten sexual arousal.

Smell, it seems, also has the power to influence brain functions that affect psychopathology. As noted, axons from the olfactory bulbs connect to the amygdala, a structure in the limbic system essential to nurturing behavior and fear-conditioning. Removing the amygdala causes animals to neglect their young and to forget the negative associations that they have previously formed with particular stimuli. Oversensitivity of the amygdala has been implicated in anxiety, panic disorder, posttraumatic stress disorder (PTSD), and attention deficit hyperactivity disorder (ADHD). The amygdala receives stimuli from every sensory modality, though none as directly as olfaction.

Olfactory projections are also found in the hypothalamus, the brain's hormonal center, which is responsible for the fight-or-flight response. Consequently, odors can alter heartbeat and blood pressure

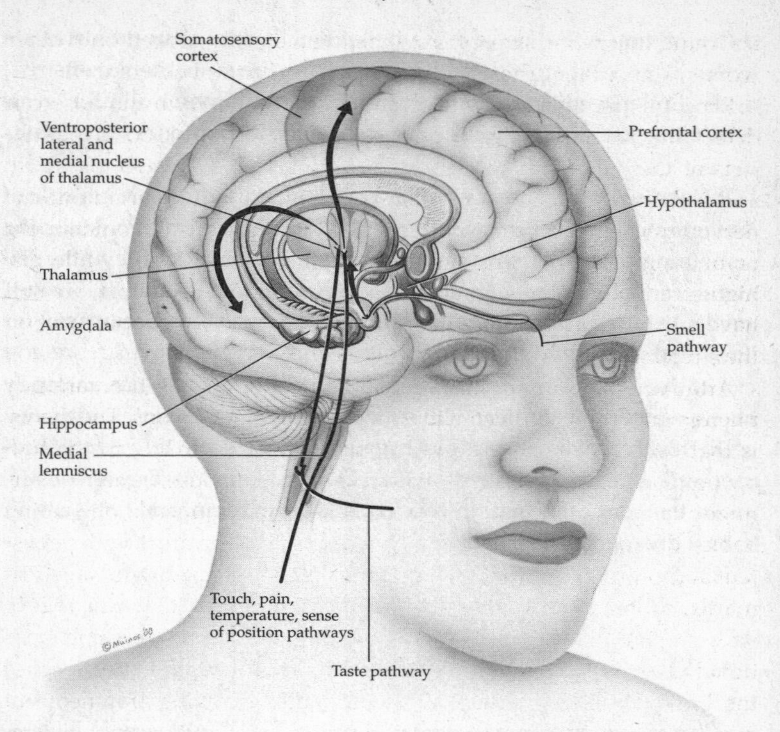

THE SOMATOSENSORY SYSTEM Sensory information received externally is processed by systems in the brain to create sensations such as smell, taste, and touch. The olfactory or smell pathway is the simplest, the only direct route of our sensory apparatus. Odors are taken in through the nostrils, processed as electrochemical information, and passed on to the amygdala and then to the cortex.

The other sensory pathways all pass through the thalamus, the "grand gate" of sensory input. The taste pathway starts with the taste buds, where information is taken in, then sent on to the brainstem. From here it travels up to the thalamus and on to the cortex. Touch, pressure, temperature, and pain sensations from the body travel through the spinal cord to a way station in the brainstem. This information is then sent to the thalamus and on to the somatosensory cortex for further processing.

directly, with very little mediation. Olfactory fibers also project into the pleasure areas of the limbic system, including the amygdala and the septal area, where dysfunctions are seen in schizophrenia, addictions, ADHD, and the ability simply to feel satisfied. Depression, interestingly, often causes a substantial decrease in patients' ability to identify different smells. Women have a better smell sense than men and are better at picking up odors during certain times in their menstrual cycle. Epilepsy that starts in the limbic area is notorious for overwhelming the sufferer with strange or overtly foul odors and tastes during a seizure.

It is worth noting, however, that very few individuals are capable of "imagining" smells very well, either by hearing a word or visualizing something particularly odorous. This inability may be due to the relatively small area of higher cortex devoted to olfaction. However, smells are strong prompters of memories, because the olfactory nerves are wired directly to the hippocampus and amygdala, which are crucial to memory. We are all familiar with how a smell can instantly evoke memories complete with sounds, images, and feelings: the smell of a freshly baked apple pie transports you back to your grandmother's kitchen; the scent of Play-Doh to your kindergarten classroom; the odor of incense to your college dorm room; and hot, stale air to that dreaded lecture hall.

TASTE

AS OLIVER IN THE famous musical put it, life would simply be very bland without taste. Like smell, our sense of taste developed early in our evolution, to protect us from ingesting poisons that could have killed a fledgling species. But today taste also serves to enrich our lives. Imagine if the food you ate had no taste? The experience of eating would lack the incredible emotional and cultural experiences we have built up around food.

As in the nose, the receptors for taste respond to chemical stimuli. We have between 2,000 and 5,000 taste buds in and around our mouths. Chickens have only 24, so you can pretty much feed them anything and they won't care. On the other hand, catfish have 175,000 taste buds, most of them on the outside of their bodies so they can sample their food without ever opening their mouths.

Contrary to popular belief, taste buds in humans are located not just on the tongue, but inside the cheeks, on the roof of the mouth, and in the throat. For the purposes of this chapter, however, we will focus on the papillae, the bumps on the tongue that give it its rough texture.

Each papilla has anywhere from one to several hundred onion-shaped taste buds with an opening on top called a taste pore. The taste buds are made up of cells that detect and process the taste stimuli into signals that travel to the brain.

At each receptor cell, a chemical reaction occurs which transduces, or changes, the chemical signal into an electrical current that is sent as an impulse along a nerve fiber through the brainstem and into the brain. There are two current theories regarding how tastes get encoded into neural firing patterns. The specificity theory says that neurons are tuned to respond to specific taste qualities. The fiber-pattern theory promotes the idea that signals are generated by the pattern of activity in many nerve fibers; substances that have similar tastes will elicit similar patterns, just as similar smells elicit similar patterns. New taste receptors are being identified by scientists and researchers for the food industry as they continue to define the process by which the chemical experience of taste is perceived.

There are four categories of taste receptors, which sense the four primary tastes: sweet, salty, sour, and bitter. Receptors near the tip of the tongue are especially sensitive to sweetness. Salty and sour are most noticeable on the sides of the tongue. Bitter is most distinctive on the back of the tongue, as well as on the soft palate. Interestingly enough, the middle of the tongue has no taste buds, and is sometimes called the tongue's blind spot. Recently, Japanese researchers have identified a fifth primary taste category, called *umami,* which is Japanese for delicious or yummy. This proposed fifth receptor may be a taste-enhancer—a receptor that makes food taste delicious.

From an evolutionary standpoint, scientists believe that each of the four basic tastes serves an important ecological function. Sweet taste ensures our ongoing search for a supply of energy, and flags food as nutritional. Salt sensitivity helps maintain the body's fluids and electrolyte balance. Bitter perceptions guard against toxins and poisons. Sour warns against spoiled food. This implies that even before higher cognitive levels of the brain were developed, our primate and preprimate ancestors had taste machinery in the brainstem. Taste is not that black and white, however; some people have a stronger affinity or aversion to one category of taste or another. Certain chemicals can also alter a food's taste. For example, sodium chloride—salt—makes certain foods taste "better" because it inhibits the action of bitter compounds on our bitter taste buds.

One interesting fact about the taste and olfactory systems is that all of the receptor cells undergo a constant cycle of birth, development, and death over an average period of 10 days for taste and 30 days for olfaction—unlike the sensory receptors in vision, hearing, and touch, which are fixed and remain protected behind the eye, ear, and skin. This regeneration cycle, called neurogenesis, serves an important function because these chemical receptors are constantly exposed to the environment. Taste receptors must endure very hot and cold liquids, harsh spices, and the constant scraping action of one's teeth. They are also bombarded by bacteria and dirt, and are under constant risk of drying out. If you burn your tongue on that hot chocolate, you can thank neurogenesis that your loss of taste from the trauma will be restored fairly quickly as cells regenerate.

As mentioned earlier, taste is highly dependent on the sense of smell. Seventy-five percent of what we perceive as taste, especially the perception of flavor, is really attributable to our sense of smell. Actually, what is most important is the combined action of the two senses. When you eat a pizza, the odors of the crust, oregano, and cheese combine with salty taste to produce the flavor you know and love. Think about the last time you had a cold; you may have noticed that you couldn't "taste" anything. The deficit was not in your taste buds, but was attributable to your stuffy nose blocking the olfactory receptors. Next time you've got a cold, concentrate on what you're actually sensing as you eat. You will perceive salty, sweet, sour, and bitter, but without the smell, you will not perceive any actual flavor of the food.

It is worth noting that flavor is not the only function of our sense of taste. As we roll food along our tongues, we process information about texture, such as the smooth surface of noodles, the rubbery feel of Jell-O, or the hard crust of pumpernickel bread—so named because it was thought to be so difficult to digest that even the devil (*nickel*) would become flatulent (*pumpern*) if he were to eat it. Our tongues can also sense temperature. Even though you cannot taste flavors when you have a cold, you can still perceive the temperature and texture of your food.

Temperature has an effect on taste receptors, too. If taste buds are cooled or frozen, the ability to sense certain tastes is greatly reduced. This insight has been turned into a con by some sectors of the food and beverage industry. For example, the big national brewers in America spend billions of dollars advertising the great taste of "ice-cold beer." But brewmasters at microbreweries and throughout Europe will tell you that the reason the big brewers want you to drink their product ice-cold is because the shock of the very cold liquid entering the mouth temporarily dulls the taste buds. So you don't even taste the beer! It's likely that you've already conducted this experiment by drinking the last ounce or two of your beer at a party after the glass has been sitting idle for a while: "This beer has gotten warm. It tastes terrible." It's the same beverage; the composition hasn't changed, but its temperature has. Indeed, microbreweries and European brewers and pubs serve their beers at only slightly cooler than room temperature, so that you can appreciate the full taste—and it tastes good.

Whether it's beer or pizza, taste signals enter the brain at the medulla in the brainstem from three cranial nerves. From the nucleus solitarius, the arrival area in the medulla, signals are sent to the thalamus and then on to the taste centers in the cortex, which sends them on parallel pathways to the hypothalamus and amygdala, and then on to other parts of the limbic system, where emotions and memory are stored and retrieved in regard to qualities of the taste. This can cause us to avoid food that tastes a certain way or seek food that can satisfy the nutritional needs of the body, such as salt. The signals traveling back and forth along these pathways also affect consumption reflexes such as salivating and swallowing.

The hypothalamus plays a key role in feeding mechanisms. Studies of brain lesions in rats show that problems in the lateral hypothalamus area cause the animals to stop eating and drinking, while lesions in the ventromedial nucleus of the thalamus cause overeating. This and other brain regions are being furiously studied to see if they can be affected in ways that manipulate our hunger and satiety centers so that the right diet drugs can be developed.

Areas in the thalamus and hypothalamus are involved in feedback patterns that maintain the body's energy balance and body weight. Decisions about whether to eat or drink, what to eat or drink, whether to continue eating or drinking, and when to stop eating or drinking—the balance between eating and satiety—result from the interchange between these areas. Since the hypothalamus is also a key player in the motor system, emotions, and memory, it is believed to control our hunger by triggering the release of dopamine, which is greeted as a reward by our reward system. When these areas determine that satiety has been reached, the dopamine is stopped, and our desire to eat wanes.

WHY WE LOVE SPICY FOODS

GIVEN THESE SORTS of controls and our general aversion to extreme tastes, it is a curious phenomenon, scientifically speaking, that people like spicy food. Nowhere else in the animal world will you find a species that willingly ingests "tongue-burning, mouth-on-fire" foods.

The sensation of "spicy" is actually the perception of irritation. Chili peppers get their heat from a tasteless, odorless chemical called capsaicin, which irritates certain nerves in the nose and mouth. Scientists rate the "hotness" of a pepper in Scoville units, which is a measure of capsaicin concentration. Plain bell peppers rate a zero, jalapeño peppers are between 2,500 and 5,000 units, and the hottest pepper of all, the habanero, rates a searing 300,000 Scoville units. Nerves in the mouth, called trigeminal nerves, register such sensations as the bite of a chili pepper or the burn of ammonia. So why do some people expose themselves to this sensation, and even enjoy it? One explanation is that people in hot climates eat spicy foods to help the body perspire and cool off. Another is that eating spicy foods heightens the appreciation of other tastes in the meal and that they cause the brain to release endorphins, which are powerful chemicals that block pain and create a sense of well-being. It may also be that the challenge of staying with the burn of a hot pepper is similar to staying with the burn of muscles in weight-lifting or running. The enjoyment is in the end accomplishment, a "benign masochism," says scientist Paul Rozin at the University of Pennsylvania. Rozin says that your body is responding as if it is in trouble, but your mind knows you are safe, and that, to some, is a pleasurable experience.

You may have noticed that food tends to lose its flavor as you proceed through a meal. Certainly, the fourth or fifth dip into the jalapeño salsa is less dramatic than the first. This phenomenon can be attributed to simple biology. Like the entire brain, the five senses react most strongly to change, so that we can survive in a constantly changing environment. Therefore, when the same stimulus is presented continuously for a significant amount of time, the receptors undergo a process called adaptation—they essentially accept the signal as routine and the messages in the brain therefore weaken in intensity. It takes about one minute of continuous stimulation for particular taste receptors to reach their maximum sensitivity. After that, the taste receptors will undergo adaptation and the taste will fade.

The best way to circumvent adaptation is to give your taste buds different foods to taste throughout a meal. Instead of eating all of your meatloaf at once, present your taste buds with some potatoes, then some vegetables. When you return to the meatloaf your taste buds will react to it anew, and it will taste more flavorful. So will the potatoes and vegetables, through each rotation. Like the rest of the brain in general, the sense of taste is always on the hunt for the novel. The new may be a threat or a new source of food or comfort. The impact of the new on our survival arouses us; it stimulates us to notice and either welcome it or be wary of it.

Adaptation occurs in all five senses. Perhaps the most recognizable example is when you see "afterimages." If you stare at a still image on a television or computer screen, then suddenly shut your eyes, you will still see the image—though the colors will be inverted, like the colors on the negative of a color photograph. If you stare at a colorful dress for about a minute, then look immediately at a white wall, you will notice that you can still see the dress—again, with the complements of the colors instead of the real thing. What has happened is that your brain, after prolonged stimulation, has adapted; it has gotten used to the image, and the weak, plain signal such as the black of your closed eyes or the white of a lone wall doesn't disrupt it too much. The circuitry in the brain is activated and this activity takes awhile to dissipate. It is the sensory hangover of the brain.

Just as too much constant taste signal weakens a taste, too little will not trigger any taste at all. There are two types of taste thresholds. The first is called an absolute threshold, the point at which one can just barely detect the presence of a particular substance. The second is a recognition threshold, the point at which one can recognize the taste substance. The two thresholds differ significantly because detecting the presence of a stimulus is very different from identifying what that stimulus is.

Thresholds also play a role in adaptation; when adaptation occurs, the threshold needed to perceive more of the same taste increases. Americans learned this in dramatic fashion in the 1970s when diet soda was widely introduced. When adults who had switched to Tab or Diet Coke for a while tried a regular Coke, they would comment that it tasted sweet—much sweeter than they remembered. That's because in the years they had spent drinking regular Coke, they had adapted to its taste; the threshold needed to perceive "sweet" had been raised. Switching to diet soda eliminated the constant sugar stimulus, and the threshold required to perceive "sweet" dropped.

There are two deficiencies that can affect a person's perception of taste. Dysgeusia occurs when a taste nerve is damaged, causing a person to perceive tastes that are not there, notably salty, metallic, or bitter sensations. Dysgeusia can also be a side effect of certain types of drugs, such as clarithromycin-amoxicillin antibiotics used in combinations for treating ulcers, and captopril and dipyridamole, drugs often used to treat high blood pressure and congestive heart failure. Hormonal abnormalities also can alter taste perception, the most widely known being the early stages of pregnancy (where there is a drop in the ability to taste with subsequent pursuit of spicy foods and bizarre combinations, which may be considered an effort to get the taste function back) and hypothyroidism, or too little thyroid hormone, which can impair both taste and smell.

The second deficiency is ageusia, which is a loss of the ability to taste. Total ageusia is rare, but it may occur following radiation therapy, which can damage the nerves from the taste buds to the brain, or result from head trauma, which can damage the cortex's ability to recognize taste signals. It is also possible to inherit total ageusia through a very rare condition in which people have no taste buds or papillae. It is much more common to suffer a partial loss of taste, including the inability to taste one specific primary substance, such as sour. Partial ageusia is often attributable to drug use or to tumors of the taste pathways, though it also can be inherited. The condition is analogous to color blindness; partial ageusiacs can taste most foods, but are much less sensitive to certain tastes.

Recently, Swiss researchers have discovered an interesting disorder called "gourmand syndrome," a brain disorder that occurs in a small percentage of people who have suffered from strokes, brain tumors, or head traumas, which causes an intense craving for fine foods. Gourmand syndrome was first discovered eight years ago in a stroke patient who was operated on for a lesion around the middle cerebral artery in the right hemisphere. After the surgery, the patient was not only unable to walk but displayed a preoccupation with food.

When more patients exhibited the same symptoms following strokes, scientists began investigating the possibility that one particular area of the brain was affected. Researchers Marianne Regard and Theodor Landis of the University Hospital of Zurich then studied more than 700 patients who had had cerebral lesions within a three-year period and found 36 men and women who met the criteria for gourmand syndrome; 34 of them had a discrete lesion in the right anterior cerebral hemisphere, and this was the only lesion site in 30 of them. Patients who are afflicted with gourmand syndrome seem to enjoy its symptoms. Prior to their strokes, they had had no inordinate preoccupation with food. Following their strokes, they indulged their newfound taste. Interestingly enough, even though they exhibited an obsession with food, shopping, and dining rituals, none of the patients became overweight.

Another interesting aspect of taste involves disgust. The same area of the brain that responds to offensive tastes—the anterior insula—is also activated when one person sees another make a face showing disgust. This is a good example of how the brain combines senses to improve our chances for ongoing existence. In this case, the coupling of taste and vision allows us to perceive the disgust of another person eating, say, a rotten food, so we don't try to eat it ourselves. Even if we were to bring the food toward our mouth, the sight of it would prepare our taste system to perceive disgust.

Furthermore, when such an event occurs, the conclusion is sent on to a higher part of the brain called the orbitofrontal cortex, where we make associations. It leaves a marker there that will guide us in the future, so that when we see food next week that looks like the disgusting food we saw last week, we'll opt for something else instead. This disgust area has been generalized by us to make it more social and human. When people with obsessive-compulsive disorder have a disgust phobia of dirty linen or of germs on their hands, it lights up the anterior insula as well as the other parts of the circuit: the anterior cingulate gyrus and the orbitofrontal cortex. Furthermore, this area is highly activated in moments of extreme anxiety, and is a big part of the worry circuit. So we see how our emotional responses evolved from our sensory apparatus.

The insular cortex illustrates the "free lunch" that we get owing to evolution, whereby we use cortex already in place to serve other functions. The insular cortex is the taste-sensory cortex, but is also there for disgust and for pain. All of this makes sense, if you consider the brain developing to help us avoid things to eat that bring us disgust and pain. Likewise, pain stimuli activate the anterior insula, a region heavily linked to the sensory and limbic systems. Such connections may provide one route through which painful input is integrated with memory to allow a full appreciation of the meaning and dangers of painful stimuli. As we evolved, the brain just went on its merry way, using whatever it had available to group stimuli together to arrive at today's pain and disgust area.

TOUCH

HUMAN BEINGS POSSESS an instinctive urge to touch and to be touched. It is part of the human drive to explore and interact with the world. Touch is unique because it is the only sense that allows us to experience the world through direct physical contact. Touch is also our most powerful and intimate form of communication. A touch can move us, and hurt us, in a way that no spoken word can, sending messages from comfort to hate across language and cultural barriers.

Touch is far more than a sensory apparatus. Our sense of touch affects the development and expansion of our brains well into adulthood. It is a key component in growing, learning, communicating, and living. Touch is the first of the five senses to develop, and is far more developed than hearing or seeing in newborns. A baby's touch is a significant factor in the development of certain areas of the brain. Edward Perl, a professor of neurophysiology at the University of North Carolina, says, "When you watch a baby touch, you are watching the development of intelligence in his or her cerebral cortex."

Pediatric research at the University of Miami's Touch Research Institute (TRI) in Florida has shown that babies first develop the sense of touch in the uterus, which may explain why they tend to respond so quickly to being held at birth. A significant portion of an infant's first few weeks of life is touch-related. The rooting reflex, the most primitive early sign of touch, occurs when a baby turns its head toward a mother's hand when she touches its face. It helps the newborn locate its mother's nipple when it is feeding. Infants respond instinctively to other forms of tactile stimulation. For instance, if you touch an infant's hand she will grip your finger tightly. Tickle her foot and she will curl her toes.

Studies of premature babies also show that touch can speed their growth and development. In order to survive, they must often be put into incubators. While necessary, this isolates newborns from human touch. PET scans of touch-deprived infants show that critical sections of their brains are barely active, stalling entire areas of development. Researchers at TRI are investigating how to counter this risky isolation. A famous study by Tiffany Field at TRI showed that premature human babies who were massaged for 15 minutes, three times a day, for 10 days gained 47 percent more weight than similar "preemies" who were given the same diet but were not massaged. (One reason the massage helped was that it stimulated the vagus nerve, which initiates the release of food-absorption hormones such as insulin and glucagon.) Episodes of apnea, a brief cessation of breathing common in premature babies, were also greatly reduced in caressed babies. There were economic benefits, too; the massaged babies were fit to be discharged from the hospital a week earlier than other preemies, reducing hospitalization costs dramatically.

Findings on the importance of touch date back to the thirteenth century in the Holy Roman Empire. Frederick II was interested in learning what language children would develop if they were raised without hearing any spoken words. He took a number of newborn infants from their parents and gave them to nurses, who fed the babies but were forbidden to touch or talk to them. The babies never learned a language, and died before they could talk. Frederick had inadvertently discovered the role of touch in a baby's initial development.

This hypothesis was confirmed in the 1990s. Tragically, when Harvard Medical School researcher Mary Carlson visited the overcrowded orphanages of Romania, she found hundreds of swaddled babies in cribs who were never touched, not even during feeding time (the bottles were simply propped in the cribs). Some of the babies had lived that way for nearly two years. Carlson's studies at one orphanage showed that the babies were stunted, acted about half their age, and had abnormal levels of the stress-fighting hormone cortisol, compared with babies who had been raised in homes nearby. These results harshly highlight how critical touch is for development.

Receptiveness to being touched varies from culture to culture. Americans do not feel as comfortable touching one another as people in other cultures do, whether it be casual touching or affectionate caressing of children by parents. The latter point could be significant, because cross-cultural studies have demonstrated that societies in which parents show more physical affection toward their infants and children tend to have significantly lower rates of adult violence. On a lighter note, one study of adults in social settings such as cafés reported that casual touch, such one friend patting another's shoulder or hand, occurred about some two hundred times in a 30-minute period in France versus twice in 30 minutes in the United States.

THE CHALLENGES OF AUTISM

PERHAPS THE BEST WAY to understand the enormous benefit of touch is to consider the challenges faced by people with autism. Extreme difficulty in communicating emotionally and interacting socially are principal features of autism, a developmental disability that affects several areas of the brain, including the cerebellum, hippocampus, and limbic system, beginning at a very young age.

Autistic children often shun many kinds of physical contact. Part of the reason is that sensory information coming to them from the outside world comes too fast for their brains to process, and they are simply overwhelmed by the stimuli around them. A typical reaction is to shut down or attempt to get away from the stimuli. This is compounded by an inability to pay attention, because the sensory information that they receive comes in fragments or pieces. For example, a healthy baby can shift his attention from the eyes to the nose to the mouth of his mother within fractions of a second. But an autistic baby often takes five to six seconds just to process the nose. Because of the delay, the infant cannot take in a whole face at once, just parts. Missing a social cue such as a smile or a frown is easy. The result is that the baby receives only partial information about the world, and the information he does receive is often confusing.

Some autistic persons have normal sensory abilities but have difficulty sorting out the important information from the noise. They cannot prioritize the multitude of sensory signals pouring into the brain. To cope, autistic children respond by exhibiting behaviors whose ultimate goal is to shut off the massive and confusing sensory overload. They accomplish this by screaming, covering their ears, or running to a quiet place—anything to block the noise. An aversion to touch adds to their social isolation from the outside world.

This kind of behavior was a significant factor in diagnosing a patient brought to me years ago. She had been in and out of mental hospitals for twenty years, diagnosed with everything from manic depression and schizophrenia to antisocial personality.

Now forty, Delores had been born the middle child of five, and her problems had begun early in life. She was a quiet, reserved toddler, and was content to be alone amid the bustle of her siblings. But she began to cry every time her shirt was changed. Sometimes she would take off a shirt she hated and run around the house half-naked. When she was four or five she got into rip-roaring fights with her otherwise agreeable mother, insisting that she would wear only one particular shirt. Her mother refused. Delores screamed, complaining that the other shirts scratched or pricked her. Her mother got furious. Delores did too.

Delores developed an antagonistic relationship with her mother over what clothes she would wear. Her mother told her she was being obstinate and demanding and that such behavior wouldn't be tolerated. This led Delores down the path of being very contrary and aggressive toward her mother and then her siblings, because they all came to see her as a problem child. Other behavior supported the family's view. For example, Delores didn't like to go into malls or stores. She'd fight going in, get loose if her mother was distracted, and run out into the parking lot, despite the weather or whatever punishment might await her.

Intellectually, Delores seemed perfectly fine. In fact, she was a good student. By age nine she was writing fascinating poetry. Her teachers viewed her as gifted, and she maintained stellar grades. But Delores continued to feel completely at a loss for who she was. She was naturally shy, and her self-image had become bruised and battered owing to her family's constant redress. As a teenager, she saw herself as a complainer, because that's what her mother and siblings and even her friends told her she was. They said she was spoiled because she had to have everything her way. She also had a hard time feeling confident, that she was okay, when there were more than just a few people around her. She couldn't stand for anyone to hug her, either. Why? She couldn't be held, and therefore couldn't feel comforted. She felt unlovable.

By her late teens these feelings intensified, leading Delores to serious depression. In college she began a series of suicide attempts. After shocking herself with her own behavior, she went to a psychologist. She and her therapist quickly focused on her bitter relationship with her mother. They reconstructed the childhood incidents and ongoing anger into a complex, Freudian scenario of hatred, dominance, acquiescence, and battling personalities.

For the following two decades, Delores's various therapies got her nowhere. She had counseling, took medication, was admitted to mental hospitals for periods of time for observation and different treatments, none of which found the root cause of her problem and all of which further reinforced her own self-image as a deviant person.

Delores was first referred to me by a neurologist who wanted me to reevaluate her medication regimen and test whether she might suffer in part from ADD. Delores gave me her history, but said nothing about her early childhood fights with her mother over her clothing. To her they were only faintly remembered episodes in a lifetime of episodes.

As I interviewed other family members, however, one of Delores's sisters recounted the fights to me. The story tipped me off. I asked Delores about her sense of touch. She didn't seem to think she had a problem. But she did comment that many different types of clothes irritated her. They scratched her skin. She wore only a few different pieces of clothing, wore them over and over, and washed them with excessive amounts of fabric softener so she could stand them against her skin.

After some other tests it became apparent that Delores had an oversensitivity to touch. What might seem to the average person like a slight brush from a rough piece of clothing felt to Delores like the swift scratch of a cat's claw. She didn't perceive the sensation simply as discomfort—it was pain. That's why she had fought so violently with her mother when she was a little girl about not wearing certain shirts. That's why she tore them off when she could—if that didn't cause even more pain—and ran around the house so her mother wouldn't catch her and force her to put them back on.

It turned out Delores also had an overly sensitive sense of hearing. Loud noises made her panic. She hated to go to the mall because the ambient noise—all the conversations and footsteps and cash registers and paper bags—overwhelmed her. She had to run out into the parking lot to escape the sensory overload. She hated parties in college for the same reason, and never went to them so she wouldn't have to confront the problem.

After more investigation, I and several of her other doctors concluded that she was also very mildly autistic. The shame of Delores's case is that an undiagnosed perception problem led her in part to a life of psychological burden. Maybe in retrospect she wasn't obstinate, she wasn't a difficult child, and she wasn't spoiled. She didn't always have to have her way. What she had to have was clothing that didn't overstimulate her sense of touch, social situations that weren't overwhelming, and understanding. Her perception problem was a huge factor in her early childhood, which if understood could have sent her down a very different path in life. Being shy and gifted, she could have matured into a smart woman who was a bit of an odd duck. Instead, she was driven to try to self-destruct and spent much of her adult life in institutions.

Delores is hugely relieved to have learned about her touch problem. However, four decades of horrible experiences will not simply vanish. At age forty, it will be hard for her to stop believing that she isn't a problem person, to change her attitudes about the abusive nature of other people and about herself.

Had the problem been diagnosed when Delores was three, had her mother known her little girl was touch-sensitive, there would have been ways to prevent the fights. The bad attitudes on both sides would not have developed in the first place. In the 1950s, physicians simply didn't know about touch sensitivity. Today, a pediatrician might say, "Oh, she keeps taking her shirt off. Well, maybe she's overly sensitive to touch. Let's explore it." With the neurological evidence for this problem at hand, Delores's mother would have believed her daughter when she complained that a certain shirt scratched her.

Delores's case, like that of Rickie, shows how perception is much more than simply sensing stimuli from the outside world. It is an enormous factor in personality development. Even the smallest ongoing perception problem can lead to a cascade of events that can result in a psychologically traumatic life.

Apart from perception, another factor may be sensory integration, a crucial process of understanding the world as a whole unit. It means that a person can process information from more than one sense at a time. People with autism may have difficulty doing this because of differences in the thalamus and its cortical connections. In her book *Thinking in Pictures,* the gifted autistic woman Temple Grandin relates the words of a fellow autistic person who cannot process more than one sense at a time: "Donna Williams described herself as a monochannel; she cannot see and hear at the same time. When she is listening to somebody speak, visual input loses its meaning. She is unable to perceive a cat jumping on her lap while she is listening to a friend talk. She often handles telephone conversations more easily than face-to-face meetings, because distracting visual input is eliminated." Many autistic persons, like Temple, suffer from a milder form of this problem, one that we may all be a bit familiar with. She relates: "I still have problems with losing my train of thought when distracting noises occur. If a pager goes off while I am giving a lecture, it fully captures my attention, and I completely forget what I am talking about. It takes me several seconds to shift my attention back." The rest of us can shift our attention back more smoothly.

Temple says that since people with autism cannot process information quickly, parents and teachers can help them by altering the environment. For example, some autistic people will hear better if visual stimuli that cause sensory overload can be removed. Temple suggests placing them in a quiet, dimly lit room that is free of fluorescent lights and bright wall decorations. They should be spoken to slowly to accommodate a nervous system that processes information slowly. Sudden movements should also be avoided. Other autistic children will do better if auditory distortions are removed. Thus, it behooves the caregiver to learn which perceptual domains are "too sensitive" and to make the appropriate environmental adjustments.

An autistic person's aversion to touch stems from deficits in the parietal lobe, medulla, and thalamus, all of which are involved in the touch pathway. Today some researchers share a hypothesis that the normal process of cell pruning during prenatal development didn't work as it should in the autistic brain, leaving too many neurons to respond to perception, so that the brain is flooded with sensations. (Evidence to support this line of thinking is that the brains of autistic people are bigger and heavier than normal.) An inability to properly control stimuli coming through the brainstem could also be a factor. The result is oversensitive skin, which makes wearing clothes, except for the softest kind, incessantly difficult. Thanks to the wonderful insight of people like Temple, we are afforded a look inside the mind of an autistic person to experience the world as he or she sees it. To Temple, petticoats were like "sandpaper scraping away at raw nerve endings." Shampooing her scalp felt as if "the fingers rubbing my head had sewing thimbles on them." Getting accustomed to new clothes was exceedingly difficult; it would take Temple two weeks to adapt from shorts to long pants. Underwear, the item of clothing closest to the skin, was particularly hard to get used to. To Temple, "new underwear was a scratchy horror" because of the developmental mishaps of neuronal pruning.

Most of the sensory problems in autism involve vision, hearing, and touch oversensitivities, but taste and smell can sometimes be affected as well. In one study, 80 to 90 percent of autistic adults and children reported oversensitivity to touch, sound, and/or vision, and 30 percent reported an oversensitivity to taste or smell. It may be that since taste and smell are the brain's most primitive senses and there are fewer brain regions involved, there are fewer sources of confusion and noise, and it is easier to adapt to noxious smells and tastes.

Temple notes that autistic children experience eating difficulties that stem from sensory processing problems. Typically, the children are finicky and are often unable to tolerate the texture, smell, taste, or sound of food in their mouths. Temple herself hated eating anything that was "slimy," like Jell-O or egg whites. Others will only eat very bland foods like oatmeal. She discussed how some autistic children are highly sensitive to the smells of certain foods, and will avoid them at all costs. One boy could not play outside with other children because the smell of grass overwhelmed his olfactory system. Ironically, many autistic persons have a strong association of smell with memory. A number of them will remember people by their smell. Another study showed an individual who associated feeling safe with the smell of pots and pans, because he associated that smell with his home.

Stochastic resonance, which is background noise that we all have in our perception systems—random clutter, static, or hiss that is too low to be directly felt, heard, or seen—is playing an interesting role in helping some autistic people. Normally, this low level of noise can serve to enhance otherwise weak, undetectable tactile, auditory, and visual signals. This noise, as I mentioned earlier, actually primes the neurons in the brain, heightening their readiness to perceive a potential incoming stimulus. Once a stimulus is perceived, stochastic resonance can even enhance the detection of the weak signal.

In some autistic individuals, stochastic resonance has proved useful in combatting sensory overload. For example, if a child has difficulty hearing or attending to people who are speaking, whispering may enhance his ability to focus on spoken words. The concept is analogous to the psychological aspect of attention. You are more likely to hear what someone is about to say if you are already paying attention to them. If you are unaware that they are about to speak, you may miss the beginning of their statement and therefore some of the meaning. Stochastic resonance is also being used to counteract the sensory deprivation that often comes with old age, stroke, or neurological disease.

It is easy to understand how the many processing difficulties of autism can lead to social isolation. If one has an aversion to being touched by another human being, if clothes feel like steel wool, and if sensory information comes too fast and furiously for one to process, a perfectly natural reaction is to avoid the overwhelming stimuli in any way possible. Sadly, this social isolation, which begins in early childhood when the brain is developing, sets up behaviors that can last a lifetime.

PLASTICITY, PHANTOM LIMBS, AND PAIN

ON ITS BROADEST SCALE, touch gives us all the tactile, palpable information about our bodies. It processes information at three sensory levels: stimulation of the skin (the body's largest sensory organ, with about two square yards of receptive surface); the position of various body parts in relation to one another, as well as the position of the body in space and whether a limb is static or moving; and the status of internal body processes such as heart rate and blood pressure.

We use our fingertips and hands most frequently to identify objects through touch. There are about a hundred tactile receptors per square centimeter on the human fingertip—three to four times more than on the palm. The density of receptors on other parts of the body, such as the back, is far less still. The signals are sent up the spinal cord and to the medulla and then on to the thalamus, which processes them and sends them directly upward to the cortex. The somatosensory cortex contains multiple maps of various tactile functions throughout the body. One such map is a representation of every square inch of skin covering the body. Other maps include plots of limb positions and joint movements. Some areas of the skin, particularly those that need fine tactile discrimination, are disproportionately represented on the cortex. For instance, the area devoted to the thumb is as large as the area devoted to the entire forearm. The lip has more cortex devoted to it than the leg. The face, eyes, nose, jaw, teeth, gums, tongue, and hands all have very large areas, while the torso is not as finely represented.

Although the maps remain fairly stable, so that the brain and body work efficiently together, they can change, giving the brain the plasticity necessary for learning. Merzenich's experiments described in Chapter 1—which examined the tactile processing of monkeys' fingers—elegantly demonstrated that there is plasticity in the adult brain. Further experiments by Timothy Pons of the National Institute of Mental Health demonstrated that neural reorganization can occur on an even larger scale. Using monkeys subjected to a variety of brain surgeries, Pons showed that sensory nerves that once responded to an arm later responded to touching the face. The neural reorganization spanned more than a centimeter, stretching more than a third of the way across the entire somatosensory map. The changes in Merzenich's experiments occurred with neurons spaced less than 2 millimeters apart.

Vilayanur Ramachandran, a neuroscientist at the University of California at San Diego, proposed that cortical reorganization occurs because two sets of signals travel along the nerve pathways, one a strong, dominant signal and the other a weak signal. Normally, the weak signal is inhibited by the dominant one. But if the nerves that ordinarily carry the dominant signal are severed, the weaker input is allowed to thrive. This is often referred to as the "filling-in" phenomenon.

Taking this a step further, Merzenich believes that this type of plasticity occurs on a daily basis, that neuronal connections are continually strengthened or weakened—a minute bit at a time—as sensory input changes owing to external stimuli. This allows certain neural connections to dominate at certain times, but as conditions change, others better suited to the changing task take over. This, he says, is how the somatosensory cortex learns—and how our brains adapt in ways that improve performance.

It may be that adjustments in the filling-in phenomenon account for some of the sensations in phantom limb syndrome, the pervasive and mysterious condition that has plagued amputees and baffled physicians for close to a century. The typical sensations reported by patients include pressure, warmth, cold, wetness, itchiness, sweatiness, and even ticklishness, but a considerable 70 percent of all amputees suffer from phantom limb pain. This often agonizing condition includes burning, crushing, cramping, or shooting pain, and can range from mild and occasional to continuous and severe. The pain often begins shortly after the amputation, and it can last for years. For the sufferer, the pain is a constant companion and a never-ceasing reminder of the absent limb.

For many years, phantom limb patients were farmed out to psychologists, who told them that they were practicing a form of "wish fulfillment." We now know that the underlying causes of phantom limb syndrome stem from the thalamus and the somatosensory cortex. During the reign of the hard-wired brain theory, medical science was often at a loss to explain the causes of phantom limb syndrome; after all, if the limb was missing, then the brain circuits for it would not receive input and would therefore shut off.

However, there is now evidence that when a limb is amputated, the area of the brain that was devoted to receiving sensory signals from that limb still activates. Ramachandran discovered that when pressure was applied to different (normal) parts of the body, a patient could feel it in precise locations of the phantom limb. Ramachandran concluded that the cortex remapped the areas of the brain once devoted to the missing appendage to a new area of skin. He asserts that these remapped pathways are not brand-new, but that the cortex utilizes weaker neuronal pathways that are allowed to come to the forefront because they are no longer being inhibited. The underlying cause of phantom limb pain could be these weaker pathways, which emit signals that are subsequently misinterpreted by the brain as originating from the missing appendage. The weaker pathways adapt themselves to respond to the area of the body that the dominant signals once responded to. Even though there is no limb to receive the signals, the sensations these pathways send still feel real.

Phantom limb syndrome is a primary focus of neurological research. Neurologists continue to investigate the nature of the erroneous signals in the hope of someday being able to provide relief for those who suffer.

The troubles of phantom limb syndrome belong to a larger category of tactile sensations: pain. The perception of pain varies greatly from one individual to another because it is influenced by a number of psychological factors, including culture (not everyone can walk on hot coals), previous life experiences, and current mental status.

Pain serves as a danger signal to warn us that a part of the body is being injured. It also helps the healing process by reminding us to take precautions in how we treat an injured area. Although feeling pain is unpleasant, lacking the ability to feel pain is a danger in itself. Clinical studies have shown that people who have a congenital insensitivity to pain often incur severe burns and cuts because their bodies do not warn them of danger. In extreme circumstances people have died owing to serious infections or a ruptured appendix because they did not feel the danger signals of pain.

The primary receptor for pain is called the nociceptor, which responds to stimulation such as intense pressure, extreme temperature, burning sensations, and more. The electrochemical signals for pain travel from the nociceptors to the spinal cord and ascend through the brainstem to the thalamus. Soon thereafter, neuronal signals project to the cortex in the parietal lobe.

Other interesting processes are at work in the pain mechanism. For example, when you hit your knee on a desk, you probably rub it to make it feel better. Why does rubbing a spot that has just been injured reduce the pain instead of aggravating it? Because the act of rubbing sends a second set of tactile signals to the brain. As the brain is finite and will have to pay attention to both signals at once, the second stimulus leads to a reduction in the perceived severity of the more intense first one. This concept is called competitive inhibition. When your mother rubbed your elbow after you fell off your bike, she was applying a scientific principle to the reduction of pain. The perception of pain also decreases when normal tactile stimuli such as rubbing, massage, or gentle vibration activate fibers in the nociceptor that send inhibitory signals to the brain. The act of rubbing your knee also produces an effect that releases morphinelike opiates. The opiates bind with and excite opiate receptors in the amygdala and the hypothalamus, resulting in signals to the medulla that feed back to the spinal cord, countering incoming signals from the nociceptors and diminishing the transmittal of pain information to the brain.

Just the same, the pain signals continue to the amygdala, home to the body's warning system against potentially harmful or life-threatening situations. If pain is going to be a threat, the amygdala, which is responsible for fear, startle, and autonomic reactions, will respond with the signal to fight or flee. This allows us to react quickly to a harmful or dangerous situation involving pain. After the relevant pain information has been processed in the amygdala, it is sent up to the frontal cortex for higher-order processing and our bodies' response.

Pain signals are sent to the brain when the nociceptors are activated. Scientists have recently discovered that this activation is triggered when a cell is damaged and releases a chemical called adenosine triphosphate (ATP). The ATP molecules bind with the nociceptors, and the alarm signals begin. The exciting aspect of this discovery is that it may lead to methods of treating pain, particularly debilitating and chronic pain. Research on the release of special neurotransmitters is also providing clues to methods for blocking the signals of persistent pain.

SOUND

DO YOU HEAR your jaws moving when you are speaking? Do you hear your heart beating while you are reading this? Certainly not. Just like our other senses, the sense of hearing groups together stimuli in terms of features—larger, more manageable units—which then determine the way we hear. In our discussion of touch, we found that we could not experience life without the sense of order brought about by features, because the noise in our external environment would overwhelm us. In similar fashion, our ears pick the features—the primary sounds we are hearing as well as what we expect to experience—and screen out the ambient background noise.

Much of this processing by the auditory system—much more than for the other senses—is accomplished long before we are conscious of it. In the way stations along the path from the ear to the point when we become aware of them, the sound signals are adjusted and refined. That is why we are not consciously aware of our jaws moving and our hearts beating. If we heard these sounds and other background noises, it would be quite annoying, upsetting, and distracting.

People who suffer from tinnitus, a condition that leads to persistent ringing, buzzing, or humming sounds in the ears, have this problem. An estimated 40 million Americans are distracted with noise that ranges from mildly irritating to incapacitating, which can cause sleepless nights, exhausting days, and frustration that can border on madness. They are not able to suppress the sounds in their surroundings that are not crucial. Background noise becomes the feature on which they focus, and it can be as loud as a roaring train.

There is no cure yet, but some people suffering from tinnitus find comfort in listening to tapes of soothing sounds, such as waves at the seashore, which allow them to focus on something other than the sounds in their ears. Silence is the worst thing they can be subjected to, says Stephen Nagler of Atlanta's Southeastern Comprehensive Tinnitus Clinic, which he started in 1994 after he was forced to leave his general surgery practice at Northside Hospital because of the distraction of his own tinnitus in the operating room.

Recently, doctors have begun implanting a small device in the ear that electrically stimulates the cochlea. This stimulation prevented tinnitus in 23 percent of test patients, and reduced it in 46 percent more. One such device may have literally saved the life of William Shatner, the famed Captain Kirk of *Star Trek.* The constant sound in his left ear, a "shhhhh-like drone," became so bad it nearly drove him mad. In September 1997 he admitted on the NBC show *Today* that he "began to actively think of what means you could use that . . . could end your life." Fortunately, Shatner battled his desperation by seeking high and low for a coping mechanism. He landed at the University of Maryland Medical Center, where he was fitted with a device like a hearing aid that fed so-called white noise into his left ear. The point was to try to make the "shhhhhh" sound so much a part of his subconscious that he was no longer aware of it, just as the humming of refrigerators and computers dissolves into the background noise.

It's not clear what causes tinnitus. Chronic exposure to loud noise may be a culprit; many rock musicians suffer from it. Sudden noise trauma also might cause it; a person standing too close to an explosion may not immediately sense hearing damage, but may wake up days later with a constant ringing in his head. Shatner believes such a scenario led to his tinnitus, and to a similar problem for Leonard Nimoy, who played Mr. Spock. They were standing near an explosive device on a set of *Star Trek* when it suddenly went off.

Unlike Shatner's device, when hearing aids were first introduced, they simply amplified the sound waves reaching the wearer's ear. This made signals perceived as weak stronger, but also increased the noise. Recent devices are much improved, but wearers sometimes complain that they can't distinguish foreground noise from background, throwing off the balance they learned to expect all their lives and making it difficult to distinguish one conversation from another. Altering the signal-to-noise ratio is helping the hearing-impaired deal with multiple conversations. People with normal hearing sort these out through "selective listening"—the brain focuses on one conversation while ignoring the others by selecting which input to listen to and which to classify as noise. New hearing aids with full-dynamic-range compression and frequency-response shaping can improve the balance of signal and noise, helping the wearer's ability to listen selectively.

Normally, in all perception, each new experience builds on previous features, and we develop categories of features through which we compare and contrast new experiences. This process is continually being elucidated for me by my dyslexic wife. Dyslexics experience fewer of these features than most people. Although most dyslexics typically score in the normal range on standard hearing and visual tests, some hear and see things that others do not. Furthermore, my wife mishears a lot of speech that is understandable and clear to most of us. She becomes frustrated by large-group conversation. On the other hand, she loves one-to-one conversation, telephone chats, and small groups in which she does not get lost in the noise. Ironically, the confusion of dyslexics when confronted with a plethora of sights and sounds might account for their ability to see the novel in the ordinary; because they tend to focus on all the stimuli, they are less apt to ignore a perception that is mildly unexpected, while the brains in the rest of us routinely lump it with "the noise" and disregard it.

Other tests have shown that some dyslexics have a superior talent for hearing slower sounds, such as vowels and bass tones, and see shapes and colors more sharply in their peripheral vision, which may explain why many dyslexics become poets, musicians, and painters.

Some dyslexics' processing problems most likely originate from a difference in a kind of cell, the magnocellular neuron, in an early way station in the thalamus for input entering the brain. They may also have structural brain abnormalities called ectopia—nerve-cell bundles on the surface of the cortex that may contribute to the brain's inability to process input from fast sounds.

Individuals with autism also often have a unique relationship with hearing. Because their perceptual systems become easily hyperaroused, or oversensitive, and because they process incoming stimuli slowly, they receive disordered, inaccurate information about their environment. Many autistic persons never learn to speak because they cannot distinguish differences between speech sounds. Further, even though they perform normally on standard hearing tests, many autistic children seem to be somewhat deaf because they shut out noises that are disturbing to them; they are unable to modulate incoming sound, and consequently minor sounds that most people ignore can be nearly impossible to tolerate. Noises can also completely distract them from everything else that is taking place around them. Temple Grandin says that her hearing is "like having a hearing aid with the volume stuck on 'super loud' . . . like an open microphone that picks up everything." She says sudden loud noises hurt her ears with the intensity of a dentist's drill hitting a nerve.

As we have noted, in perception the brain breaks down sensory information into the smallest of elemental units, tiny parts that are seemingly unrelated. The brain then distributes these bits of information and somehow reassembles them, according to a person's memories, past experiences, and possibly even wishes. In hearing, the pressure of sound vibrations impinging on the eardrum is transduced or converted into energy that moves the ossicles, the three tiny bones located in the middle ear. They stimulate the cochlea, a spiral-shaped, seashell-like cavity, causing a sort of rumbling within it. The cochlea contains on the order of 15,000 cilia, fine hairs that bend in one direction or another in response to vibrations. Particular hair cells are sensitive to particular frequencies of sound at particular levels of loudness.

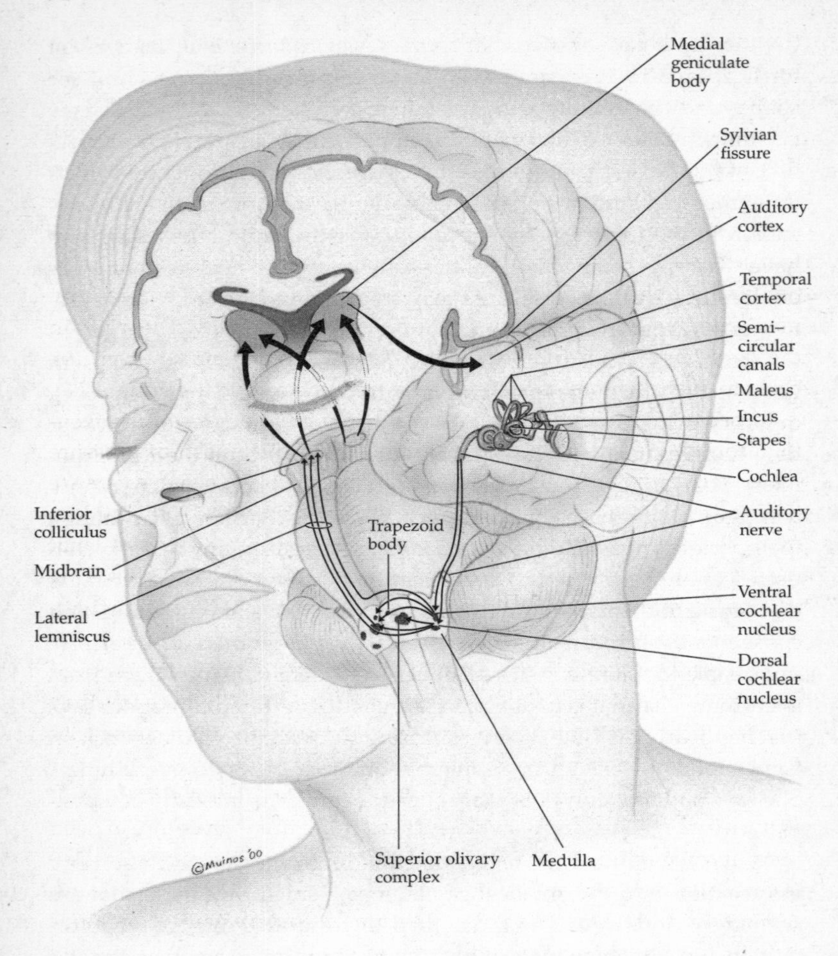

THE AUDITORY SYSTEM Hearing, a complicated affair, begins when our eardrum responds to air-pressure changes caused by incoming sound waves. These are converted into electrochemical information through the workings of the inner ear and sent to the brainstem, where the trapezoid body and the superior olivary complex process the information and position the head and adjust pressure in the inner ear to optimize picking up sounds. The pathway to the top of the brain begins with the inferior colliculus and then goes on to the medial geniculate body of the thalamus. From there the information is sent to the auditory cortex of the temporal lobe for further processing.

(Loud sounds can cause hair-cell loss, which is permanent and leads to hearing loss.) The motion of the cilia is converted into electrical signals that fire neurons along pathways to the brain.

However, as with information from the other senses, we transform this auditory information into meaning along the way. The process of hearing occurs even earlier than with the other senses, right at the ear itself. Our brains immediately begin to adjust the pressure in the inner ear and to make other changes in the ear and head so as to maximize our ability to hear what it is we want or need to hear in our environment. Evidence that our brains continually shape what we hear lies in the fact that there are more neuronal networks extending from the brain to the ears than there are coming from the ears to the brain.

To perceive sound, our brains engage in a great degree of monkeying around. We receive sound as an incoming mishmash of pressure waves—not just one vibration but layer upon layer. That we make sense of this cacophony is the miracle of hearing. Part of our ability to make sense of it all is due to the fact that we develop models of what we expect to hear: phonemes, words, music. As we perceive sound, it either fulfills our expected models or surprises us. Dyslexics with auditory processing problems are continually being surprised, because nothing they hear seems to fit the models. They must guess or intuit a lot more than most of us about what they hear. Eventually, they come to hear predominantly what they expect to hear, excluding or reshaping what doesn't make sense to them.

Like all of us, some dyslexics learn to tolerate this inability to fit auditory information into models by becoming more inventive in deciphering the sounds around them, perhaps by asking probing questions all the time that help make sense of what is being said and heard.

Once the ear processes incoming sound, the auditory information is sent to the brainstem through the auditory nerve, which has only about 25,000 nerve fibers—very few compared with the billions of neurons involved in touch or vision. Consequently, information must be assessed in the ear before it is sent to the cortex. So these fibers are more efficient, the cells and connections involved are active, on ready alert, even in the absence of incoming sound; they prime themselves to act, just like soccer players running along the sidelines getting ready to come into the game.

In the brainstem, sounds are sorted according to tone and into units defined by timbre, or the quality of the sound. Sounds are also sharpened by deletion of the legions of echoes surrounding us—such as the sounds of voices echoing off the walls, ceiling, and floor—before we are even aware of them. Cleverly, the brainstem preserves the sound from which the echoes originate, and does not delete new or strange sounds. It also begins the process of comprehending speech by identifying specific sets of sounds as phonemes, the units of speech that form a person's native language, even though they do not carry meaning. Consequently, a Chinese man listening to English may not "hear" particular phonemes because his brain has not learned to recognize them as units of language.

Information sent to the medulla is also analyzed for spatial characteristics. The ability to detect the spatial location of sounds is a profound evolutionary advancement. Animals that were able to "localize" sound—to fix in space where a sound originates from—had a crucial advantage. Today this ability is not as necessary for human survival—although we are still quite good at it—because we have evolved in our cortices, further upstream in the auditory pathway, an ability to identify sounds rather than simply localize them. In our cortices, we also have detailed maps of our surroundings to guide us, unlike more primitive animals that must move their heads to focus on sounds. Vestiges of our history remain, however; we still possess small muscles to move the pinna, the external ear, although we don't use them much.

Neurons in the brainstem called the superior olivary nuclei also adjust to the fact that sounds should be louder in the ear nearer to the sound. These two pairs of nuclei send messages to the midbrain, which then coordinates the body's reflexes and reactions to this input. At the superior colliculus in the midbrain, input from the ears, eyes, and skin (touch) begin to come together, almost reflexively directing us to orient toward the specific stimuli. The superior colliculus is crucial for integrating sensory information from the visual, auditory, and somatosensory systems—for creating the unified, coherent surroundings that we experience.

From the superior colliculus, the auditory neural impulses run up to the thalamus and then on to the primary auditory cortex, which links with the secondary auditory cortex, the structure that has connections to other parts of the brain needed to coordinate hearing with memories, the other senses, and awareness.

In the medial geniculate bodies of the thalamus, signal processing is divided between two types of nerve cells, the parvocellular and the magnocellular neurons, the latter being responsible for the transfer of rapidly incoming sounds into the auditory cortex. People who lack these cells may have trouble making fast discriminations and could be dyslexic. Their auditory processing difficulties include problems with discerning "fast" phonemes, such as certain consonant combinations.

The lower levels of the perceptual system are in a state of constant response; they are focused on the demands of survival, unable to discriminate between the novel and the habituated. But the cortex contains columns of neurons that process different pieces of sensory information and bind them all together somehow into seamless, integrated perceptions. The cells in these columns are very sensitive to specific differences in sound frequencies, and changes in frequencies cause different columns to fire. In order to come up with the tremendous range we hear in sounds, columns fire together in deliberate mixes. The cortex then compares the patterns of firing to stored patterns, or features, with which it is already familiar.

The auditory cortex does not function in isolation. The senses interact to create a world that, well, "makes sense" to us. Researchers have found that linguistic visual cues, such as the shape of the lips, activate the auditory cortex, while facial movements that are not identifiable as speech do not. The activation of the auditory cortex during lip-reading suggests that visual signs impact on the perception of heard speech even before the sounds themselves are processed into phonemes.

A recent study at Brandeis University even found that we sometimes "see" with our ears, not just our eyes. In order to test the connection between vision and hearing, researchers assessed the impact of sound on the perceived movements of objects on a computer screen. When a sharp "click" sounded as two round images on a computer screen moved toward each other and then away, people thought that the objects had collided. When no sound was played, far fewer people thought that the objects had collided. We tend to think that when we play Ping Pong or baseball, we react based on where the ball looks as if it's headed, but the sound of the ball striking off the opponent's paddle or bat is an important clue to its trajectory. Subjects who have tried to play these games while wearing gear that blocks their ears perform less well.

Most of us use cues to combine vision with sound to help us process what we are hearing. One woman who was being interviewed by a doctor while being admitted to a hospital told him, "Wait a minute, I can't understand you. I need my glasses." She was perceived with a bit more suspicion as to the intactness of her mental faculties. Yet she needed to see his facial expression to cue her in to what he was saying. Some people who are great talkers in person hate to talk on the telephone, in part because they can't see the face of the person they are talking to. The reverse phenomenon is also common, evidenced by a person who is driving a car and demands that the radio be turned down when he's negotiating a busy road; he is trying to hear audio cues that help him see what's going on. I do this all the time.

Both sides of the brain work together in the tremendous project of discriminating complex sounds. The right hemisphere is more concerned with relationships between simultaneous sounds, such as harmonies, and relationships between close sounds. The left hemisphere contains the "language centers," which are concerned with the ability to use and understand language. The auditory cortex reports to the language center of the left hemisphere. But interestingly, scientists have discovered that the ability to hear sounds is not crucial for the development of the language area. MRI studies with deaf persons who use American Sign Language have shown that watching ASL activates neurons in the left-hemisphere regions that have classically been associated with language processing (as well as stimulating the right side of the brain, which handles spatial and visual skills).

Other recent research has found that the processing of language in the brain is much more complicated than traditionally recognized. For instance, Paula Tallal at Rutgers University has studied children with language-based learning impairments (LLIs). She has found that impairments in the left hemisphere—which is a faster processor than the slower, more ponderous right hemisphere—lead to difficulties in identifying specific, "fast" consonant combinations, such as "br" and "pr." Using computers to exercise the hearing circuits in LLI children, Tallal is finding that their ability to discriminate these speech sounds can be improved by retraining the neural networks—by presenting stimuli in slower, extended forms and gradually speeding them up. We will investigate her techniques more closely in Chapter 7, "Language."

Tallal has also found that fast processing of speech takes place in Broca's area of the left hemisphere, rather than in Wernicke's area, behind the ear, which is typically assumed to be the auditory region of the brain. Since Broca's area is usually thought of as the controller of the motor cortex (controlling the tongue and voice box), and not the receptor area, these results suggest that speech has a great deal to do with the movement regions of the brain.

The interdependence of both hemispheres is particularly evident in the processing of music. The left side is better at targeting the succession of sounds—the rhythm. Particularly fascinating are PET studies by Henri Platel of the University of Caen, France, of non-musically-trained men who listened both to brief excerpts of well-known pieces of classical music and to random sequences of musical notes. He found that Broca's area was activated when subjects listened to well-known pieces of music. Platel suggests that this area may involve the recognition of all familiar sounds, not just speech. Changes in the rhythm of the unfamiliar pieces also activated this area. The only musical quality that predominantly activated the right hemisphere was the quality of the sound itself—the timbre. Platel concluded that music appreciation is amazingly complicated, involving memory, the recognition of sequences of musical components, and the coordination of specialized regions of the brain.

Furthermore, music appreciation, just like the overall process of perceiving sound, is influenced by a person's experiences. Our memories or symbolic factors concerning different pieces of music in the same key, for instance, can affect how music sounds to us. No piece of music—or sound—is inherently "happy" or "sad." Yet a melody backed by major chords may sound pleasantly bright; the same melody played in a minor key can sound dark. We add emotional content to what we hear, further evidence of the sophisticated nature of perception and its many levels of processing.

The process of hearing underscores a phenomenon that recurs throughout the sensory system: the components of the system overlap. The incoming sensory information is sliced apart and repackaged, then recombined to form a final "perception."

VISION

I MET A PSYCHOTHERAPIST from the West Coast named Rolf at a conference in Aspen, Colorado. It was autumn, cool and overcast, yet Rolf was wearing yellow-tinted sunglasses. I just thought, Oh, it's the California thing. But Rolf, age sixty-eight, had discovered only two years earlier that he had a visual-processing problem. He had begun to work with dyslexics when he retired from active practice, and in studying all he could, he learned about a technique called the Irlen method for helping a small subset of dyslexics.

Certain dyslexics have difficulty reading because as they move their eyes from left to right across a line of type, the letters seem to shimmer—they move. The affected individual can't keep track of the words, and so has to struggle mightily to read. The Irlen idea was that if such a person looked at written material—or any fine details—through a certain type of filtering lens, the shimmering would stop.

Rolf had been tormented all his life with the idea that he was not as smart as he thought he was. It had taken him much longer than other students to study. He was smart enough to get by, and got his medical degree by forcing himself to listen well and ask lots of questions. Indeed, his first love was neurology, but that required much more detailed reading than psychology, which relied more on talking and listening, so he ended up becoming a psychiatrist. He had always loved literature, but just never read it because it was too much of an ordeal.

Upon discovering that different-colored Irlen lenses helped certain dyslexics, Rolf drove to his neighborhood pharmacy, picked up a magazine, and began trying on different-colored sunglasses. He tried blue, then brown. Nothing happened. But then he put on a $5 pair of yellow-tinted lenses, and began to read the magazine. The words stood still! He read it more easily than anything he had ever tried to read before in his life. He was elated.

Rolf was already wearing glasses for common farsightedness. He hurried to his ophthalmologist to explain his discovery, and together they ordered a pair of Irlen lenses. Today Rolf is a voracious reader.

It's important to note that Irlen lenses help only a small fraction of people who suffer from dyslexia, which, as we will see in Chapter 7, can be caused by many different perceptual or brain-processing problems. The shimmering of letters is not a problem that can be diagnosed with routine eye exams. Rolf happens to be in the small group of dyslexics who can be helped by Irlen lenses, was aware enough to apply what he was learning about dyslexia to himself, and was clever enough to find some ready evidence for a possible cure at his local pharmacy.

Once again, however, as was the case with Rickie and Delores, Rolf needlessly spent much of his adult life with a poor image of himself. Despite his outwardly successful career, he had been in analysis for years trying to understand why he thought of himself as inadequate and lazy—why he had to study so hard to achieve what others did routinely, why he didn't read the journals as his fellow psychiatrists did, or keep track of the news in the papers. His struggle had nothing to do with an intellectual deficit or a motivational problem. It was pure perception.

Just for a moment, look up from this book and examine the scene around you. Be it a sterile office, cozy bedroom, or beautiful park, allow yourself to sit back and really "see" the world that surrounds you. In the amount of time that you averted your gaze from this page, your eyes meticulously dissected the image cast upon your retina into approximately 126 million pieces, sent signals for every one of these tiny elements to a way station in the thalamus, which then fired neuronal networks to and within the visual cortex, then sent the information to the frontal cortex, and somehow you put the pieces back together into a seamless pattern perceived by you as a sterile office, cozy bedroom, or beautiful park.

To add to this complexity, recent physiological findings suggest that all this processing takes place along several independent, parallel pathways. One system processes information about shape, one about color, and one about movement, location, and spatial organization. If you look up and see a clock, the image of its face and the action of its sweeping second hand are being processed independently, despite how unified the image appears. It may seem bizarre to think of vision as functionally subdivided. But how otherwise could a person who has perfect focus and tracking of moving objects be color-blind? Some "blind" people who cannot see colors or objects can still see movement.

As humans, our highly convoluted cortex enables us to combine visual messages with other sensory messages and past experiences to give unique meaning to particular visual situations. The sight of a fresh bouquet of red roses will probably have a different effect on me than on the florist who works with roses every day. Most other species do not have cortical convolutions, so the greater part of their visual processing occurs as pure sight. Humans have evolved to process most visual information "upstream" in the visual cortex.

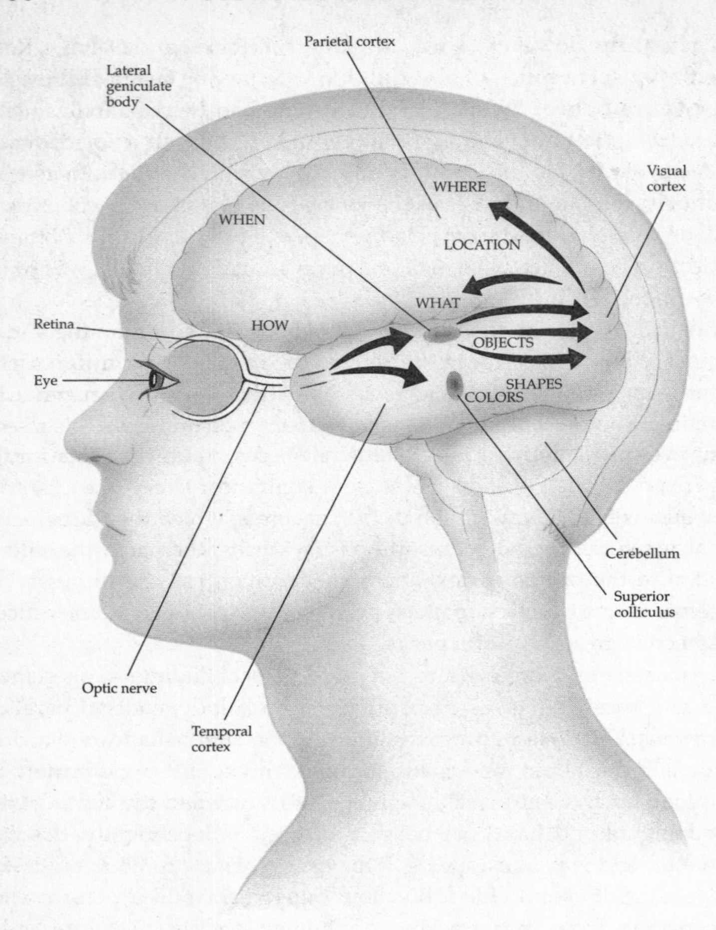

THE VISION SYSTEM Vision begins when light rays hit the eye. There the input is transformed into electrochemical energy and sent to the brain via the optic nerve. The first stop is the lateral geniculate body of the thalamus, with a small segment that goes to the superior colliculus, helping adjust the head and the eye to maximize information input. From the lateral geniculate the information is sent to the occipital or visual cortex and then dispersed to regions close by to be processed. The information mainly goes along two distinct routes: the HOW pathway of the parietal lobe and the WHAT pathway of the temporal lobe. Vision, like most brain functions, is distributed—sent to various brain regions to be processed—and that information is somehow bound together again, and we SEE.

Imagine that you are staring at a colorful painting hanging on a white wall. The entire wall is well within your field of vision, but your gaze is inevitably drawn to the painting. The white wall seems to "fade out" of significance while the painting seems to "fade in." Why do you have a visual preference for the painting and not the wall? The answer is what psychologists call salience: the brain pays particular attention only to specific images in its visual field.

Learning how salience occurs tells us a great deal about just how much executive control our brains have over what we ultimately see. Our visual experience begins when light bounces off the painting and the wall and passes through the lenses of our eyes to a thin sheet of highly specialized neural tissue in the back of each eye known as the retina. Photoreceptors there—the famous rods and cones—sense the light's wavelength and intensity. They convert this raw data into neural impulses, a language the brain understands. The retina is very selective and allows less than 10 percent of the light entering the eye to pass through to the photoreceptor cells. Otherwise, our brains would be overwhelmed with too much light.

The human retina contains approximately 120 million rods and 6 million cones. The cones are responsible for color vision and for most of our visual perception in normal and bright light. Different cones are more receptive to long-wavelength (red), medium-wavelength (yellow), or short-wavelength (blue) light, the primary colors from which other colors are made. We see an object as having a given color because that signal from the retina is the strongest; the wavelengths coming in from a red object synchronize the light reception of the red cones and reinforce their firing, resulting in a stronger signal to the brain that says "red." The brain compares this strong signal to the weaker ones for yellow and blue and concludes that the object is red.

The rods have poor visual acuity, but are much more sensitive to low levels of light and provide the best part of our vision when it is dark. Rods were of significant benefit to our ancestors, who hunted prey after sundown. Most animals, on the other hand, are essentially colorblind. They are sensitive to some colors, but not the full range—they have poor cone vision. However, they have far superior rod (night) vision than humans. Rods depend to a great extent on a photopigment known as rhodopsin, which consists of vitamin A, the nutrient found in spinach, tomatoes, fruits, and of course carrots (which explains Mom's claim that eating carrots will help you see better).

Our rods and cones see the painting simply as a field of independent dots of light, dark, and color. (The poor ability, or complete lack of ability, of some specific types of cone to distinguish colors is what causes partial or total color blindness.) The resulting neural signals created by the rods and cones leave the back of the eye through the optic nerve en route to processing centers in the brain. There, a tangled mass of neuronal connections fine-tunes the overwhelming amount of visual information that the retina passes on from the environment.

Our eyes take a snapshot of the outside world and chop the image up into millions of tiny bits of information segregated according to the color and the direction of the light, all without the help of the brain. What occurs next is a fascinating aspect of human vision that a number of laboratories have devoted copious time and energy to trying to understand.

There are two major pathways from the eyes through the brainstem to the cortex: the geniculostriate pathway and the more overlooked tectopulvinar pathway. The latter is believed to be responsible for orienting the eyes toward a particular stimulus. It provides the mechanism by which we shift our eyes—for drawing our attention to the vibrant painting instead of the dull wall.

Experiments involving a phenomenon known as "blindsight" support the theory that the tectopulvinar pathway plays a crucial role in visual processing. Patients in these studies had suffered damage to the primary visual cortex as well as to regions of the geniculostriate pathway, resulting in blindness in one half of their visual field. Interestingly, when they were shown a pattern in their blind field, their eyes shifted toward the image despite not being able to see the pattern itself. This means that the tectopulvinar pathway was able to shift the patients' attentional gaze to novel stimuli despite damage to the geniculostriate pathway. They weren't able to see, yet they saw. They were not conscious of what they saw.

What does this have to do with you and me? The answer is: attentional specificity. Imagine looking for a pen on a cluttered desk. Your eyes are being bombarded with visual information—piles of paper, disks, books. Still, you find the pen amid this visual clutter because your tectopulvinar pathway ignores everything except for what it craves: the features of a pen that are novel compared with the features of paper, disks, or books, such as its cylindrical shape, point, and color.

Once the tectopulvinar pathway senses the novel object, the geniculostriate pathway allows us to actually see it.

The geniculostriate pathway ultimately begins in the lateral geniculate bodies of the thalamus. It is this pathway, researchers now believe, that contributes to the reading problems of some dyslexics. Like Rolf, many dyslexics say the words on a given page tend to "shimmer" or "jump around," making the task of reading frustrating and laborious. Several perceptual studies have confirmed that dyslexic individuals process visual information more slowly than other people. For instance, they have problems distinguishing the order of two rapidly flashed visual stimuli, but can see the same stimuli perfectly when they are presented slowly. Margaret Livingstone and Al Galaburda of Harvard Medical School found related anatomical evidence when they examined sections of the lateral geniculate bodies in autopsies of five dyslexic and five nondyslexic subjects. Neuronal networks in the lateral geniculate bodies were more disorganized in the dyslexic brains, and the neuron cell bodies appeared smaller. There were fewer magnocellular cells that bring in rapidly changing information. According to one theory, this would inhibit the brain's ability to clear out one image before the next one comes in, so that the images seem to overlap, or to come and go without clean breaks between them, which might explain how words could appear to shimmer or jump around. Recent research by Guinevere F. Eden of the National Institute of Mental Health provides additional evidence. She used MRI scans to compare brain activity in the motion-detection region of the visual cortex of eight nondyslexics and six dyslexics, and found that the output of this area was much less active in the dyslexics, meaning the region was doing less to control the perception of the movement of images.

There is another step in visual processing that may also contribute to dyslexia. It is best explained through a personal story told to me by one of my researchers. Her words follow:

> As a teenager, I dreaded the last Thursday of every month. It was Book Club night, which meant Mom would whip together a quick nutritious meal while frantically skimming the chapters of her required book in search of a simplified theme. What good is being in a book club, I remember wondering, if you don't read the books? But despite her hectic schedule, my mother never missed

her monthly meeting—and in the ten years she has been a member, she's never finished a book. Not until recently did I discover why.

While gathering information, I stumbled upon a detailed explanation of the anatomy of the visual system, which maintained that the human visual system can be divided into independent parallel pathways with remarkably different functions. Imagine that you are standing at a busy intersection watching cars drive by. Your eyes send information to two peanut-size clusters deep in the brain known as the lateral geniculate bodies of the thalamus. This structure is segregated into parvocellular (parvo) and magnocellular (magno) divisions [as noted in the hearing discussion earlier]. The parvo, or slow-processing system, appears to process information about color, while the magno, or fast-processing system, appears to process information about movement, location, and spatial organization. The parvo system "sees" what color a fast-moving car is, and the magno system "sees" how fast a car is moving. It is then the responsibility of the cortex, working with the cerebellum, to piece this information together and give us the seamless perception of a red car whizzing past us.

I then read an article by Margaret Livingstone and her research involving developmental dyslexia, the selective impairment of reading skills despite normal intelligence, motivation, instruction, and sensory acuity. Livingstone examined sections of the lateral geniculate bodies of autopsied dyslexics and nondyslexics. While the parvo layers appeared similar, the magno layers were smaller and more disorganized in the dyslexic brains. Livingstone hypothesized that this magno "chaos" could cause the difficulty dyslexics have in focusing on words on a printed page. MRI scans by Guinevere F. Eden supported Livingstone's conclusions.

One evening, while this research was still fresh in my mind, my mother called on the phone. Her voice was anxious. She asked if I had possibly read *The Accidental Tourist,* and if so, could I quickly summarize it for her? It must be the last Thursday of the month, I thought. When I told her that, no, unfortunately, I hadn't read *The Accidental Tourist,* she began describing how her exhaustive efforts to complete this book (like all the others) were utterly fruitless. As I listened to her frustration, certain words grabbed my attention. "It's so hard to focus on the pages," she said. "The words seem to jump all over the place. It takes so much longer for me to finish a book than anyone else." No, it couldn't be, I thought to myself. My mother couldn't possibly be dyslexic. She had graduated at the top of her class, she's a perfectionist, and she absolutely loves to learn. How could she of all people be dyslexic?

Unfortunately, in today's society, reading ability is taken as a proxy for intelligence and most people (not excluding myself) assume that if someone is smart, motivated, and educated they will have no problem learning to read. This is simply not the case. Consider some of society's more famous dyslexic individuals, including Thomas Edison and Albert Einstein. Dyslexia clearly does not equate with a lack of intelligence. While it is true that some children fail to learn to read because they are poorly taught or lack the motivation to study, some 5 to 10 percent of children have difficulty learning to read despite the best possible circumstances. As Livingstone and Eden suggested, these individuals may simply have biological deficits in their magno, fast-processing centers.

In fact, the more I thought about my mother's life, the more her personality seemed to be shaped around a magnocellular deficit. My mother hates crowds. More accurately, she hates being in unfamiliar places where there is "too much going on"—cities, flea markets, amusement parks, malls. When she is in control of her environment, for instance at a mall she knows well, she is assertive and headstrong with an agenda in mind. In an unfamiliar mall, however, it is as if she regresses in age, becoming timid and almost scared. I can remember going with my family as a child to the Statue of Liberty Centennial Celebration in New York City. There were thousands of people in Liberty Park, packed in like sardines, desperately trying to get a glimpse of the festivities. I remember feeling suffocated and glancing up at my mother's face for assurance, only to find that her eyes were nearly glazed over in fear, and to realize that her skin was moist and pale. We left soon after that.

In these situations, she is limited by a magno deficit. Her brain cannot fast-process visual information properly; it cannot properly process a changing and unstable visual scene quickly enough. The unfamiliar malls and the crowded park were "too much to handle." She became overwhelmed. And rightly so.

My mother is an extraordinary artist. When I was young I would quietly watch her steady hand paint surrealistic panoramas on canvas. Years later, when the game *Pictionary* came out—where players have to draw pictures to convey phrases—she would crush the competition with her effortless talent. I also know that I developed my love of nature's beauty from listening to my mother's passionate accounts of the landscapes she had seen in her lifetime. I'm convinced that she sees shades of colors most others do not, and that she sees the world from a truly different perspective than most—through the eyes of an artist.

With research in hand, I finally came to understand that my mother has an exceptional parvo, slow-processing system; like most artists, she expresses an enhanced perception of color. She also expresses a mildly deficient magno, fast-processing system. I also became aware that her brain "compensated" for its deficiency; and finally, I put to rest the notion that perhaps my mother had a slightly abnormal brain.

Is there such a thing as a "normal" magno system, or a "normal" parvo system? Is there a prototypical brain to which others can be compared? Should my mother's brain be considered "below average" because her magno system is not up to snuff? Does her enhanced parvo system bring her back up to par? It is impossible to quantify such complex brain functions into a "normal" equation. Consider this exercise: Visualize a house. Are there two bedrooms? How big is the kitchen? Your house will likely be different from mine, but they're both legitimate, are they not? There is no such thing as a "normal" house, only variations of houses. Similarly, there is no such thing as a "normal" brain. Brains are dynamic and ever-changing. While a brain may be weak in one area, it likely has a corresponding strength in another. This is why I consider my mother's magno-deficiency an extraordinary gift.

My research assistant had not only found an underlying cause for her mother's reading difficulty, she had also hit upon a central conclusion about perception, and indeed all brain function. Each brain is different, and each is more efficient at certain kinds of processing than others. For most people, their plastic brain attempts to reorganize to compensate for its deficiencies the best it can. The more we learn about how this occurs, the more we will be able to help the brain retrain itself.

All the visual processing we have been discussing with reference to her mother has taken place in the thalamus. Once the thalamus does its job, the processed visual information is sent to the back of the brain, the occipital cortex. Here the signals arrive at the visual cortex, an area known as V1. It acts as the secretary of the visual system, assigning signals to each of at least thirty separate areas dispersed throughout the brain that specialize in one kind of feature discrimination or another—color, shape, size, or orientation. Most investigators believe that the primary visual cortex is organized into modules and that neurons within each module are devoted to the analysis of specific features of one small portion of the visual field. Some modules process individual pieces of the perceived image, while others respond to features such as orientation or movement within the visual field. For instance, a neuron might respond to a vertical stroke of paint on a picture simply because it is oriented in a vertical direction.

We still cannot perceive the colorful painting on the white wall until information from the individual modules has been combined. The puzzle pieces must be fitted together to create a seamless pattern. This process begins in the visual association cortex, where we add data about "what" an object is and "where" it is located. Information about color, texture, and shape are handled by the temporal cortex, and spatial details by the parietal cortex. These areas ultimately pass the results on to higher regions of the brain—the frontal cortex—for further analysis. Once the visual information is there, we become conscious of our bodies, standing in a room, and staring at a colorful painting on a white wall.

That is not the end of the story, however. The visual pathway is not a one-way street. Higher areas of the brain can also send visual input back to neurons in lower areas of the visual cortex. Try this: Create an image of your favorite vacation spot in your mind. Is the sun shining? What shade of blue is the sky? As humans, we have the ability to see with the mind's eye—to have a perceptual experience in the absence of visual input. For example, PET scans have shown that when subjects, seated in a room, imagine they are at their front door and starting to walk either to the left or the right, activation begins in the visual association cortex, the parietal cortex, and the prefrontal cortex—all higher cognitive processing centers of the brain.

There is yet another crucial reason why our brains send substantial projections from higher cognitive processing centers to lower areas of the visual cortex. Again, it relates to attentional specificity. This feedback system provides the brain with an intrinsic mechanism for shutting down inputs when they are repetitive, unnecessary, or should be ignored. Think about how incredibly overwhelmed our brains would be if we had to relentlessly take in every single visual detail from our environment. Without a filtering process, the neurons of our visual pathway would likely be in a continuously overwhelmed state. The higher regions of our brains remind the lower areas of the visual cortex that the white wall is not of interest, and to pay attention to that which is of interest—the colorful painting.

Our other senses operate under similar feedback mechanisms. Think about your sense of touch. Do you feel the clothes on your body right now? You probably didn't until you actually paid attention to the perception after I just suggested it. The higher regions of your brain are helping you ignore the repetitive feeling of clothes on your skin, thus enhancing your sensitivity to other aspects of touch needed to read this book, scratch your chin, or pet the dog.

The brain has an amazing capacity to filter our incoming stimuli so that we can negotiate our environment without being overwhelmed by it. Without such intricate neural circuitry, our olfactory, culinary, tactile, auditory, visual world would be intensely chaotic and disorganized. So the next time you drive down the highway grasping the steering wheel, watching the traffic, and listening to the radio, or the next time you smell and taste that beautiful dinner, remember that your brain is managing an incredible flow of sensory information. Remember, too, that the next time you examine a colorful picture on a white wall or search for a pen on your cluttered desk, what you are seeing is actually much more than meets the eye.

THE SIXTH AND SEVENTH SENSES

AS IF THE FIVE SENSES of perception aren't interesting enough, science has debated for years over other, nontraditional senses. Not extrasensory perception or other paranormal feats, but a sense of direction and a sense for sex.

There are countless reports in history about people with an extraordinary ability to know where they are going: pathfinders, guides, mariners, pioneers. Certainly, ancient peoples found their way across continents and oceans without compasses, sextants, radar, or the global positioning satellite system. We know, too, that species such as migratory birds and salmon have an instinctive sense of direction. Certain cells in the heads of honey bees and of homing pigeons contain crystals of magnetite, a natural magnetic material. The crystals align in the earth's magnetic field much like the hands of a compass, which is somehow used by these species as a frame of reference in navigating.

Joseph Kirschvink and researchers at the California Institute of Technology have identified the same kind of magnetite particles in human brain tissue. They don't know what function the particles actually serve, but the possibilities are intriguing.

Some scientists believe our seventh sense may be a sex organ . . . inside the nose. Many mammals have a vomeronasal organ (VNO) for sensing pheromones released by potential mates. The VNO is located just inside the nostrils, above the hard palate that forms the roof of the mouth. Anatomists have identified a similar structure in humans—a pair of pits on either side of the septum, which separates the nostrils.

Whether this "human VNO" plays any role in sexual attraction, and therefore mate selection, has yet to be shown. The debate is reaching near-rage proportions in the medical community, because recent evidence indicates that a VNO structure does develop in the fetus, but may disappear before birth. Even if it does persist in some minute form in adults, the next question is whether it can actually send signals to the brain. Some scientists think that the two pits have lost their ability to function during human evolution, and are a mere remnant of our heritage. But until evidence can conclusively say that the VNO is not active, we will continue to hear the hawkers of certain aftershave lotions and perfumes containing synthetic human pheromones claim that their products can help attract the opposite sex.

3

ATTENTION AND CONSCIOUSNESS

YOU'RE DRIVING IN the tight left-hand lane of a fast metropolitan expressway. It's a sunny Saturday morning and you and a friend seated next to you are laughing hysterically about what happened at a party the night before. She makes a crack, you turn to look at her, burst out laughing again, and reach for the tissue box in the back seat so you can dry your tear-filled eyes. Meanwhile, the guard rail, a mere five feet from your left elbow, is screaming past your car at 55 mph. You've paid no attention to the road whatsoever, yet you've managed to avoid the metal barrier.

Or perhaps it's late on a Thursday night. You're exhausted from fourteen hours of work and weak from four days with little sleep. You've been driving a long time. The lonely two-lane highway ahead of you is dark. The tires drone. Your eyelids get heavy. Your eyes glaze over . . . Suddenly you grip the steering wheel in a panic. Your eyes focus sharply, your heart races, and you break into a profuse sweat. You had started to drift onto the narrow shoulder and an overpass abutment is just ahead. You correct your position. Thank God, you somehow managed to stay on course.

How is it that you were spared in both these situations? You perceived the sights and sounds of the road even though you were not paying them any attention. Something in your brain remained tuned in even though you were not aware of it.

For decades scientists studied attention and consciousness as separate disciplines. But in recent years both theorists and experimentalists have begun to regard attention and consciousness as different levels of the same brain activity. In the newest view, our brains are always in "ready mode," always tuned in to the never-ending incoming stream of perceptions. Then an event takes place that causes the brain to deliberately pay attention to a stimulus—to put a spotlight on it.

Attention and consciousness are inexorably intertwined, and some scientists now believe that they are actually the same thing. Despite the volumes that have been written on consciousness, we still don't know how to define it, or what brain activity gives rise to it. But we can't talk about the brain, much less present a user's guide to the brain, without including some working models of consciousness that we can understand and discussing where the controversies lie between competing models. After all, without consciousness little else that the brain can do would matter.

Understanding consciousness is fundamental to understanding ourselves, and it is fundamental to diagnosing patients with any type of psychological, psychiatric, or neurological problem. It is not enough to assess a person based on the question "How do you feel?" The key questions are "How do you perceive and comprehend the world? How do you attend to it and become conscious of it? How do you know?"

Attention and consciousness are the foundations on which we create an understanding of the world. Together they form the ground upon which we build a sense of who we are, as we define ourselves in relation to the myriad physical and social worlds we inhabit. They also are the basic functions that give rise to "the mind"—a real kettle of fish.

SALESCLERKS OF THE BRAIN

UNTIL RECENTLY, THEORIES about attention and consciousness were largely that—theories, even if they were based on observational evidence. But we are now beginning to learn what, physically, in the brain creates attention and consciousness, thanks to painstaking research by people such as Rodolfo Llinas, Chief of Physiology and Neuroscience at the New York University School of Medicine, who has spent forty years investigating how neurons work.

For a century, the dominant view of the brain has been that of a simple "reflexive" organ. In this view, human brains are little more than input-output machines. Although they make very complex responses to the world, they still essentially respond to outside demands as they come in. But Llinas and others have shown that neurons don't simply sit around quietly waiting for incoming data to rouse them. They are always active and choose whether and how to respond to stimuli.

Llinas uses the analogy of going into a store (the brain). Before you enter, you (the stimulus) look through the window and see a group of clerks (neurons) talking to each other. There is activity, communication. You enter the store, making yourself now a new incoming stimulus. At first the clerks keep chatting; they may have sensed you, but are not paying attention to you. But if you clear your throat loudly enough, one of them will look at you. You now have part of the brain's attention. If you say, "I want to buy that overcoat," suddenly all the salesmen pay attention. The brain is now conscious of you. Having become conscious of the stimulus, a few of the clerks take action: one goes to the rack to get the coat and another goes to the cash register to ring up the sale. Some of the remaining clerks continue chatting; others rearrange shirts on a shelf. Still others, who have been in the front of the store arranging a window display, continue their activity without noticing the other clerks at all. In a similar way, the brain is attending, responding, deciding.

If neurons are always communicating and acting in the background, we begin to see a physical system that is always in a ready state. It does not sit idle, waiting to respond; it is talking, rearranging shelves, building displays, and directing some of its neurons to respond to a stimulus while others continue with their ongoing tasks. This state of activity, itself, may be consciousness.

If the brain were simply reflexive, it would never be able to plan a future action. The brain is a powerful prediction machine, continuously making elaborate mental maps of the world that are reliable enough to enable us to predict what lies ahead, both in space and in time. All animals that move must have some predictive power—at the very least a simple image of what they are moving into and a sense of how they are moving into it. Building these navigational aids forms the basis for ongoing activity in the brain. As time passes and the body moves, this conscious brain—or mind—experiences the world, adding new information and updating and revising its maps.

Although it hasn't been proven, this model makes intuitive sense. How else could you speed down a highway without paying close attention to the road and still cheat death?

In a great case of serendipity, recent research about how people keep track of objects supports the theory about the physical nature of the brain's consciousness. Quick, where's your wallet? You can answer that question, and perhaps even reach for your wallet, without your eyes ever leaving this page. Why? Your brain keeps mental maps of nearby objects. Michael Graziano and his team at Princeton University found out how by inserting a tiny electrode into the ventral premotor cortices of two monkeys. Further, their findings dovetail beautifully with current brain theory.

For an animal to move and interact with its surroundings, it must coordinate its muscles with what it sees, hears, and feels. In primates, the command center for this activity is in the ventral premotor cortex. Graziano found that when the monkeys' eyes locate an object, groups of neurons start to fire, and a subset of them continue to fire even after the object is out of sight.

His team placed a plastic tube in the monkeys' field of view and recorded the response from the electrodes. When they turned off the lights, one group of neurons continued to fire, as if the monkeys were still seeing the tube. These neurons appeared to be preserving the brain's working memory of the tube's location. The researchers then quietly removed the tube and turned the lights back on. When the new sight data told the neurons that the object was no longer there, they stopped firing.

Graziano's findings back up the idea that the brain constructs maps of its surroundings, and that some subset of neurons, like the clerks, keep talking in the background to maintain those maps. When new data come in, such as the displacement of the tube, the neurons reconfigure their maps. It is likely that the brain draws maps based on past experiences and memories, too. Sit on a carpeted living room floor and toss a small, soft ball behind a couch that's almost against the wall.

Watch the ball's path as it disappears from view. You can predict pretty closely where it will land, even though you don't see or hear it hit the floor, because maps from experiences and memories of hundreds of other trajectories you've witnessed in your life let you predict what will happen.

AROUSE, ORIENT, DETECT, EXECUTE

REACTION TO STIMULI, ongoing talk between neurons, and predictive maps of the world all work together to create consciousness. Before we can be conscious of something, however, we have to pay attention to it.

Attention is much more than simply taking note of incoming stimuli. It involves a number of distinct processes, from filtering out perceptions, to balancing multiple perceptions, to attaching emotional significance to them.

Imagine yourself at a cocktail party. You can attend to many features of this environment simultaneously: sip a drink while listening to a friend talk and watching a colleague dance with a new partner. Or you can process the many stimuli through a filter that lets the numerous insignificant bits of the environment pass through, leaving you with the big, meaty pieces; you might focus in on your friend's face and words as he tells a captivating anecdote, while seeming not to hear the dozens of other voices or see the other visual images in the room. You also combine perceptions—say, sight and sound—to identify the abstract notion that a fight might be brewing across the room.

The intensity with which you attend to such stimuli is determined by your own level of interest, alertness, and anxiety. The cognitive process of assigning an emotional weight to perceptions affects attention as well. If you sensed a possible fight, your limbic system, which controls your fight-or-flight response, would ensure that your attention system monitored the target until the situation had been resolved. If you suddenly saw the would-be combatants shake hands with a laugh, your limbic system would subtract the emotional valence it had assigned to the perception. Your attention system would feel free once again to ignore the situation, illustrating yet another aspect of attention: that the brain continually decides when to stop paying attention to something. If all this weren't enough, the brain also has to attend to stimuli coming from within itself, such as memories and thoughts, as well as from the body. Attention is a complex system.

Scientists have identified four distinct components within the attention system, which together create the brain's overall ability to monitor the environment: arousal, motor orientation, novelty detection and reward, and executive organization.

At the lowest level of monitoring, the brainstem maintains our vigilance—our general degree of arousal. At the next level, the brain's motor centers allow us to physically reorient our bodies so that we can immediately redirect our senses to possible new villains or food sources. Then, the limbic system accomplishes both novelty detection and reward. Finally, the cortex—especially the frontal lobes—commands action and reaction and integrates our attention with short- and long-term goals.

Arousal is the ability to suddenly increase alertness, inherited from the days when sleep left one vulnerable to predators. Nowadays, fear is still a good arouser, though it may stem more from missed deadlines than from almost being eaten. As *Homo sapiens* learned to think with greater abstraction, novelty from within the forum of his own thoughts also began to excite arousal.

Arousal is controlled by the reticular activating system, which connects the frontal lobes, limbic system, brainstem, and sense organs. Incoming information from the senses, or thoughts, can arouse us, and depending on the startle value it alerts the rest of the arousal circuit. The hippocampus—a key player in long-term memory—also communicates with the reticular activating system. With its store of knowledge, the hippocampus is the way station to our memories, able to compare the present with the past and thus monitor events as either novel or ordinary. That way, if the reticular activating system startles over something harmless, the hippocampus works as a protective filter to conserve the brain's energy by inhibiting the system.

In the story of survival, the obvious second step after startling is to orient the body—and specifically the body's sense organs—toward the novel object in question. This allows us to process the new information in as short a time as possible, which increases our chances for safety. Like the arousal system, motor orientation is basically involuntary. We do not need to think before we perk up our ears, turn our heads, focus on the approaching pit bull, or lean over the enticing salsas for a better whiff. We do it without effort or awareness.

Michael Posner and Steven Petersen of the University of Oregon suggest that motor orientation be broken down into three steps: disengage–move–engage. First, the posterior parietal cortex helps us disengage from a stimulus. For example, if you are staring at this book and the telephone rings, it allows you to stop visually attending to the book and disengage your attention. Now your brain can prepare your motor pathways to do something new. Next, the basal ganglia and the frontal parietal attention circuits shift the focus of attention to the new stimulus, in this case directing you to move your eyes and ears (your head) to focus on the phone. Finally, a group of neurons in the thalamus engages attention by focusing the brain on the new stimulus and inhibiting other noise and therefore distraction, allowing you to pick up the phone and concentrate on the caller.

Damage to the posterior parietal cortex can cause a classic deficit called Balint's syndrome, in which patients are unable to attend to multiple objects simultaneously; they cannot see the forest for the trees. The damage limits a person's ability to shift attention from one location to another, and perhaps from one sensory modality to another. People with autism share some of this characteristic.

Eric Courchesne at the University of California at San Diego has taken this idea a step further, proposing that a congenital defect can cause an inability to shift attention between behavior and social cues. It might be the problem, for example, for a baby who cannot smoothly shift its attention from its mother's face to a tree on the sidewalk when the mother points her hand at one. The baby does not adjust its behavior (gaze) to follow the mother's social cue (pointing). This could hamper the child's ability to learn how to develop social sharing, and therefore social relationships. There will be more on this in Chapter 8, "The Social Brain."

Once we are aroused and oriented, the reward and novelty system kicks in, governed by the mesolimbic pathway (a group of dopamine-containing neurons), which is a key driver of the limbic system. This system is integral not only to attention but to many other brain functions, notably the emotional and social brain.

Detecting novelty and seeking reward are the two primary forces that direct the selection of where to focus our attention. The novelty system takes note of new stimuli. The reward system produces sensations of pleasure, assigning an emotional value to a stimulus, which also marks it for memory. If, later, the same stimulus reappears, the memory of these visceral emotions provides a response, from joy to disgust, which then directs the individual to seek out a plan of action. The mesolimbic system also functions as the novelty detection system, which is sensitive to unexpected scene changes.

The nucleus accumbens is a central player in the reward system. This small cluster of cells is located in the forebrain and is well connected to the amygdala and other parts of the limbic system. It has long been identified as the principal pleasure center of the brain. It contains one of the highest stores of dopamine in the entire brain, and is sensitive to other pleasure neurotransmitters such as serotonin and endorphins. These chemicals are key factors in feeling satisfied and rewarded, and therefore in providing motivation.

The nucleus accumbens has been studied extensively. Several addictive stimulant drugs such as amphetamines and cocaine, for example, achieve their pleasurable effect by changing the concentration of dopamine in the nucleus accumbens. Several of the primary drugs used to treat ADHD, including Ritalin, Dexedrine, and Cylert, also enhance the concentration and action of dopamine, which in these cases allows ADHD patients to better sustain attention and complete tasks. In fact, most of our addictive substances, such as chocolate, coffee, nicotine, and marijuana; external endorphins such as heroin, morphine, Percodan, and Dilaudid; and moderate levels of stress (which we call stimulation) act to increase the availability of dopamine in the entire system. A key area of action is this little group of cells, the nucleus accumbens.

Monkeys with lesions in this region are unable to sustain attention, which hinders them in performing tasks that are not rewarded immediately, therefore affecting motivation. ADHD can be thought of as an addiction to the present. Patients are often impulsive, lacking in inhibitions, and quick to act because they are hooked on immediate feedback. They tend to prioritize tasks according to which offers the most immediate gratification. As a result, they tend to not interrupt current activities in order to rehearse skills or evaluate the consequences of their actions.

For example, in the monkeys with lesions, the immediate pleasure of eating a peeled nut often overcame the future, though greater, benefits of hoarding. The long-term goal of hoarding, ultimately more important, was unable to compete with the overpowering and disinhibited reward system instructing the monkeys to go for pleasure and immediately eat the peeled nut. The monkeys' freewheeling or impulsive reward system drowned out the signals about how best to survive. The disinhibition is also related to memory problems, contributing to the monkeys' inability to maintain long-term goals.

We see the effects of dopamine in psychiatry when we use antipsychotic drugs like Thorazine and Haldol. They block the action of dopamine. In a psychotic patient this has a calming effect and helps restore the balance in the person's psyche. However, often patients also experience bizarre effects on their movements. They are less able to learn and remember, and often are chronically less motivated. Blocking dopamine in one area of the brain may help control hallucinations and wild paranoid ideas, but it may also compromise other brain functions, so that the patient is less able to pick up his life and go forward. The newer antipsychotic agents offer much promise because they seem better targeted toward the areas of the brain that are disturbed in psychosis.

Many addictive substances such as nicotine, narcotics, and alcohol have been shown to disrupt the nucleus accumbens. The addict's inability to inhibit drug consumption and the obsessive-compulsive person's inability to restrain himself can both be interpreted as a disinhibition of desire. The faulty novelty and reward system causes the addict and the obsessive-compulsive to feel that every new drink or situation is novel and exciting no matter how many times the stimulus has been encountered. The obsessive-compulsive brain errs further by treating the novel event as a danger that must be immediately grasped, washed, fixed, checked, or otherwise attended to. What results is a constant feeling of unease, which breaks down the establishment of any daily routine. Haunted by a feeling that something is not right, individuals with obsessive-compulsive disorder look for relief by checking up on the "novel" stimuli over and over again.

The fourth system of attention—executive organization—commands our actions and integrates our attention with short- and long-term goals. The CEO of this executive function is the frontal lobe, which plays a major role in the ability to sustain attention by blocking out irrelevant stimuli. It allows you to ignore the other conversations at the cocktail party so you can focus on the story your friend is telling. The caudate nuclei and basal ganglia, deep within the temporal lobes,

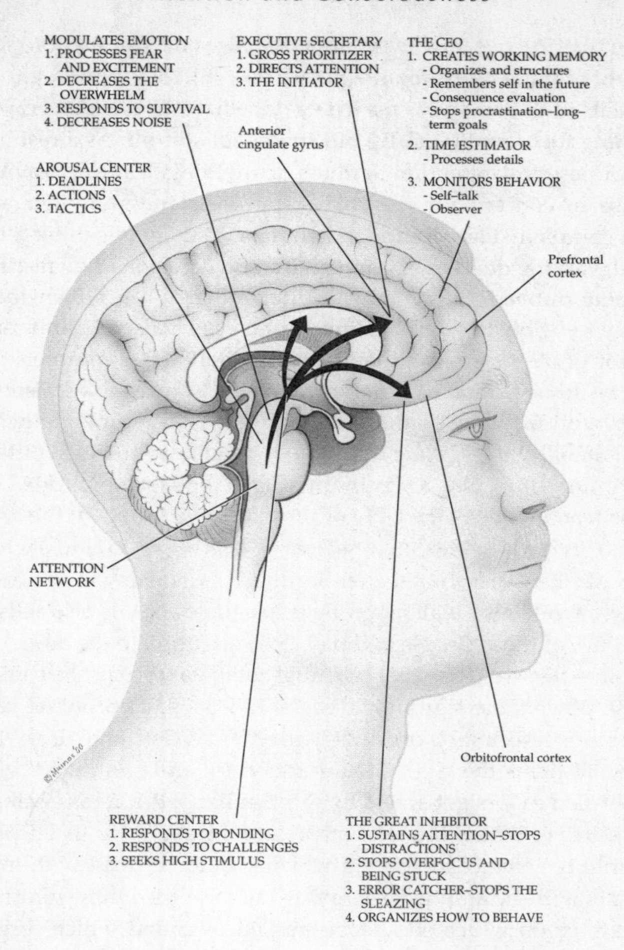

THE ATTENTION SYSTEM AND ITS DYSFUNCTIONS The attention system is a complex one. It begins at the arousal center in the brainstem and travels up through the limbic system and into the cortex, ultimately connecting to the frontal and the parietal cortex.

To understand this system from an ADHD investigator's view, we have joined neurogeography with clinical pictures. This allows one to look at the attention system and the many problems that may result from dysfunction in the identified areas.

help by filtering out unwanted internal noise and stimulation, such as your worry at the party about whether you still look stunning.

Patients with damage to the frontal lobe have tremendous difficulty sustaining focus and blocking out irrelevant stimuli. As a result, they may not pay close attention to much of anything, leading to a marked decrease in concern for the social appropriateness of their actions or to a complete disinhibition of behavior. They may exhibit a loss of arousal, varying degrees of impulsivity, and occasionally, a fascinating syndrome dubbed "environmental dependency," in which they are drawn irresistibly to external stimuli, often compulsively imitating the behavior of others around them. Another deficit is a complete loss of curiosity: some frontal patients seem to lack any desire whatsoever to interact with the world around them. There is also growing evidence that an inability to regulate frontal-lobe function—to block out irrelevant stimuli—may play a significant role in the symptoms of ADHD.

If the frontal lobe is the CEO of the executive function, the anterior cingulate gyrus is its executive secretary. This region is active when we need controlled, distributed attention, such as listening to our friend at the party while also watching our colleague dance. It also tells us to forget both of those people and pay close attention to the other side of the room when we sense that potential combatants may start a fight.

These powers are significant. In the old days of corporate culture, the executive secretary organized the CEO's day and desk, prioritized what items the CEO should attend to, and controlled whether unscheduled callers got access to him. Similarly, the anterior cingulate gyrus decides which bits of sensory information are to be granted entry into the frontal lobes, and which should be dealt with for how long and with how much energy. It tags the incoming information with ratings on which should be mulled over and which forgotten, allocating the frontal lobes' resources accordingly.

Since the 1930s, studies have linked the anterior cingulate gyrus to attention, emotion, memory, somatic and autonomic motor responses, motivation, and even responses to painful stimuli. One of the keys to its widespread power is that it can regulate its own dopamine levels, which enhance the reactivity of neural networks. It also has extensive neural connections to regions throughout the brain, helping it regulate other regions involved in attention. The complex system of arousal, emotion, and motivation feeding the attention system seems to be coordinated through the anterior cingulate gyrus.

EMOTIONAL TAGS

THE PRIMARY EMOTIONAL signal the anterior cingulate gyrus receives comes from the amygdala, at the core of the limbic system, which influences attention by assigning emotional significance to incoming information. Even before a sensory perception has reached the frontal lobes, where it enters conscious awareness and undergoes fine categorization, the amygdala has already branded it with a raw emotional valence somewhere along a continuum from mildly interesting to "oh my God!" It activates the body and the rest of the brain in response to how significant it deems the stimulus to be to survival. If the stimulus seems threatening, it activates the alert centers of the brain and notifies the hormone system and brainstem to get ready to rock and roll.

The amygdala provides a preconscious bias of intensity to every stimulus you come into contact with, even before you actually pay attention to it. It can, and does, operate outside consciousness. One example is its ability to immediately prepare us to flee when we perceive an exploding noise, long before the cortex has begun to make sense out of what the noise is. People without an amygdala owing to infection, stroke, or surgery have what we call the Kluver-Bucy syndrome, a bizarre set of symptoms with a tendency to react to all stimuli in the environment without discrimination or learning. They also have a marked indifference to people and lose their emotional attachment to family members.

By anticipating events, the amygdala's emotional tagging, which occurs in consultation with our memories, allows us to instantly judge and then react to the world. If a stimulus is deemed dangerous, the amygdala adds impetus to the attention system to "keep arousal going." However, strong emotions can also be a hindrance if they cause us to prejudge the environment. People with depression, for example, may not respond to attempts by others to help them feel optimistic because they are constantly tagging stimuli with the label "sad."

How does limbic tagging occur? Dopamine may again be the key. It is abundant in the amygdala, and both the amygdala and the closely linked striatum, the home of the nucleus accumbens, have been shown to release dopamine in response to pleasurable rewards or painful punishments. Whether the experience is registered as painful or pleasurable, however, seems to be less important than the intensity of the pain or pleasure. In this fashion, a person learns to associate positive or negative emotions with an event.

As we learn more about the various brain regions, we are beginning to get the picture that attention is not a function limited to a few "centers" in the brain. Responsibility for attention is widespread. There is even evidence, now, that the cerebellum is involved. The cerebellum has always been thought to be involved solely in coordinating motor function. However, it has recently been found to be crucial in coordinating and timing cognitive function—which would help you attend to the rush of stimuli you receive when you suddenly "awake" and react to drifting off the highway.

THE ROLE OF NEUROTRANSMITTERS

THE STRUCTURES OF the attention system make use of various neurotransmitters, as does all of the brain. Serotonin and norepinephrine are important, but the system is most heavily dependent upon dopamine. Although the specific functioning of dopamine remains somewhat of a mystery, evidence suggests that it has two seemingly conflicting duties: it decreases spontaneous neural firing, which slows random noise, but it also has a depolarizing effect on neurons that makes them more ready to fire. Thus it has been called the "learning neurotransmitter."

The depolarizing effect of dopamine is especially interesting because it may represent a link between the reward, novelty detection, and executive functions, and also between the overall attention system, learning, and memory. It is essential to the formation of long-term memories. Dopamine strengthens the prolonged chemical firing of messages between neurons and allows for unfettered communication between neurons. Conceptually, this results in the ability to formulate new concepts by forging links between old elements of knowledge—memories—which allows one to identify novel objects as opposed to those previously encountered in a given environment.

Dopamine may be the link between rewarding sensations of pleasure and long-term memory. This "learning neurotransmitter" may also be the link between the motivational reward and motor systems; problems with working memory (a form of short-term memory) correlate highly with dopamine deficiencies, and it is working memory that enables us to maintain continuity in our attention from one moment in our daily lives to the next.

THE EFFECT OF GENES

WE ARE BECOMING increasingly aware that genes too play a role in attention. For example, the attention system involves distinct contributions from dopamine, the enzymes that make it in the nerve cells, its binding sites on the postsynaptic side, its transport mechanisms, its metabolism. Each of these components has its own gene to guide the production of the proteins involved in these functions. One of these, the D2R2 receptor gene, which codes for the D-2 receptors on the postsynaptic site for dopamine that is mainly concentrated in the area of the limbic system, has an allele, or alternative gene, that has been linked repeatedly with a variety of psychiatric conditions such as alcoholism, ADHD, cocaine abuse, nicotine addiction (better known as smoking), compulsive gambling, and other addictions. These findings remain, but a number of studies now show that there is a relationship—this is an association, not a causal connection—between the D2 receptors and problems with reward and attention. It is believed that an abnormally low density of D2 receptors in the nucleus accumbens reduces an individual's ability to experience pleasure. This diminished capacity would almost inevitably drive that individual to seek external forms of self-gratification. During the course of a life, such strategies could settle into pathological patterns of reward-seeking, from substance abuse to sexual conquest to problem gambling. This has led Dr. Kenneth Blum of the University of Texas and others to define a new syndrome called the "reward deficiency syndrome."

Recognizing that a reward you are receiving is "enough" is in part a function of memory, and sustaining attention is the primary gateway to the encoding of memories. Dopamine is central to both attention and reward. What's more, substances such as nicotine, cocaine, chocolate, marijuana, carbohydrates, and alcohol increase the level of dopamine in this area of deficiency, so searching for and ingesting these substances may be in part an attempt to compensate for individual differences in dopamine levels. These levels may also be increased rapidly by engaging in high-risk behaviors, or by constantly confronting novel and challenging situations. Putting yourself at prolonged risk, say, by climbing a rock wall or driving a dirt bike at high speed through unknown woods demands that your attention system remain on full alert for long periods of time, forcing the sustained release of dopamine.

The D2R2 allele, although present in only 20 to 25 percent of the American population, was found to be present in 70 percent of severe alcoholics dying from cirrhosis of the liver. These findings, presented to the public in 1990, stirred up a flurry of debate in the media, which immediately jumped to the conclusion that the "alcoholism gene" had been discovered. Despite the researchers' insistence that they had in no way found a gene that "causes" alcoholism, a series of independent replication studies were performed to further investigate the findings. Although these studies have thus far yielded mixed results, fourteen independent studies have supported the findings.

When alcoholism is seen as a complex disorder affecting a person's body, thought processes, and behavior, it appears very unlikely that a single genetic factor could give rise to it. Instead, it may involve the interaction of various environmental factors with a group of aberrant genes. Each gene alone may not have the power to cause alcoholism, but the group, acting together, has a stronger genetic influence, increasing the likelihood of the onset of alcoholic behavior. There are also associations with other dopamine genes, such as the novelty-seeking gene—the D4 receptor gene. Perhaps the most exciting news is the finding replicated by two independent groups of investigators—a rarity in behavioral genetics research—that the dopamine-transporter gene allele, or alternative copy of the gene, is significantly correlated with ADHD families. This is important because in treating ADHD, the stimulants block the action of the dopamine transporter. So this seems to be a ready-made drug-to-gene match. However, as is so often the case with research covered in this book, we are still on the first step of a very long staircase.

The mounting evidence about neurotransmitters and genetics has become too compelling to ignore. People suffering from ADHD and other stimulation and disinhibition disorders can no longer be simply described as individuals who "like high stimulation." Instead, these

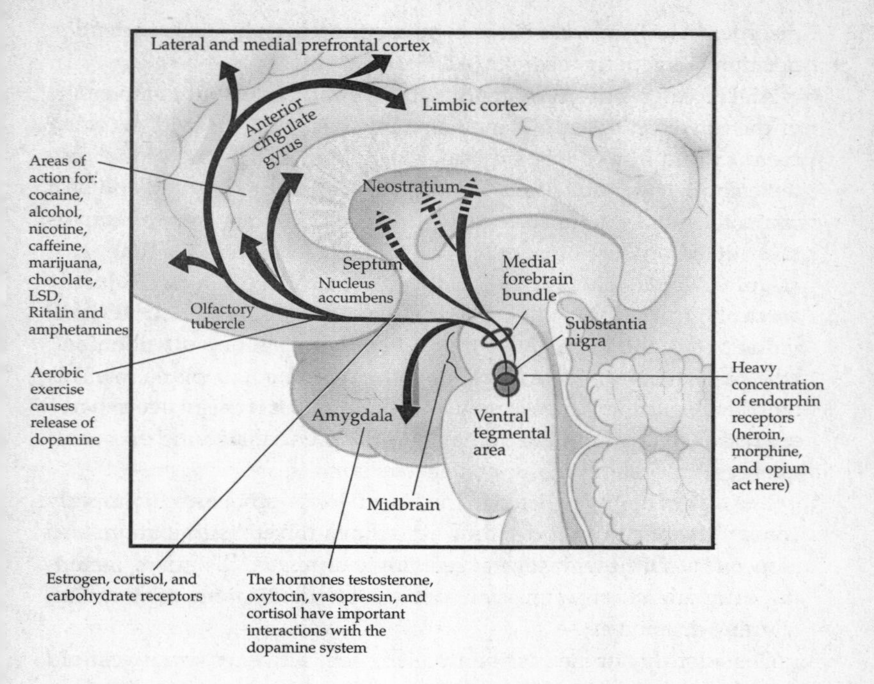

THE ADDICTION PATHWAY The ventral tegmental area and the substantia nigra are the key areas of the dopamine system, and this illustration shows where the many addictive agents and activities play a role in maintaining and altering the attention/motivation system.

individuals can now be said to possess a genetic and neurochemical desire to boost the dopamine system in their brains. Hopefully, this knowledge will open the door to a new era of medicine—and understanding.

ATTENTION-DEFICIT HYPERACTIVITY DISORDER

INDEED, NO DISCUSSION of attention would be complete without addressing the high-profile illness associated with it: attention deficit disorder (ADD), or what has become more accurately labeled attention deficit hyperactivity disorder (ADHD).

ADHD can result from a malfunction of any of the four components of the attention system. It may also involve problems with working memory and frontal-lobe arousal. Most often, ADHD individuals are deficient in the motivational sensations of pleasure or pain, and as a result they struggle to sustain the drive required to complete important but tedious tasks that only reward after a long period of time, such as doing well in school to eventually take on college or a career. Just as an alcoholic feels he or she must drink in order to survive, ADHD individuals seek the intensity of the present because their attention and reward systems are fueled by the pursuit of immediate pleasures. This neurobiological imperative may be so strong that it easily overpowers the reasoned advice of the frontal lobes, such as considering the consequences of pursuing an impossible love interest.

The lack of ability to listen to the frontal lobes' advice, or to properly consult memory when deciding on actions to seek stimulation, also associates ADHD with several seemingly unrelated disorders, including drug addiction, aggression, depression, tic disorder, anxiety, and learning disability.

Consider this in the case of a patient, Jeff, a twenty-seven-year-old who has attempted to control his drug addiction since adolescence. His parents both had problems with smoking and drinking, and Jeff began smoking at age eleven and drinking at fourteen. Over the years he attended several twelve-step programs, including groups for alcohol and marijuana abuse. He also followed Edward Khantizan's "self-medication hypothesis," utilizing interchangeable heavy dosages of exercise, alcohol, marijuana, cocaine, food binges, daredevil skiing, basketball, soccer, caffeine, and nicotine throughout his life. During college, Jeff could only induce the sustained attention necessary to complete papers through the stress of chronic procrastination and by snorting cocaine.

Jeff certainly had life's advantages. His IQ scores were extremely high. His academic parents encouraged his intellectual development and sent him to a high-powered prep school. Still, he experienced extreme difficulties at every educational institution. Constantly procrastinating, he attended to his schoolwork only sporadically and was poorly organized. In his yearbooks he was consistently voted "class clown" or "the bad boy." Though handsome, Jeff had never been able to maintain a meaningful long-term romantic relationship. Unfortunately, although family members agreed that Jeff clearly had an attention disorder, therapy was consistently limited to his drug abuse. Unable to fulfill or truly express his academic potential, Jeff was often scolded for being narcissistic, lazy, self-indulgent, and even a "spoiled brat" by many of his teachers. For much of his life, Jeff suffered from low self-esteem.

This vignette illustrates not only the close relationship between ADHD and addictive or thrill-seeking behaviors, but also the ease with which people attach severely critical labels to people without seriously considering the driving force behind the objectionable behavior. Jeff's actions clearly developed from entangled problems with the attention, motivation, and reward systems in his brain. Unfortunately, his critics reduced his story to a single aberrant behavior within the context of one aberrant incident at a time, and judged him by it. This approach is not only demeaning but also ineffective. Perhaps Jeff's disinhibition was caused by his attention disorder, which in turn contributed to his poor motivation to complete the banal yet necessary tasks of schoolwork. Negative comments from teachers or peers contributed to poor self-esteem and discouraged hopes for improvement. Most important, Jeff's heavy drug use appeared to be directly related to his problems with attention and motivation. Jeff had stated explicitly and repeatedly that the drugs helped him "focus." In college he would regularly resort to cocaine in order to complete a paper. What a plain clue!

There have been suggestions for some time that ADHD be considered a motivation deficit disorder. But it is only recently, owing to our growing knowledge of the dopamine system, that there has been neurochemical evidence to show that ADHD is indeed a motivation deficit or, more technically speaking, a reward deficiency syndrome, which results from a deficiency of pleasure neurotransmitters, principally dopamine but also serotonin and endorphins, in the reward systems of the brain. Examples of reward deficiency syndromes include not only ADHD but also alcoholism, smoking, and Tourette syndrome. Deficits in attention are not due to one or another area being out of whack, but to the entire system being dysregulated. The attention system has many parts and they all work together; a problem in one area will affect the entire operation. The deficit is throughout the system. It may have a common resting place that is highlighted by our fancy scans and tests, and that may be the sleepy executive; but the entire system is driven by dopamine, and it is this entire system which is dysregulated.

As a result of the deficiencies, reward-deficient individuals may be more susceptible to the temptations of dopamine-boosting activities, including substance abuse and other addictive ventures. Many drugs and thrill-seeking activities seem to achieve their pleasurable and addictive effects by inducing the excessive release of dopamine in the hippocampus and nucleus accumbens. This is confirmed by studies on addiction dating back to the 1980s that reveal that dopamine blockers reduce the pleasure of stimulants by inhibiting dopamine receptors.

These insights explain why dopaminergic medications like Ritalin, Dexedrine, and Cylert are used to treat ADHD. They work by both exciting presynaptic dopamine receptors and inhibiting molecules that clear synapses of dopamine. The antidepressant Wellbutrin, also used to treat ADHD, similarly increases dopamine levels in the brain.

There is another link between reward deficiency syndromes and motivation deficit: disinhibition. Individuals suffering from reward deficiency syndromes have difficulty inhibiting interest in irrelevant actions that reward immediately but interfere with their long-term goals; for example, a student who skips class even though he wants to graduate. ADHD distractibility is like the addict's lack of motivation to abstain. Indeed, addiction, which is the sum of motivation deficit and disinhibition, may be the common factor behind all the reward deficiency syndromes. Even the earliest studies seem to support addiction as a problem of inhibition. For the same reason, the disinhibited ADHD individual or addict is insufficiently motivated to abstain from counterproductive activity.

A different, though equally often misunderstood, aspect of ADHD is the hyperactivity part of the disorder. Ironically, the hyperactivity is due not to excessive movement but to a problem with inhibition. Faster movements simply involve good reflexes or perhaps well-trained, athletic muscles. The problem is disinhibition, which leads ADHD individuals to devote less time to analyzing a situation before acting, and alerting themselves that their behavior is in error. For example, tests in which images are flashed on a computer screen show that the reaction times of ADHD people are no faster than those for other people, but are indeed slower when a signal demands that they inhibit a response.

ADHD individuals are susceptible to distracting novelty, especially thrill-seeking, immediately gratifying activities that tend to waste time that might be spent reaching for the greater benefits of a long-term goal. The truly valuable tasks—like studying for the math test in high school rather than racing cars in the parking lot—are often ignored simply because the reward can't be held in mind long enough.

WHAT IS CONSCIOUSNESS?

WHETHER IT FUNCTIONS smoothly or not, the ultimate purpose of our attention system is to help our brains tune in to the world, including our own minds. Tuning in opens the door to that most fascinating aspect of our lives: consciousness.

People love to debate consciousness. Considering that we don't know what it is or how it works, the fervor with which it is debated can be embarrassingly presumptuous. The exploration of consciousness involves many fields: philosophy, psychology, psychiatry, neurology, biology, and quantum physics. It is heartening in this age of overspecialization that consciousness is being approached from all angles. The problem that plagues any interdisciplinary subject, however, is that experts from one domain may feel justified in dismissing the criticisms of experts from another. Alas, for now, this is the best we can do.

There is a great analogy—one that I've used before—that is useful in thinking about consciousness: that of a symphony orchestra. At any moment the brain is receiving and generating all sorts of signals. Like musicians tuning up randomly on stage, the signals are constant but haphazard. However, when the conductor strikes his baton on the podium, the musicians suddenly all pay attention. When he then gives the opening downbeat, they suddenly create signals in wonderful harmony. The musicians, in working together, create consciousness.

We can test this analogy by considering a few familiar experiences. When we sleep, the conductor is on break. Without his direction, some musicians stop—our sight and reason turn off. Other musicians may continue to play quietly in the background—our breathing and digestion. Occasionally, though, some musicians might play random notes that for brief periods come enough into harmony to sound something like a song—our dreams. But without the conductor the song soon falls apart again.

What if someone is knocked unconscious, say, by falling backward while skating and hitting his head on the ice? The blow physically rattles the conductor and musicians, making it impossible for them to play the right notes or coordinate. The song stops until the conductor and musicians can restore order and get back on track.

Too much damage from a bad blow or too much alcohol can put a person into a coma. In this case, the musicians are injured and cannot right themselves to play. If enough are hurt, the orchestra simply can't produce the song. Only if they can heal—or if substitute musicians take their places—can they rebound.

ATTENTION, MEMORY, AND CONSCIOUSNESS

LIKE A LIGHTING technician directing a powerful spotlight onto different characters in a Shakespeare play, most humans can easily manipulate their own focus of attention. In a simplistic sense, when we turn our attentional beam onto something, we are conscious of it. Objects outside the periphery lose their distinguishing features, falling out of our consciousness.

Whether we continue to pay attention to and thus remain conscious of our surroundings from one minute to the next also depends on working memory, which keeps our attention advancing forward. Attention, memory, and consciousness build upon one another to give us higher-order cognition. An impaired attention span, the culprit in ADHD, can make life seem incomprehensible, indistinct. Indeed, many ADHD patients describe their conscious experience as a blur, or as filled with static.

But how do we swing our attentional spotlight from one object or thought to the next? Paul Churchland, a professor of philosophy at the University of California at San Diego, suggests that this ability is driven by the relationship between working memory and long-term memory. Working memory is a significant part of the executive functioning of the prefrontal cortex. The systems that handle working memory are located in the frontal lobe, right in front of the areas concerned with motion and process. They hold data, motivations, and ideas all in mind for a bit, and then count on the long-term memory system to encode the information in the hippocampus and other parts of the cortex.

Working memory holds small amounts of information for only a few seconds at a time. It gives us the ability to remember a telephone number taken from the phone book long enough to dial the phone. Without mental rehearsal of the information, we lose the contents of working memory within a few seconds. Information in long-term memory, however, remains reliable over extended periods. We use working memory to conceptualize immediately occurring events and long-term memory to direct the present and plan for the future. Working memory is our brain's RAM, or rapid access memory, and is somewhat similar to the computer version. Our brains, just like our computers, need a space where many things can be held together and manipulated, so we can process them, evaluate them, rehearse them, make decisions about them. We in a sense can run programs on the information held in our working memory.

Early theories of cognition failed because they explained consciousness only in terms of long-term memory. They held that at a given instant, sensory data entering the brain are given to long-term memory processors, which try to make sense of them by looking for information that fits what the brain already knows.

In this model, sensory data of, say, four-leggedness, brown color, and angular shape are fed into long-term memory and compared to data already stored there. We identify the object as a chair. One's experience of the chair does not usually end at mere identification, however. Perhaps you recall an incident in which, as a child, you banged your knee against the back of a chair. Or perhaps you recall a newer memory, from yesterday, when you sat in this very chair and were uncomfortable because the seat was hard.

The criticism of feed-forward theories is that they miss the crucial role of working memory. Because of this one-way stream, your present thoughts would have no means by which to influence your immediate or future actions. If you remember the chair as being hard, you will probably not want to sit in it again today, but you may decide to anyway because you are aware that your legs are tired. Without working memory, this thought pattern and action plan would not be possible. Both working and long-term memory are necessary for consciousness.

This model is also useful because it explains why we tend to focus intensely on a minute portion of our environment and put the rest into the background. As you concentrate on this text, you are totally unaware of the mild squeezing sensation in your feet caused by your shoes—until you turn your attention to that perception. Your interest in the book prevails, facilitated by working memory, while the sensation in your feet—even though it is equally perceivable—fades into the background, allowed to go there by the long-term memory that the perception doesn't matter much.

Working memory and long-term memory allow us to prioritize certain stimuli over others by keeping the less important issues circulating in the background, though at the ready to be called upon. In this way, a mother who is worried about her sick child in the next room can actually sit on her couch and read this book, without having to think constantly about whether the child is okay, and yet remain ready to spring to her feet if the child suddenly starts to cry. This is true because the mother's past thoughts about her sick child have primed "future memories" of thinking, "I need to be on the alert for crying sounds." Although these future memories may never achieve the status of conscious thought—the mother may never say to herself, "I need to be on the alert for crying sounds"—past memories in this energized state ready her to pay immediate attention to a cry if she hears it.

The importance of memory systems to consciousness is especially clear when memory malfunctions. Bizarre changes in a person's conscious and unconscious awareness often result from memory trauma or artificial factors such as drugs that affect memory. One fascinating example of breaks in consciousness owing to memory problems involves people who are given anesthesia during surgery. It was always assumed that an unconscious patient could not perceive any information about his surroundings. As a result, surgeons were careless about comments they made about patients they were operating on. Beginning in 1964, however, D. B. Cheek of the University of California at San Francisco studied patients who had undergone surgery but had given their surgeons problems prior to the operation. The patients, while under anesthesia, were exposed to unflattering comments by their surgeons. Cheek found that some of the patients could, under hypnosis, recall their surgeons' conversations at a verbatim level.

In 1965, Levinson staged a mock crisis during the surgeries of a number of patients, during which surgeons made statements that the procedure was failing and the patient might die. When asked about their experiences on the operating table, some of the patients became extremely upset. Other studies tested the effects of statements made to anesthetized patients, which indicated they would have quick postoperative recoveries. These individuals did indeed spend less time in the hospital than patients who were not given such positive suggestions.

Taken together, these studies show that attempts to distinguish clearly between attention and consciousness are fraught with difficulty. Most people would say patients under anesthesia are "unconscious," and certainly that they are not "paying attention," yet if these patients can hear conversations, remember them, and even repeat them, the labels don't hold up. We can know and remember things even when we are not paying attention to them, or are not even "conscious" of them.

CONDUCTING THE ORCHESTRA

RATHER THAN DESCRIBE individuals under anesthesia as unconscious, it might be better to say they are in an altered brain state. We experience two different brain states each day—waking and sleeping—and on occasion others, from drunkenness to trance. Attention, memory, and consciousness collaborate to create different states.

Obviously, our waking state predominates. Following that is sleep. What happens when we sleep and do not actively focus the spotlight of our attention? What do our brains do when we are in this "unconscious" state? Using modern imaging technologies, various scientists have eavesdropped on the electrical activity of the brain's neurons in this state. What they have found is a revelation in explaining consciousness.

During both waking and sleeping, there is an ongoing din in the cortex. Neurons are constantly interacting with each other, even if they are not currently being called upon to perform a specific duty. This "noise" is not random, however. Rodolfo Llinas at New York University made the remarkable discovery that all areas of the cortex emit a steady level of noise, or oscillation, at a frequency of 40 cycles per second (40 Hz). Furthermore, some areas of the cortex, humming along at 40 Hz, were phase-locked, meaning their sound waves all oscillated in unison; they kept the same beat. How could this be occurring on such a widespread scale? Llinas, Churchland, and others suggest that the neurons perform in synchrony because they follow a kind of conductor in the brain.

The prime candidate for the conductor's job is the many intralaminar nuclei, located deep within the thalamus. These nuclei receive and project long axons to many areas of the brain. They take in information, reply to it, and monitor the neurons' response to their replies, creating an elaborate feedback loop. The information flowing back and forth between the intralaminar nuclei and the rest of the brain modulates itself, setting up a regular loop of electrical activity oscillating to a synchronized beat of 40 Hz.

As we might expect, during waking there are large bursts of electrical activity in every brain region, in addition to the steady 40 Hz oscillation. The bursts correlate to changes in our environment and activity. During nondreaming sleep the intralaminar nuclei are inactive; there is no 40 Hz oscillation. But during dream sleep (REM sleep), the 40 Hz background hum returns, and it is again accompanied by heavy regional activity, similar to that which occurs during waking. A host of neurons operating on automatic pilot create imagery; they're just less constrained in the absence of sensory input, and they feed you fantastic images and stories as they work to make sense of and store the prior day's events. Unlike in waking, however, the bursts do not correlate in any way to changes in the sleeper's environment, not even to deliberate efforts by researchers to create moderate noise or gently move the subject. In dreaming, the cortex activates itself purely from within.

Such findings provide a strong clue as to how our brains allow us to perceive a unified external world. This is a puzzle because our brains register each sensory input (such as smell, sight, or sound) separately, while internal inputs such as thoughts come from everywhere in the brain. The question is, then, Where in the brain do all these data meet to create a meaningful story of an event? That is, where in the brain does sensory input go to become conscious experience?

The answer is, nowhere and everywhere, but the intralaminar nuclei are a crucial part of the system. The myriad inputs are the sounds of the orchestra musicians tuning their instruments. The noise is a cacophony until the conductor gives the beat and the inputs synchronize. They subsequently stay together as long as the conductor pro-

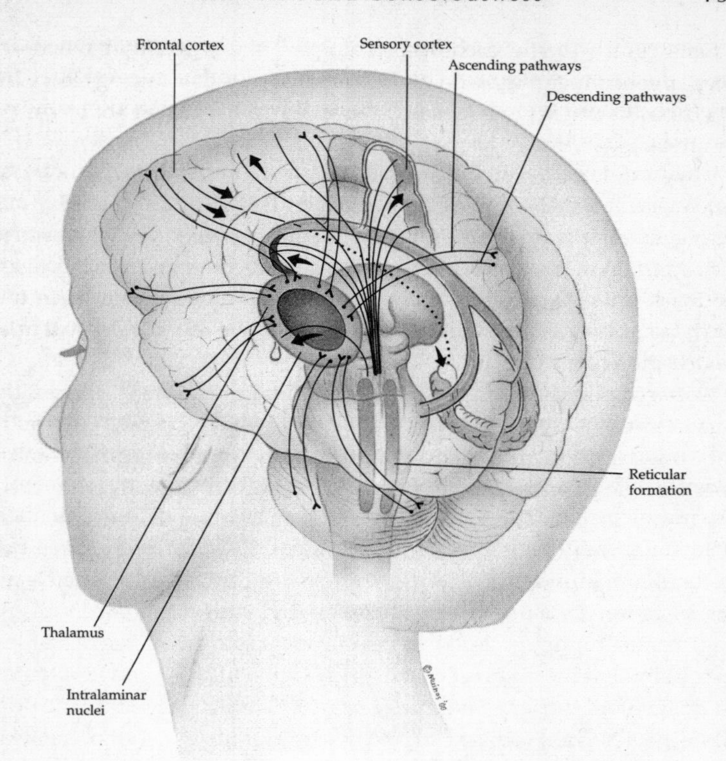

CONSCIOUSNESS One of the most appealing explanations of consciousness is the proposal that the recurrent network set up between the thalamus and the cortex is the neurology of consciousness. The thalamus is connected to the cortex by the intralaminar nuclei, which project long axons to all areas of the cerebral hemispheres. These areas in turn send back projections to the same intralaminar nuclei, and when this circuit is humming with a steady oscillation, consciousness may result.

vides the beat they are to follow, which is the 40 Hz oscillation set up by the intralaminar nuclei. It is only when the conductor can synchronize the brain's neural networks that we become conscious. When this happens with enough networks, the oscillations become ordered. They then spread their influence, coopting more networks to join them, and consciousness arises and widens.

Let's return to the orchestra analogy. I suggested that when we sleep, the conductor is on break. Research shows that during sleep, the 40 Hz oscillation is gone. Indeed, the intralaminar nuclei are on break, and there is nothing to keep the musicians together.

If we are knocked unconscious, the blow rattles the conductor as much as it does the musicians. It disturbs the electrical synchrony of the feedback loop among the brain's neural circuits, and the system falls apart. Until the intralaminar nuclei can recover and get the 40 Hz feedback going again, the brain cannot regain consciousness. In this view, consciousness itself is the sustaining of the 40 Hz electrical hum among the brain's circuits.

Research also shows that if the intralaminar nuclei are damaged, the person enters a deep and irreversible coma. If the damage occurs in only one hemisphere, the individual does not become comatose, but does lose the powers of awareness of half his body. He fails to perceive any events on one side of his visual field; and although one entire side of his body may be paralyzed, he dramatically fails to perceive this and often rigorously denies that he is disabled in any way. Some patients even deny that the paralyzed limbs belong to them.

THE FRINGE

ALTHOUGH THE SPOTLIGHT metaphor for attention and the orchestra metaphor for consciousness help us understand a great deal, there is more to the story. To a varying degree, we are also aware of, and think about, what is happening just outside the beam of the spotlight, the area which William James called "the fringe." It's not that the activities there are fuzzy; it's just that they are not commanding our direct attention.

This idea extends to a more complex level, where physical and mental attention meet. While the sight and texture of a baseball in your hand is a physical awareness, you are also aware of your feelings of familiarity or pleasant anticipation of making a pitch. In James's words, your sense of the baseball is in the spotlight of your awareness, and your anticipation belongs to the fringe. Other examples of fringe awareness include background emotions, feelings of knowing, feelings of being on the right track, and having a name on the tip of your tongue. It may be that the fringe is where we evaluate the relationship of what is in our spotlight to our current goals and interests.

Both the spotlight and the fringe aspects of attention and consciousness are crucial to our experiences. One case of lost fringe awareness is that of blindsight. A person with blindsight cannot "see" objects clearly, such as a ball rolling quietly across a table in front of her, owing to damage in the primary visual cortex. But if pressed to guess at the ball's location such an individual can correctly point to it with reasonable accuracy, and can also determine which way it is rolling. However, despite a capacity for visual processing, such a person has no sense of seeing anything in the visual field. If a person with blindsight is asked "What do you see?" she will often deny that she sees anything. She must always be coaxed by the experimenter to point at where she thinks the object might be. When she "guesses" correctly, she shows the most genuine amazement. It seems that the visual information absorbed in blindsight never reaches the status of conscious information. It is not processed in the primary visual cortex, but must be picked up by fringe parts of the visual input system.

This kind of dissociation provides evidence for an "unconscious" ability of the brain to contribute to perceptions and behavior. For most of us, conscious and unconscious systems are inextricably tied, but in the case of blindsight, the conscious and nonconscious visual processing are dissociated. Unconscious abilities may be what keep us aware of things happening in the fringe.

How the conscious and unconscious abilities of the brain work together is not known. Some researchers think it is due to processing between the brain's two hemispheres. Evidence for this comes from patients whose corpus callosum, the bundle of nerves that carries traffic between the hemispheres, has been cut to treat severe epilepsy. They often have dissociative experiences after their operations. As the left hemisphere controls the right side of the body and vice versa, incoming sensory information has lost its bridge between the two. Neither side of the body can know what the other is experiencing. This problem is compounded because, in most people, language is a tool of mainly the left hemisphere, and so information entering the right hemisphere from the left side of the body is left inexpressible through language.

Victor Mark, a split-brain researcher, describes the fate of one young woman. She had attained a two-year college degree and worked as a clerk, but by the time she was thirty-three her epileptic seizures were so severe that physicians decided to sever her corpus callosum. The seizures declined dramatically.

During rehabilitation therapy, however, the patient suddenly developed unusual problems. In one test she was asked whether her left hand was numb. She was told to answer by pointing either to a piece of paper with "No" written on it or to one with "Yes." Her body was not numb and her left hand responded by jabbing at "No," but her right hand pointed to "Yes." Understandably, she got upset and "furiously and repeatedly tried to indicate which thought was the 'correct' answer." Ultimately, the left hand forced the right aside and covered the word "Yes."

Why did this happen? It seems that the two hemispheres of the woman's brain had different assessments of the world and wanted to express contradictory opinions. Part of her firmly believed that her left hand was numb while the other part did not. More specifically, the right hemisphere, because of its connection to the left side of the body, knew that the left hand was not numb. However, the right hemisphere had no means of sending this knowledge to the patient's conscious awareness because it lacked the left hemisphere's capacity for language. Because the left hemisphere lacked connections to the left hand as well as the right hemisphere's awareness, it sensed nothing about the left hand and therefore concluded that it was numb. The interesting twist, however, was that although the right hemisphere was voiceless, it still tried to indicate its knowledge that the left hand was not numb by pointing to the "No" sign. As a result, the patient felt torn.

The emotional trauma for the woman was high. Mark writes:

> Despite attempts to correct herself, she had great difficulty accepting what she had just said or indicated, and sometimes she withdrew from the conflict by sighing deeply and declining to answer further. She often became upset and either pounded the table or slapped herself on the arm or leg. Her distress ceased when the topic or task was changed. During these conflicts the patient never appeared to undergo a change in personality or to be inattentive to me, and she did not adopt any consistent body or gaze orientation. During these episodes she often asked, "Why do I lie to

you?" My reply that her surgery was in some way responsible never completely reassured her and did not prevent further conflict.

Mark's startling conclusion was that in split-brain subjects, "each hemisphere could be taken to be considered 'conscious,' " thereby refuting long-standing arguments that the right hemisphere cannot be considered conscious because it is incapable of language. Mark found that his patient's right hemisphere could be interviewed: through pointing, the right hemisphere answered "No." Still, the validity of claiming dual consciousness is controversial.

What the findings also tell us is that "normal" consciousness arises only when the two hemispheres work together, and that the proper coordination of conscious and unconscious awareness also depends on this cooperation. I had a male patient who, upon testing, showed language abilities in both hemispheres. He showed many signs of slight malfunction of his corpus callosum, and thus had his hemispheres working rather independently. He was concerned about himself his entire life, because he often observed himself dissociating. For example, he would utter mean or sexually provocative comments in social settings and feel shame for saying them, but say them anyway. The language seemed beyond his control. Sometimes he heard himself saying these things as if he were listening to someone else talking. At other times, he sensed that he was overhearing himself say these things, when he was not actually speaking. The split between his conscious and unconscious abilities allowed him to observe himself as an independent person, yet also caused him to believe that he was saying things when he was not.

THE DEVELOPMENT OF CONSCIOUSNESS

PEOPLE TEND TO think of attention as a finite act of looking or listening to something, but it is an ongoing activity. We become conscious of something when we decide to pay attention to something we are already attending to. Francis Crick said that consciousness is "attention times working memory." For us to be conscious of something, we first have to attend to it, and then we have to keep attending to it from one moment to the next. Working memory is what allows us to continue to attend. That, according to Crick, constitutes consciousness.

In this theory, "the mind" emerges when the consciousness system hooks up with long-term memory. Once the data being continually tracked in working memory are compared with long-term memory of events in the past and with memory of the future, thought begins and the mind has suddenly come into existence.

Is this definition acceptable? Who knows? We can make up whatever definition we want. That's what everyone else does who argues about "the mind." While this may sound flippant, a quick survey of theories of the mind shows that they run the gamut from one extreme to the other. At one end is the opinion that the mind is the same as the brain, and at the other, that the mind is an entity completely separate from the brain, the result of a soul or some other attribute, and actually runs the machine called the brain. Somewhere in between these two positions is the idea that the mind is an emergent property of the brain—it is what results when the brain runs.

Crick's basic view of consciousness as attention times working memory is echoed by other leading theorists, including Gerald Edelman at the Scripps Institute. His view is that we are always perceiving things, but when we suddenly relate what we are perceiving to our internal categories of experiences, we become conscious. We judge stimuli against references of the world we store in our long-term memories. The act of categorizing, or judging, is what makes us conscious of the perception.

Consciousness is desperately elusive. A conference on consciousness is held every two years in Tucson, Arizona, and each year the theories to explain consciousness multiply by leaps and bounds, with more diverse players showing up each time. One reason many theories fail is because some are modeled after computers, the "neural networks" that were thought might grow so sophisticated one day that supercomputers could mimic and exceed the capacities of the human brain. These computers have proved terribly inadequate at achieving even the simplest cognitive tasks of the youngest child, such as understanding the meaning of everyday speech. These models fail because they assume that cognition is the result of a series of preset, preprogrammed rules devoid of both meaning and context. If consciousness is a set of programs, then how can you explain an organism's capacities for learning that involve adaptation and development? You can't. It might be clearer, as suggested by Steve Massaquoi at MIT, for the artificial intelligence field to use the term "neuroid network" to describe its instruments, because "neural network" describes what is going on in the brain, which is simply a different game.

Living organisms see significance in the world, develop meanings, use memory, grow from learning, and organize perceptions. A better alternative to viewing the brain as a program is to understand the brain as a selection system. This modern approach may be the key to formulating a biological understanding of what is true to the being of living creatures.

Edelman and his colleagues have been developing such a theory, which he refers to as "neural Darwinism" or the "theory of neuronal group selection." In this model, an organism's ability to categorize and adapt to the world is the result of processes of selection among neurons in the brain. It is competition between cells that is responsible for cell growth, death, strength, and weakness. Consequently, neuronal groups that benefit the organism's survival thrive and develop strong interconnections while those that are unused die. This "evolution" continues over a person's lifetime.

In this model there are two types of evolutionary selection: developmental and experiential. Developmental selection occurs before birth and is the reason that all organisms, even identical twins, are born with different brains. Although each organism is genetically constrained to develop certain features characteristic of the species, these genetic codes cannot dictate the exact destination of each developing neuron. As explained in Chapter 1, millions of neurons grow and die during this time, travel great distances, and forge connections quite unpredictably. So even though the overall pattern of brain regions is similar from person to person, each individual is born with a structurally unique brain. This prenatal level of selection organizes the brain into "primary repertoires"—singular groups of neuronal connections that create the first marks of individuality at birth.

At birth, an individual immediately enters the second stage, experiential selection. Experiential selection occurs because each person's experiences and behavior cause a strengthening or weakening of neural connections that alter the primary repertoire, although they do not cause gross changes in anatomy. Through everyday life, certain neuronal groups are selected to thrive while others die owing to lack of use. If a person is inactive mentally, that individual loses brain cells. In accordance with evolutionary principles, brain areas exercised often are presumably those that are most crucial to survival. Brain areas that are unused are seen as a waste of fuel, unnecessary to survival. In this way, experiences and behavior serve to direct nervous system development by organizing the brain into even more intricate "secondary repertoires," the neuronal groups that have been selected to thrive because they were used most often.

Values are also important for the evolution of consciousness. From developmental research, we know that infants show selective attention and preferences from birth. All organisms are born with basic values essential for adaptation and survival, such as eating over not eating, or looking over not looking toward a noise. From a broader perspective, values are the basic biases and tendencies that allow an individual to give meaning to experiences. This divides the world on many levels, from "food or not-food" to "friend or foe" and "good or bad."

The key thing to remember is that each newborn must create its own particular understanding of the world, and that each individual throughout life continually reinforces, adds to, and sometimes changes his view of the world. It is not uncommon, for example, for people who have had a heart attack or other trauma to throw out their old view of life and take on a new one. A math professor who becomes increasingly fascinated with art and increasingly bored with his job may one day throw out his books and eventually become an accomplished artist. The development of values that drive the evolution of consciousness may also explain how two people can view the same circumstances so differently: a citizen sees the gunning down of a crime boss in a restaurant as terrible; the gunman sees it as an act essential to his crime family's survival.

It is important to realize that the basic unit of this selective process is not the individual neuron but the neuronal group. There are perhaps 100 million such groups in the brain, and they range in size from 50 to 10,000 neurons. Neurons are only effective in groups working toward a single goal, such as discriminating color or producing emotion. By virtue of their size, neuronal groups can compensate for individual cell deaths. Nonetheless, changes in individual synapses will affect the entire neuronal group, which may in turn trigger alterations in a larger system.

Even with primary and secondary repertoires, however, cognition is still only partially accounted for. One must still explain how an organism develops categories, as well as synthesizes disparate pieces of data into a unified, meaningful experience. The key to answering this question concerns Edelman's most radical and important concept: that of reentrant signaling.

Reentrant signaling is the communication between maps that allows us to construct complex perceptual concepts such as "chair." It is necessary because, although we are born with rudimentary abilities such as detecting color and movement, our perception of objects must be actively created. Our ability to recognize an object according to its value, labeled name, and meaning is not innate. Recognition of objects and events does not consist of one-to-one relationships that divide the brain into a sort of one-cell–one-concept organization. For example, in the 1970s it was mistakenly thought that there was a "grandmother cell" that corresponded to one's perception of one's grandmother. In the new theory, the perception of a chair or one's grandmother relies upon reentrant signaling, which combines the activities of several mappings of brain regions devoted to sensory perception. The different pieces of the concept are transported back and forth between the regions that house them, until they resonate with each other—sustained at the 40 Hz oscillation—and lock in the idea of chair or grandmother.

The many maps that are created are inventoried by the cerebellum, basal ganglia, and hippocampus. These three areas keep track of the maps everywhere else in the brain and order the brain's output. Together, they form a kind of supermap, which contains multiple local maps. This creates a system of connections for whole categories of information, as well as patterns of motor activity. The end result of this complex value system of loop-within-a-loop layers of maps is the infinite variety of each person's thoughts and behaviors. Your concept of "chair-hood" or "grandmother-hood" in the brain is a global rather than a localized affair. Each region of the brain contributes to the recognition of a chair or grandmother, which explains why recognition can be triggered by any number of different sensory elements: the smell of mothballs; the taste of paprika; a woman with gray hair, a figure crocheting in a rocking chair, an aging female voice. Each one of these occurrences will trigger different maps in the brain. The ability to make the leap from these different sensory elements to the global concept is what cubist painters depend on in their fractured images.

Overall, Edelman's biological theory of consciousness seems to succeed where past models have failed. It both accounts for and coincides with our experience of the world as conscious beings—an experience that is provocatively and irreducibly subjective. With this theory of mind, we are free of deterministic preprogrammed laws, and instead can embrace a framework of consciousness that revolves around value and meaning in the world. This is indeed the essence of consciousness: the feeling we have of owning our actions and being able to develop our self-conceptions through experience over time.

THE EASY AND THE HARD PROBLEMS

WHILE ATTENTION IS the foremost director of consciousness, explaining it is one of the "easy" problems. Other easy problems include the ability to discriminate, categorize, and react to environmental stimuli, report on our mental states, control our behavior, and describe the difference between wakefulness and sleep. These problems are "easy" because there is no controversy as to whether these phenomena can be explained scientifically.

The "hard" problems are explaining the experiential "what-it-is-like" to be a conscious organism. Humans are undeniably subjects of qualitative experience, and yet we still do not know why and how experience arises from physical surroundings. "Consciousness" is an ambiguous term.

As we have seen, many scientists turn to physicalism to explain the relationship between consciousness and the brain. They maintain that all aspects of consciousness can be explained in terms of physical laws. Churchland, Crick, and Edelman are all "reductive materialists" who assume that consciousness arises from the communication among neurons. In this perspective, all biological and mental events are reducible to properties of matter and energy. Supporters of other views suggest that there is more to consciousness than just the interaction of brain cells. What about reason, faith, the soul, God? Even Edelman admits that reductionism can become "silly" when the mind is described only in reductionist terms. Nonetheless, for now this approach is the best we can muster, and it is teaching us a great deal about what consciousness is—or at least, may be.

The next frontier in the quest to explain consciousness is the hard question of subjective experience. We all know what a subjective experience is, but it's hard even to explain what we experience.

The most central subjective experience we have is "what it is like to be me from the inside." Examining consciousness from the inside reveals a whole new set of "data" to be explained—the qualitative aspects of our experiences, or, for short, qualia.

The term *qualia* is currently in vogue in the field of philosophy of the mind. Qualia are the phenomenological properties of experience; the "what it is like" of consciousness that are elements that can only be known from one subjective standpoint. For example, you cannot experience another person's pain. You can infer what the other person is going through, but there is no direct transfer of the experience. Other examples of qualia might be, for instance, déjà vu, a chilling dive into a cold river, or the smell of burnt rubber.

The current trend in consciousness research is to include some attempt at accounting for these qualitative elements. In 1994, for example, when the first Tucson Consciousness Conference convened, one of the most significant group conclusions was that study of consciousness will be "radically empirical." It will include subjective experience as primary data. No reported phenomenon will be written off because it "violates known scientific laws." Thus, the participants said, consciousness is not a "thing" to be studied by an observer who is somehow apart from it. Consciousness involves the interaction of the observer and the observed.

Unfortunately for us with finite lives, we will never know whether human beings as creatures on this earth are in evolutionary pursuit of more and more consciousness. That may be what our genes want us to do, and perhaps our distant descendants will find that to be true. As we attend more and are more conscious about what is going on around us, we have more freedom, while at the same time we are more bound to the reality of the world. We can think of an increasing consciousness as an expanding playground for creativity, where we can learn in new ways how the world is put together. Altruism and consciousness are the steps that we're walking through that will define us more in the future.

Becoming "more" conscious would certainly improve our abilities as social animals. It would help us focus more on decisions and consequences, on associations, so that we are more keenly aware of our connectedness to others: where we are, where other people are, and what we are doing with each other. This is where attention and consciousness come together. As we gain more attention and consciousness we can better evaluate actions and consequences and be less impulsive than our current selves.

4

MOVEMENT

YOU ARE SITTING ON the couch in your living room. Your friend, seated next to you, asks, "How many shelves are there in that tall cupboard in your kitchen?" You visualize yourself walking into the kitchen, turning your head toward the cupboard, opening the door, and scanning the objects inside, up and down, to help you focus on how many shelves there actually are. You answer, "Four."

To reach this answer, you didn't move a muscle. But you used motor programs, which re-created the movements of going into the kitchen, opening the cupboard door, and scanning up and down. Your brain created motor images—mental simulations of movement—without actually moving.

Research is now showing that you used the same brain regions to re-create this experience as you would have used had you physically gotten up off the couch, walked into the kitchen, and opened the cupboard: the occipital, parietal, and frontal cortex. The purely cognitive process you used to reach an answer was carried out by the regions of the brain responsible for actual movement.

Although it seems to be common sense, this statement is heretical. For centuries man has defined himself as "above" the animals because he can "think," whereas animals just "act." Action—movement—was thought to be a "lower" brain function, and cognition a "higher" brain function that only humans have evolved. Until the last several years, most people didn't think any portion of the "motor brain" did anything but react to incoming stimuli and instruct motor functions. But we are rapidly finding that regions such as the parietal and frontal cortex play a large role in activity related to planning, calculating, and forming intentions.

Clearly, catching a ball involves the brain's motor function. But making a mental calculation does too. Most people associate motor function with arms and legs and physical activity—a mechanical brain function that causes a toddler to crawl, Michael Jordan to leap for a slam dunk, or the inaction of the right arm of a friend who's had a stroke. But mounting evidence shows that movement is crucial to every other brain function, including memory, emotion, language, and learning. As we will see, our "higher" brain functions have evolved from movement and still depend on it.

Neurologists are finding evidence that the cerebellum, which coordinates physical movement, also coordinates the movement of thoughts. Just as it orders the physical movements needed to catch a ball, it plays a role in the sequence of thoughts needed to visualize the kitchen, make an argument, or think up a tune. As we are finding again and again in this book, the old view that each brain function is isolated in a particular region of the brain is just not true. Spatial guidance, language, emotion, and many other functions share parts of the same brain systems, bringing different regions into play, in different ways.

Motor function is as crucial to some forms of cognition as it is to physical movement. It is equally crucial to behavior, because behavior is the acting out of movements prescribed by cognition. If we can better understand movement, we can better understand thoughts, words, and deeds.

WE THINK AS WE WOULD ACT

PLANNING, DELIBERATING, PONDERING, and acting are all about behavior—translating thoughts into deeds. In most cases, we think in terms of action sequences—formulating plans and then executing them. The deeds themselves certainly rely on motor function, but so does the thinking that precedes them. Even emotions are intertwined with the brain's motor abilities; the very root of the word *e-motion* means "to move."

Consider what occurs in your mind when you have to make a decision. You get inputs from various functions of the brain: facts, opinions, thoughts, memories, predictions of consequences. You sequence the pieces, add logic, test their outcomes, and then direct a response. The steps in this process are all grounded in motor functions: sequencing, adding, testing, directing; and the neural networks that fire during the process are the same ones that fire for a motor act.

The parallel goes further. When you first learn a motor act, such as riding a bike, you use the cortex. But as you master it, the activity becomes automatic, and responsibility for it is shifted to neurons in lower parts of the brain, freeing up neurons in the cortex for new learning. The same happens with cognitive acts. At first, you use the cortex to learn 4 times 4 or how to formulate a grammatically correct question, but as these tasks are mastered they are shunted to lower parts of the brain and become automatic.

When we activate the thinking process, we take bits and pieces of data and actions and behaviors and string them together to fit a new set of demands or circumstances, creating a novel plan of action. We reshape the raw material into a properly timed sequence. This is done by the prefrontal and frontal cortex, precisely the brain regions that guide what is commonly called the motor cortex. The brain circuits used to order, sequence, and time a mental act are the same ones used to order, sequence, and time a physical act.

Even the process of contemplation is the same. Try to formulate a response to the question "How does a rose mean?" You visualize a rose, consider smells, sights, tactile sensations, emotional responses, notions about beauty and thorns, and your personal experiences of giving or receiving a rose. You mentally walk around all these pieces to see how best to approach the subject, impose order, weigh various factors, and formulate a response. All of these steps use action terms, for good reason; the action circuits in the motor cortex are "doing the thinking process." Nature is a frugal tinkerer. It has adapted the circuits that were honed to sequence actions so they can be borrowed to sequence, adjust, and decide about thoughts. Likewise, it shifts the sequencing ability used, say, in shooting a basket or in figuring out the best way to organize this chapter.

Running the thinking process is the job of the brain's executive function, which has already appeared several times in this book. Recall that the executive function is rooted in the frontal cortex. This function is all about action—how to approach the customer or the poem. The whole front half of the brain is devoted to organizing action, both physical and mental. It is also home to the great brain centers for working memory, motor planning, and the ability to inhibit competing stimuli, thoughts, and actions. At their base, higher cognitive processes, as we like to call so many of our brain activities, are about organizing actions.

The executive function also allows us to stop and consider, and not rush to navigate clumsily through life. The executive function—the CEO of the brain—weighs consequences and decides upon correct and precise strategies. To execute properly, we need to plan, to be aware of how we are affecting the environment, to monitor ourselves as to how we are doing, and to update the plan with information coming at us from our initial actions. The executive function and motor cortex do it all. Just as they direct Michael Jordan, as he leaps into the air, to shift his body and transfer the ball from his right hand to his left to beat a defender to the basket, they tell us, as we enact our first decision, how to shift and transfer parts of our plan as the world reacts around us.

As it does for attention, the anterior cingulate gyrus in the frontal cortex acts as the executive secretary, deciding which incoming information is passed on to the prefrontal cortex for planning and deliberation. There is no center of decision or center of the will to act. These result from a confluence of activity from all over the brain that ends up in the frontal cortex, which is more extensively interconnected to all the brain regions than any other area. It is here at this confluence that the inputs compete and cajole, bump up against ethics and impulses and consequences; that memories are consulted and held or lost; that we struggle, combine and recombine ideas, work things out, consider goals and the steps to achieve them, reach above where we have been, and grow to reach a higher plane. Some ideas and inputs are boosted while others are inhibited, until the whole tangle is sorted and we come up with a new creation. All this computation, taking place across the cortex and coordinated by the motor neurons of the executive function, is the very definition of many forms of cognition.

The brain then returns to what might be termed "classical" motor function in response to the determined outcome, raising or lowering blood pressure, quickening or slowing breathing, turning hormones up or down, and instructing the muscles to act or not act.

It may even be that our sense of self-awareness is driven in part by motor neurons. New research indicates that some motor and cognitive tasks are processed in another region separate from the cortex: the cerebellum, at the back lower area of the brain above the spinal cord, heavily influences the cortex.

Until recently, it was thought that the cerebellum's role was solely to regulate the speed, intensity, and direction of movement. The cortex sent signals through the cable of fibers to the cerebellum, which processed and coordinated them and sent them back to the motor cortex, which controlled movement. But while dissecting human brains, Henrietta and Alan Leiner, both retired computer scientists in their eighties, found that the cable of fibers was far thicker, in proportional terms, in humans than in monkeys. Henrietta then discovered that a small structure within the cerebellum called the dentate nucleus—the last processing stop before signals are sent back to the cortex—was relatively larger in apes and humans, and that the most evolutionarily recent part of the dentate nucleus, the neodentate, is present only in humans.

Given the evidence, Henrietta concluded that the cerebellum might play a role not just in movement but also in cognition. If this is true, its outgoing information would have to be sent to regions other than the motor cortex. Neurologist Robert Dow of Good Samaritan Medical Center in Portland, Oregon, was the first to provide clinical support, showing that a patient of his with cerebellar damage had subtle cognitive problems with planning. Other studies were started, which have since linked damage to the cerebellum with problems in word selection, judging the shape of objects, and creating properly proportioned drawings. In one test, when a person with a normal cerebellum was asked to randomly slip rings of different sizes on a pole, the cerebellum showed normal activity. But when the person had to slip the rings on in the order of their size—in other words, in a proper sequence—the cerebellum's activity increased.

MIND OVER MOTOR

FURTHER INDICATIONS OF how intricately motor activity is linked to supposedly pure "mental" functions are being found in some classic conditions with major motor components, such as obsessive-compulsive disorder (OCD) and Tourette syndrome, as well as in some unlikely ones, such as rheumatic fever and toe-walking.

On the face of it, OCD seems to be a purely emotional and cognitive problem, one of anxiety and repetitive worry and rumination. But brain scans show that when this behavior occurs there is a "locking" of the neural circuits from the basal ganglia up through the anterior cingulate gyrus and the orbitofrontal cortex, a part of the frontal cortex that lies just above the nose. The anterior cingulate tells the orbitofrontal cortex what it should pay attention to, while the orbitofrontal cortex itself identifies what seems to be an error in behavior. It says, "Error, error, this action is a mistake." When the signals about attention and error conflict, motor programs get caught up in the turmoil. A panic message results, telling the brain to activate to get out of danger or to correct the problem by taking an action, such as returning to the house for the third time to turn off the stove that is already off. The typical OCDer is a perfectionist who is interminably searching for error. He or she explodes with worry and gets caught in a never-ending do-loop of concern and rumination. Did I make the right move? Is the sequence correct? All of these concerns have roots in the motor system of the frontal cortex, the anterior cingulate, and the basal ganglia.

In Tourette syndrome, there is a combination of obsession and compulsions with tics and rituals. Neuroscientists now think that a key area involved with the troubling tic, which is clearly a motor dysfunction, is the caudate nucleus, part of the striatum, which acts like a gear shift for the attention system.

Further insight into the link between motor and cognitive dysfunction comes from the study of children who have had the unusually named condition Saint Vitus' dance—kids who, after a strep infection, developed rheumatic fever. Rheumatic fever is an autoimmune reaction to the body's own tissue. Often when we hear of this phenomenon, it is related to an attack on the heart valves; the body sets up an antibody response to the streptococcal bacterium, and then the antibodies somehow begin to chew up the heart valves. But the antibodies can attack brain cells, too. When this happens, it can cause strange tics and the bizarre involuntary muscle movements of chorea. Susan Swedo and her colleagues at the National Institute of Mental Health, who looked more deeply into profiles of children diagnosed as classic OCD, found a number of them whose symptoms had begun after a strep infection. The proper name for their condition is Sydenham's chorea, and it is a neurologic accompaniment of rheumatic fever. After a strep infection the body's antibodies cross-react with brain tissue and set up a variety of symptoms. The patients may begin to have involuntary movements. They are often described as having changed overnight. All of a sudden they seem clumsy. As this develops further the children begin to make sudden, nonrepetitive movements and prominent facial grimaces. They also have a sudden onset of classic OCD symptoms like excessive handwashing, nighttime rituals, checking behavior, and obsessions about the death of their parents. The movements can become fantastic, and look like a choreographed dance piece. Swedo and her group found that this was definitely related to the strep bacterium when there was a sudden onset. This led them to look at Tourette disorder, which is all about tics and sudden uncontrollable movements, grunts, and facial grimaces, and at OCD symptoms, and suggested that Sydenham's chorea, Tourette disease, and OCD are related disorders. The group recently identified a genetic marker called D8/17. In a large group of children with OCD, 85 percent had this trait, and 89 percent of those with Sydenham's chorea had it, while only 17 percent of control children had the trait.

It came as a shock to the professional world when NIH researchers first announced the discovery of a group of children, affectionately dubbed the PANDAS (for "pediatric autoimmune neuropsychiatric disorders associated with streptococcal infections"), who had developed OCD as a direct result of strep throat. In many cases the children were completely normal before the infection. In others, they already had well-controlled cases of OCD that sharply deteriorated immediately after the child was taken ill. In the most dramatic cases, a psychiatrically normal child might develop strep throat on a Saturday and present full-fledged OCD behaviors by Monday.

One ten-year-old boy, T.J., had no history of psychiatric or neurological problems. The weekend after several family members had the "flu," he had a sudden onset of severe obsessions about viruses and chemicals and began compulsive hand-washing. After a month of continuous illness, psychiatric treatment was sought and T.J. started drug therapy for OCD. After two months his symptoms were only partially relieved. By then his forearms and hands were chapped and red from repeated washing, and his extreme worry about picking up germs prevented him from fully opening his mouth, so much so that he would not eat in the hospital or allow anyone to take a throat culture.

Cases like T.J.'s stunned practicing psychiatrists, virtually all of whom had been schooled in the classic Freudian view that OCD somehow stemmed from a disturbance in toilet training. Acute-onset cases like T.J.'s totally upset the traditional psychodynamic, interpretative framework for the disease, even the updated, biologically aware versions that put some blame on an imbalance of serotonin. These children were virtually "catching" OCD, or at least tics, the way a person catches a cold. There was nothing social about it.

MRI scans revealed that in these children the caudate nucleus, the area implicated in OCD, had swollen to as much as 24 percent larger than normal. What's more, the degree of swelling directly correlated with the severity of the OCD symptoms. Researchers surmised that the antibodies created to attack the strep bacteria were attacking the caudate neurons of the children's brains.

The way to counter such autoimmune responses is to suppress the harmful antibodies. Some of the PANDAS were given immunosuppressant medications. Others had their blood plasma removed, treated, and then restored to the body, free of strep antibodies. So far these approaches have been successful. T.J. had six plasma exchanges over two weeks, and his symptoms declined noticeably. After the fourth exchange he was willing to eat at the hospital and allowed a nurse to take a throat culture. He was reported to be doing well several months later.

Both T.J.'s OCD symptoms and his tics and motor irregularities left at the same time. We see here the confluence of the motor, behavioral, and cognitive aspects of the brain, and can appreciate better that the cognitive functions, which we tend to view as somehow distinct and elevated from the rest of the human experience such as movements, are actually directly linked to them. The brain uses the same machinery for many overlapping functions, in this case the cognitive behavioral functions of the OCD symptoms as well as the motor behavior that was purposeless and uncontrolled.

Toe-walking is another condition that suggests a strong link between motor and cognitive development. Several limited studies have indicated that 75 percent of preschool children who habitually walk on their toes may have significant delays in their speech and language development. In one study of 799 developmentally disabled children, toe-walking was found to be more frequent in those with severe language disorders. Although toe-walking can be a sign of cerebral palsy, various brain disorders, spinal injury, muscle weakness, and autism, the children in this study had none of these problems. They exhibited what is called idiopathic toe-walking, which simply means that there is no known cause for the condition. Further research is needed to clarify whether there is a link between this motor dysfunction and difficulty with speech and language, and what that link might be. Some doctors are now suggesting that idiopathic toe-walking in young children be viewed as a marker for possible developmental problems, and they recommend that any child with this condition be referred for a developmental assessment.

The point of the findings concerning all these conditions is that not including the motor function with the other, higher brain functions is like building castles on sand. The fact that there is constant activity in our brains and throughout our bodies tells us that movement is the ongoing life force without which we could not survive. Whether the activity is maintaining body temperature, dancing like Mikhail Baryshnikov, or learning to read, movement cannot be separated from other brain systems. A lot of brain function is, essentially, movement.

WHAT KEEPS OUR MOTOR MOVING

HUMAN EXPERIENCE BEGINS with information about the world that flows in through our senses, but depends on how that information is combined with internal states to produce action. The motor system extends throughout the body, from neurons in the spinal cord to neurons in the brainstem and motor cortex. While a lot of brain talk has concerned the interaction between the left and right hemispheres, we really should be thinking more of the interaction between the front and the back of the brain—the sensory and motor divisions.

To the novice, the study of movement may seem straightforward, but it is highly complex. Many of the internal structures involved in movement are deep within the brain, making them difficult to examine. Boundaries between the structures are not sharp, and structures and functions overlap. Sometimes the brain isn't even consulted; tapping the patellar tendon just below the kneecap triggers a reflex in about 50 milliseconds that takes place across a circuit with only a single synapse to and from the spinal cord; the brain never comes into the act. If you trip while walking, though, a host of brain circuits that command balance and posture are invoked. Other stimuli initiate sequences of movements that have been previously learned. The sight of food can cause us to eat. When we see someone we love, we may be "moved" to hug or kiss that person; and when we are engaged in a debate about national health care, we manipulate a range of multidimensional, higher-order thoughts to build our argument.

Movement is fundamental to the very existence of a brain. Interestingly, only an organism that moves from place to place requires a brain. Plants enhance their chances for photosynthesis by turning their leaves to face the sun, but this is done through the growth of cells, not by changing their position. A tiny marine creature known as the sea squirt illustrates the point. In the early part of its life, the sea squirt swims about like a tadpole. It has a brain and a nerve cord to control its movements. However, when it matures, it attaches itself permanently to a rock. From that moment on, the brain and the nerve cord are gradually absorbed and digested. The sea squirt consumes its own brain because it is not needed anymore.

Over time, the natural selection process has reconfigured the landscape of brain territory. The primary motor cortex and premotor cortex are both located in the frontal lobe, one of the most advanced parts of the brain, which is also responsible for the higher executive functions such as thinking and planning. It allows us to ponder, judge, and make decisions about consequences and alternative responses before taking action. This master planner—which Karl Pribam called the executive and which I renamed in Chapter 3 "the CEO of the brain"—runs the show. In doing so, it receives a convergence of inputs from other areas of the brain and uses it to plan movements. The primary motor cortex then directly controls particular movements.

While the sensory cortex, located just behind the primary motor area, provides a significant source of input to the motor cortex, a great deal of information about our thoughts, past experiences, emotions, and stored memories also floods into the motor areas, contributing meaning, depth, and complexity to our movements and actions.

WHILE A NEW TASK is being learned by specific circuits in the frontal cortex, many neighboring neurons drop whatever they are doing to assist in the process. The learning territory is spread out to accommodate the barrage of input. Once the task has been mastered and executed a number of times, the firing patterns become established and the behavior becomes automatic. It no longer requires conscious attention. That is why you can get on a bike and ride it, even if you haven't ridden for twenty years. The ability to ride becomes second nature and lasts a lifetime. The information is condensed and routed to the subcortical structures deep in the brain, where it is stored for future situations. The neighboring cortical neurons are freed to return to previous duties, or to be available to take on new learning.

Where do these programs go? The findings of Henrietta and Alan Leiner and of Michael Merzenick, whose monkeys learned how to grasp food from successively smaller cups, suggest that they may go to the cerebellum and basal ganglia. Becoming a super athlete or piano player may require an efficient mechanism for the transfer and storage of these programs. A person who can push down more and more intricate motor sequences can be engaged in complex motion and still have a quiet frontal cortex. His higher brain is not so busy "doing" and is more available to observe and make adjustments—to the converging defenders on the basketball court or the upcoming notes and dynamics of a concerto.

Acquiring numerous repertoires of automatic actions is essential to survival. We would not be able to do much, if any, pondering or considering, or even focus on anything as simple as what to wear for the day, if we had to consciously attend to all of our actions all of the time. Your morning routine alone would exhaust the cortex! Just imagine: The alarm goes off. You wake up. You have to tell your body how to rotate and lift your arm, how far to extend it, how to move the index finger, how much pressure to apply—all just to shut off the buzzer. Then you have to consciously instruct all four limbs and your torso in how to coordinate a series of movements just to get your body out of the bed and stand up. You have to concentrate mightily, as a one-year-old does, on simply putting one foot in front of the other to walk without falling over. Fixing breakfast, brushing your teeth, driving a car? Forget it. Your cortex would sag under the load of motor instruction.

The same dynamic holds true for cognitive processes. Assuming you made it to the breakfast table, you would have to deliberately sequence through the thousands of linguistic constructs that took you years to learn in grade school just to read the morning newspaper. If the many cognitive routines you invoke each hour of the day were not automatized, you would never advance past the abilities of a child.

The frontal cortex learns, routinizes, and processes motor and mental functions in parallel. Movement becomes inextricably tied to cognition. The issue of how the connection is made between motor programs and reasoning is complicated yet simple. Processes that are fundamental and mastered are stored in and executed from the brainstem, basal ganglia, and cerebellum in the lower brain. Actions and cognition that are increasingly more complex, or very new, are managed further up in the brain, increasingly toward the frontal cortex, so that more brain regions are employed along the way that can offer input or provide delay for consideration. This allows more neurons to be involved in readjusting and sculpting a more precise final action or cognitive process.

The parallel handling of motor and cognitive functions also helps us when we have trouble mastering one or the other. For example, when we have trouble with a cognitive act, we can invoke a physical one to help out. How many times have you, or someone you know, been unable to solve a problem until you take a walk or go for a drive—distinct motor acts governed by largely automatic programs? Dyslexics often mouth the harder words as they read, totally unaware they are doing it; many of us talk out loud when we come across a complex or a foreign word we don't recognize. In these cases we are activating numerous motor centers, automatic and deliberate, to achieve the cognitive function. For the most complicated behaviors, we may well use many levels of the brain, coordinated by the smooth integration of layers that motor function provides.

As life has gotten more complex, the interconnection of movement and cognitive processes has become stronger. The seventeenth-century

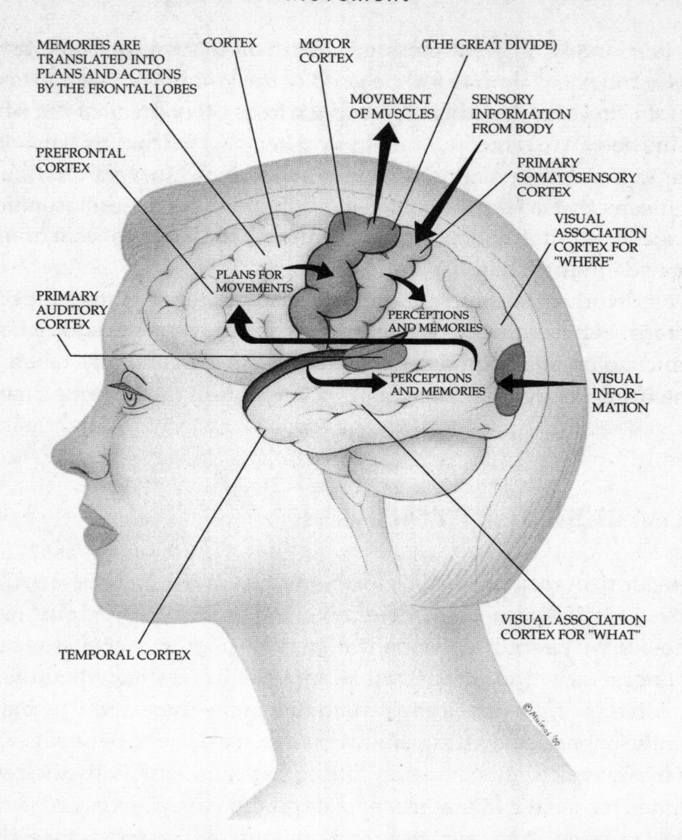

THE MOVEMENT BRAIN Here we see the gross division of the brain into two halves, the front and the back, divided by the central sulcus. The back of the brain is the sensory or input half, which receives input from the outside world and sorts, processes, and stores all of our sensory representations. In the front of the brain, the cortex is devoted to the processing of motor programs or output—we use this area to react to the input data. It is here we plan, strategize, and sculpt our responses to the world, and it is this area that has been adapted for use in abstract thinking and planning.

French philosopher René Descartes once stated, "I think, therefore I am." Many people still believe in the separation of mind and body. However, more and more scientific evidence is showing that this is an artificial distinction. What the brain communicates to the body depends largely on what messages the body is sending to the brain. Together they collaborate for the good of the whole organism. Almost all brain function depends on feedback from other areas of the brain and the body. We forget we are physical beings. We learn by behaving, either in mind or in reality. The brain's dictum is "survival first," and that means that when we confront anything, even the philosophical, we take in the information, then walk around the issue, touch it, mull it over, and then act on it.

When working properly, the motor system allows us to shift back and forth between deliberate and automatic movements and deliberate and automatic cognition. This ability, which is largely taken for granted, allows us to perform many different tasks at the same time.

FROM BASEMENT TO GARRET

ALTHOUGH DIFFERENT BRAIN AREAS are involved in movement, the connections between them determine our activity. One convenient model for the hierarchical organization of the motor system is a house with a basement and three floors. Each level has different equipment for operating and maintaining certain functions. Some of the functions can only be performed if signals are sent up or down a two-way staircase that connects all the floors. Shifting back and forth between deliberate and automatic functions depends on this communication.

The basement contains the brainstem and spinal cord, which have the hard-wired neuronal networks responsible for internal fixed actions, such as maintaining heartbeat and reflexes. The first floor is home to the basal ganglia and cerebellum, which direct movement and operate the apparatus in the basement, as well as providing input about the status of the body to the higher floors. On the second floor are the motor and the premotor cortex, which receive large amounts of information from other brain areas and send out instructions to the musculoskeletal system and organs. The third floor is the eagle's nest, the garret, the command center in the prefrontal cortex, where the CEO sits, deliberates, makes choices, and sends out the signals that inhibit or excite the lower floors, which affect, among other things, how much we feel our gut feelings and how quickly we react to them.

The flow of communications between the floors is constant and reflexive, feeding back on itself as each moment advances, so decisions and actions are as appropriate as they can be.

A BRAIN-MAPPING EXPEDITION

NUMEROUS TESTS SHOW that several brain centers work in concert to orchestrate a particular function, just as the house model would predict. Which centers activate on which floors depends on the type of movement required, the intensity of activation, and the way in which the movement was sparked, although some types of functions do seem to be more localized in certain parts of the brain than others. For example, vision is processed in the occipital lobe and speech in the temporal lobe.

We are also discovering that areas of the brain heavily associated with executing particular functions are actually just way stations in the neuronal transmission process, places where inputs converge from other brain areas, especially emotion, cognition, memory, and perception, before the brain determines which actions and behaviors it will order. It is the particular context of information that surges throughout these areas and influences, if not determines, the action and behavior that follows. As the flow of information is fed into these way stations, circuits get primed, a threshold is crossed, and a particular movement or behavior is executed.

Say you are walking in the woods and a snake suddenly appears. The snake image is processed in the visual cortex and compared with memories of other snake images. When "snake" is identified, the amygdala is alerted and a reaction is initiated to go into full-scale alert. This activates the basal ganglia, cerebellum, and motor cortex to get all body systems fired up. The heart pumps harder, and the leg muscles contract. The eyes stay fixed on the snake as the head turns to follow its movement. You may start to run, and when the frontal lobe finally gets involved, you can determine whether to panic or relax.

All systems have to be acting in concert with one another. The absence of one of them will affect the whole train of associations, and the decision may never get made. So to say the decision to flee is "made" in the frontal cortex is not really true, since it is only the final step, dependent upon calculations done in many other brain regions.

We therefore have to be very wary when we are presented with a nice, neat map of the brain. For six decades investigators have created brain maps that reflect the "latest" research. The maps have evolved from well-defined point-to-point grids to messy configurations showing complex overlapping areas. The phrenologist and the geographer live within us: we want to know where and how. But this is just not the way the brain is organized.

One of the first brain maps of motor functions was drawn in the 1930s by a Canadian neurosurgeon, Wilder Penfield, who studied epileptics. He identified precise areas of the primary motor cortex that were directly responsible for specific parts of the body; the feet and legs lay at one end of the map, followed by the torso, arms, hands, and face. This picture has come to be known as the motor homunculus. Today, however, rather than a spatially organized map, there is a map of combinations of muscles arranged in useful ways, like stops on an organ that can be pulled open when you need a certain combination of sounds. In such a scheme certain neurons might control a coordinated set of muscles to produce a common component of many movements, such as reaching, while others would be added to provide the fine-tuning necessary to make that movement unique, say, grasping, so you can pick up a paper clip.

Even though specific areas of the brain are not solely responsible for executing a particular movement or behavior, and even though several different brain centers work in concert to orchestrate a particular motor function, there are still basic brain structures that are crucial for basic functions on and between the four floors of the house model.

The basal ganglia and cerebellum on the first floor provide basic movement control and store many of our primitive reactions, as well as many of the learned programs that have become automatic. Right alongside them is the limbic system; hence the close relationship between emotion and movement, and the feelings of emotional consequences of our actions. This explains why emotional changes may accompany certain movement disorders; for example, it is not uncommon for depression to accompany Parkinson's disease.

The cerebellum on the first floor is primarily responsible for balance, posture, and coordination. Because the cerebellum allows us to rhythmically shift our attention, it is heavily involved with most if not all

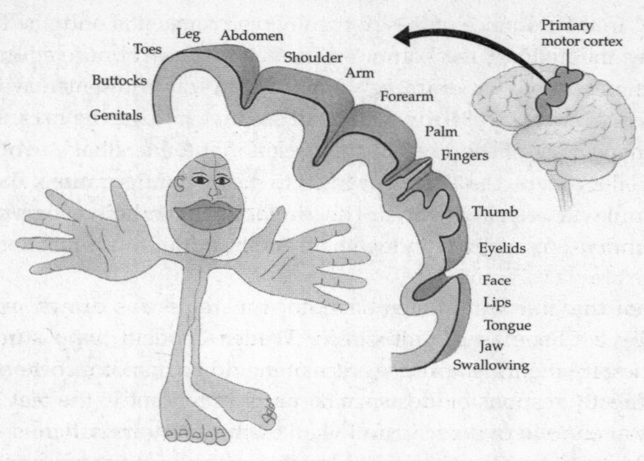

A MOTOR HOMUNCULUS Neuroscientists have employed a variety of homunculi, or "little men," distorted figures of the human body, to show how much of the cortex is devoted to specific body areas. The motor and the sensory systems roughly parallel each other, and the "little man" pictured here illustrates how much cortical space is devoted to the hands, lips, mouth, and feet, all of which have assumed far greater importance in our lives compared to our primate ancestors.

systems. Cerebellum is Latin for "little brain," and at times it seems to have a mind of its own. Information about body movement and position enters the cerebellum, where it is processed. Instructions are then sent out to modify posture and coordinate muscle movement. This is more crucial than it may sound. For movements to be performed, the brain must know the position and speed of your body and of each limb and where you are in space and time. Spatial orientation and posture are essential to knowing "where you stand." The only reason you remain upright and don't fall down because of gravity is constant monitoring by the cerebellum. It adjusts postural responses at the brainstem, which sends messages down the spinal cord that control muscles that straighten and extend the torso and limbs, fighting against the downward force. This incredible feat is being accomplished all the time, without our being aware of it.

On the second floor are the motor and the premotor cortex, which control things such as specialized movements of the face and limbs, particularly manipulative movements involving the arm, hand, and fingers. Good evidence of the extensive interconnection and feedback between the floors of the house can be seen by studying the effects of the motor cortex. For example, when we are happy we smile, and when we smile we feel happier. One of the major emerging principles in the neurology of the 1990s is the notion that the feedback between layers or levels of the brain is bidirectional; if you activate a lower level, you will be priming an upper level, and if you activate a higher level, you will be priming a lower level. So smiling can improve our mood.

Some fundamental movements do not require communication across levels. For example, although PET scans indicate that the motor cortex is active during walking, climbing, and swimming, these were possible after removal of the motor cortex in rats. Some levels may also substitute for others in certain cases; the same rats who had no cortex were also still able to eat, drink, nurse their young, and reach for food, but they were not able to hold food in their paws very well and could not move their fingers one at a time. Even though the second floor was removed, the first floor and basement took over, although the movements under their control were less refined.

The well-connected motor cortex guides complex actions that require the coordination of several muscles. In some ways it functions—in our now familiar analogy—as the conductor of an orchestra. Without the conductor, the musicians could play together, but at times there would be chaotic or random sounds. What transforms the music into a symphony is the conductor, who maintains a connection with all of the musicians and oversees the whole orchestra. The conductor also controls who will play and when, how fast or how slow, and how loud or how soft. Feedback is important, as the conductor listens for and receives information about how the whole orchestra is doing in order to make adjustments, such as slowing down or speeding up different players. The conductor makes sense of the whole system.

The motor cortex oversees multiple muscles that create movements ranging from the simple pressing of an elevator button to performing a gymnastics exercise or tying a shoe. It provides the organization of smooth, timed, and rhythmic movements among the many brain structures and spinal cord. Monkeys that have lesions in their motor cortex can be rehabilitated to do many simple movements that depend largely on single muscles, such as flexing and extending their wrists.

But they never regain their ability to make smooth diagonal movements, which requires the use of multiple muscles at the same time. Instead, the movements must be carried out in two parts, using two different muscles, one at a time. The result is a chopped-up motion, like that of a classic Charlie Chaplin routine, in which he raises a leg to take a step, but then stops it in midair, holding it there until he decides what to do with it next.

If injury-free, the motor cortex has amazing plasticity. It is different, for example, in a professional violin player's brain than in ours. Playing a stringed instrument involves considerable manual dexterity and sensory stimulation of the fingers of the left hand, which continuously press on the strings. MRIs reveal that the neuronal region representing the digits of the left hand of string players is substantially larger than that in other people—in the primary motor cortex as well as the sensory cortex. The corresponding representation for the thumb, which is not as actively involved in violin playing, is normal.

Motor activity is affected by many factors, including some we may not immediately think of, such as motivation. The anterior cingulate gyrus, also on the second floor, appears to play a crucial role in initiation, motivation, and goal-directed behaviors. It is well interconnected with the amygdala and other structures of the limbic system that regulate our emotions, the fight-or-flight mechanism, and conditioned emotional learning. In short, it assesses just how important something is, determines an appropriate response, and decides how quickly the response will be executed.

PLANNING AND MOTIVATION

PLANNING MOTOR ACTIONS from moment to moment is obviously crucial to daily life. Motor planning involves having an idea about what to do, planning an action, and finally executing it. New actions are planned using knowledge of sensory information, and past experiences and the sensations that accompany them. When motor planning occurs, a person is able to deal with a new task by organizing a new action, such as when a preschooler who encounters a riding toy for the first time is able to figure out how to get on and off.

Psychological and contextual factors affect the physiology of the motor brain. We find some understanding of how this works by looking at ideomotor apraxia, or IMA: making errors in producing movements needed to work with objects as tools, when there is no underlying disorder of movement or language. It is associated with many diseases of the central nervous system and is often seen in patients who have suffered stroke or degenerative dementia such as Alzheimer's disease. For example, when people with IMA are asked to pantomime using a pair of scissors, they may move their fingers as if they were the blades, instead of positioning their hands as if they were holding the scissors. They use their body parts as the tools themselves.

Intention and attitude can affect motor function, too. When Antony Marcel of Cambridge, England, asked IMA patients to pick up a cylinder about the size of a drinking glass and imitate drinking, they could not do it. However, if patients were asked to do the same task while eating a real meal, they were more successful. The more meaningful the task or the intention, the better the patients' performance; the context and meaning of the situation in which a motor act is learned affects how well we learn it. The more the act relates to things that are important—eating real food versus picking up an imagined glass—the easier it is to learn and sequence the movements or behavior. The simple implication for all of us is that if we can see challenges in everyday events, then add cheering, coaching, and support to the mix, we make more neurons—whether in ourselves or in others—available for learning.

MOVEMENT DISORDERS AND REPAIR

SCIENCE IS LEARNING more about motor function from movement disorders resulting from injury, which have recently received a great deal of attention. The tragic equestrian accident that left Christopher Reeve paralyzed from the neck down brought an increased public awareness of the devastating effects of brain and spinal-cord damage. Reeve has become a public spokesperson, urging Congress to allocate more funds for research into repairing crushed nerve tissue, and has inspired foundations and organizations to fund such work. His courage and persistence have also inspired neuroscientists with the hope of finding new and innovative ways to repair damage to the motor system.

Here again, traditional beliefs about the structure and function of the brain are being challenged. It was once thought that if a particular area of the brain was damaged, then a particular function would be lost. It was also believed that brain damage was permanent—that the brain could not be repaired or repair itself.

There are many documented cases in which patients who have sustained brain or spinal-cord damage have shown significant improvement. This goes for lost motor function as well as other brain functions such as language. These incredible recoveries, like those we have encountered in people who have had significant portions of their brains removed, are attributed largely to coping strategies that patients' brains have developed to compensate for the damage.

Researchers are exploring the brain's own natural capacities to repair nerve damage. Neuroplasticity is a new term that describes the ability of nerve cells to change and modify their activity in response to changes in the environment. This capacity may be instrumental in the brain cells' ability to resist or overcome injury and disease. Recent research suggests that although a given neuron may display a particular function, it is likely to take part in other functions as well. The ability of neurons to perform more than one function may have significant implications for helping people with brain and spinal-cord damage to reverse motor loss.

One controversial treatment for brain damage involves the transplanting of healthy brain tissue from aborted fetuses. The first transplants a decade ago prompted a media frenzy. Journalists reached for their Frankenstein clichés in a way that anticipated the recent reaction to the cloning of sheep. As President Clinton would later do for human cloning, President Bush banned the new technology using fetal tissue, or if that were impossible, cut off federal funding for it. Then as now, researchers defended the value of what they were doing, and continued their work at universities in Europe and privately funded ventures in the United States. In 1993, presidential disapproval of the experiments was lifted.

However, partly as a result of that start, no conclusive clinical trial of the process has ever been completed. Since the first grafts were performed at the universities of Lund, in Sweden, and Colorado, at Denver, only a few hundred transplants have been carried out, so no one group of patients has been big enough to yield meaningful results. As a consequence, the real value of the treatment is only just emerging.

The target has primarily been patients with Parkinson's disease, whose substantia nigra, at the base of the brain, stops producing dopamine somewhere around age fifty. Dopamine is essential for carrying signals between these neurons. Normally, the substantia nigra supplies packets of dopamine to the striatum to coordinate movement. No dopamine means no smooth muscle movement. One way of restoring the dopamine supply, the theory goes, would be to replace the faltering substantia nigra cells with ones that work; and a good source of such cells would be the brain of an aborted fetus.

According to Curt Freed, one of the pioneers at the University of Colorado, roughly two-thirds of Parkinson's patients improve with fetal grafts. Half of these can abandon their medication altogether while keeping up normal appearances. A study by Olle Lindvall, who helped to start the work at Lund, suggests that remission can last for up to six years.

The figures are still open to debate, however, because a standard technique for doing the transplants has not been used. There is disagreement about how much fetal tissue to graft (up to eight fetuses are used for some patients), how best to scatter the material in the adult brain to ensure a successful outcome, and why one-third of patients fail to improve. Large-scale U.S. trials begun recently may soon provide answers.

Researchers are also trying to tackle Huntington's chorea (now known as HD, or Huntington's disease), which also involves the striatum and has even nastier effects than Parkinson's disease. Parkinson's often can at least be checked with the drug L-dopa, but there is as yet no relief or cure for the progressive course of Huntington's disease.

Even if transplants prove effective, there are major ethical concerns to consider. Legislators are wrestling with how to assure that a woman's decision to have an abortion is not influenced by payment, or by the idea that her fetus's tissue may help a victim of Parkinson's disease. A number of U.S. states still prohibit the procedure.

New techniques may make the debate moot. Researchers at Harvard Medical School have shown that nerve cells taken from fetal pigs can survive and mature in the brain of a Parkinson's patient much as human grafts do. Only a dozen Parkinson's patients and a dozen Huntington's sufferers have had such transplants, but their recovery rate is similar to that for human transplants. Over half of them showed improvement six months after surgery, regaining some of their motor control. Transplanting pig cells into people is still controversial, however, since it brings the risk of transplanting pig diseases (which may be contagious) at the same time.

There is another option: generating replacement cells from a patient's own brain in a tissue culture. Recent research has overturned the old neurological dogma that adult brains cannot renew themselves. It used to be thought that neural stem cells—which divide to produce nerve cells in an embryonic brain—shut down in adulthood. But Brent Reynolds and Sam Weiss at Neurospheres, a Canadian biotechnology company, have shown that stem cells are still alive and well in adults. They just need to be switched on again. This prompting is done with growth factors—molecules that stimulate tissue growth by turning genes on and off and then maintain mature organs. Reynolds has shown in controlled experiments that stem cells treated with growth factors can be persuaded to produce new nerve cells. If these can be successfully transplanted back into the patient's brain, they might solve the problem, with no ethical issues or worries about rejection or disease.

Yet another alternative is being pursued by Ontogeny, a company in Cambridge, Massachusetts, that is working with the potent, if improbably named, protein "sonic hedgehog." This is one of the growth factors that—in a lab setting, at least—transforms stem cells directly into mature dopamine producers. Ontogeny is betting it will be able to do the same *in vivo.* Its researchers are now shooting sonic hedgehog directly into the brains of mice to see what happens. Human testing will come later.

Amgen, one of the oldest and largest biotechnology companies around, has begun clinical trials of a growth factor called GDNF. The trials, however, require regular injections of GDNF straight into the brain. So, presumably, will Ontogeny's. This means that a sufferer has to have a hole drilled in his skull, and a catheter fitted.

It would be easiest of all, therefore, to install the means of producing the growth factors directly on site. Fred Gage, of the Salk Institute in La Jolla, California, and Mark Tuszynski, of the University of California at San Diego, have been experimenting with unassuming skin cells. They removed some skin cells from their experimental animals, added a few new genes to encourage them to make neuronal growth factors, and then let them multiply. Then they implanted the daughter cells back into damaged areas. Rats with severed spinal cords that received these grafted cells have been able to regrow neurons and regain some function.

A different approach for spinal-cord repair is being taken at the University of Florida in Gainesville, where Douglas Anderson and Paul Reier are gearing up to transplant fetal spinal cord into patients whose own cords have been crushed, as happens in over two-thirds of all spinal injuries. In animal experiments, such transplants mature into adult nerve cells and insinuate themselves into the damaged cord, restoring at least some movement to an otherwise paralyzed body. While promising, this technique brings back the abortion issue.

The progress in transplants and regeneration may prompt even wider discussion of issues because the techniques are already being applied to other conditions. Bone marrow has stem cells that replenish red and white blood cells day in and day out for decades. The skin, liver, intestines, and perhaps other organs are also thought to have their own stem cells, which replace injured and dead cells. Researchers are looking into how transplants and regeneration might help to alleviate ailments that affect all of these systems. Research is also under way by Professor Jeffrey Gray at the Institute of Psychiatry at London's Maudsley Hospital to see whether transplants or regeneration might replace brain cells that die from lack of oxygen during stroke or heart attack.

There is hope, too, in understanding more about how the brain can reorganize itself around damaged areas. William Jenkins and Michael Merzenich at the University of California at San Francisco recently performed an experiment on monkeys to determine whether or not reorganization could occur in response to training or rehabilitation. The animals were trained to maintain hand contact with a rotating disk in order to get bananas. The training caused a clear reorganization of the brain regions, a reorganization not found in untrained monkeys. And the differences lasted for a significantly long time after the training was discontinued. The research demonstrates how training can produce functional reorganization in the cerebral cortex and has important implications for rehabilitation therapy. It also provides additional support for the notion that when a reward is involved, especially one basic to survival (food), the brain can quickly recruit extra neurons to help it adapt.

THE INFLUENCE OF ATTENTION AND EMOTION

THINKING ABOUT PEOPLE who struggle to perform even basic movements reminds us that despite all our grand ideas and plans, the brain is first the supreme survival organ. Taking in information, processing it, and responding are driven by what is needed to ensure that the brain's owner survives. Therefore, motor function takes place under the influence of attention and emotion, which have evolved to rapidly size up and respond to imminent danger. Attention and emotion are the primary processes that our bodies and brains use in the combined effort to thrive and survive in the face of continual challenge. Our brains use attention to constantly survey our internal and external environments to determine what is important and what is not. Emotion provides a quick, general assessment of the situation that draws on powerful internal needs and values. Clearly, these systems heavily influence the motor system and the motor system heavily influences them.

The most fundamental attention system involves the fight-or-flight response. Understanding how it works illustrates how attention, emotion, and motor systems work with and on each other.

Suppose you are walking down an avenue in a busy city, following directions to an apartment you've never visited. A bit lost, you venture down a dark, cluttered alley. Suddenly you hear a loud, crashing sound behind you. Signals from the ears head toward the cortex. On their way, some of the signals take a short side route to the amygdala, which checks whether an immediate response is needed. This tiny cluster of brain cells shouts, "What is that? Search for it. Find out. Alert!" to the autonomic nervous system.

The autonomic nervous system oversees the body's vital functions through subconscious signals that originate in the anterior cingulate and are relayed to the hypothalamus and the spinal cord. It has two reciprocal and complementary branches: the sympathetic and the parasympathetic nervous systems. These send out neurons to regulate the internal organs, such as the heart, lungs, stomach, and genitals. They balance and offset each other to keep the body in just the right tone automatically and without our being aware of the changes. This frees up the cortex to pursue the conscious services of sight, speech, hearing, thinking, emotion, and voluntary movement.

The instant after the amygdala shouts emergency, the parasympathetic nervous system ever so briefly suppresses the heart rate, breathing, and other internal functions. It quiets all systems so you can fully take in information and focus on perceiving and evaluating, and creates a bodily delay so the cortex can efficiently assess what that sound might be before you respond to it. An instant later, however, the sympathetic nervous system is driving up your blood pressure, pulse, and breathing, and producing adrenaline so that your muscles can spin you around and prepare you for fight or flight.

As this is happening, the perception and alert signals reach the frontal cortex, which evaluates the situation and decides whether or not there is danger. If it determines that a cat has tipped over a metal can, it calms the amygdala down, saying, "There's nothing to fear." Signals from the sympathetic nervous system reverse. Your blood pressure comes down and your heart rate returns to normal. The lower brain surrenders some control to upper portions of the brain. You begin to "think" about what is happening rather than just responding.

If a wild-eyed man is waving a gun in the air, signals are sent immediately to the hypothalamus. Unlike other responses that require a decision-making process, this response bypasses the upper cortex, so immediate action can be directed. Because it plays a crucial role in the regulation of body systems, the hypothalamus is often referred to as the brain of the brain. CRF, the brain's own stress hormone, is released, which heightens anxiety and vigilance and eventually sends instructions to the adrenal glands to release epinephrine (adrenaline) and cortisol, the "stress" hormone, to prepare you for action. The hypothalamus also directs the pituitary gland, the body's master gland, which secretes hormones affecting every major gland of the body. These systems reactivate the amygdala and the brainstem, triggering the sympathetic nervous system to put the body into overdrive so that you can run like you've never run before.

How these systems respond in the case of surprise in an unfamiliar alley or in situations that are not life-threatening depends to some degree on a person's history. Individuals who are generally calm will spend fractions of a second more time in the parasympathetic response, to better gather information and evaluate before they react. But people who have a history of overresponding may flee, or at least hurry off, even if they do not perceive an obvious danger. People who are anxious develop an overly efficient neural pathway from the amygdala to the sympathetic response. If this pathway is too well hewn, the brain is instantly hyperaroused to "Attend! Attend! Attend!" This excessive signaling makes it impossible for the frontal cortex to focus, because the noise and the strong panic signal commandeer the bulk of the frontal cortex's resources.

This kind of "hyped up" response mechanism is what can cause performance anxiety. If people who have histories of being anxious in front of a group have to step up to a podium on stage and give a speech, they may get thrown into a hypervigilant response mode. Their anxiety arouses the sympathetic nervous system. This, in turn, starts hormonal and cellular reactions that cause their muscles to tense. Heart rate and blood pressure go up and breathing gets shallow. These physiological changes send an attention signal back to the brain saying, "You're tense, you're tense, you're tense. Oh boy, this is bad news, this is bad news. Check everything." An upward spiral of anxiety is initiated; the feedback loop amplifies itself and becomes louder with each cycle. Once again, the strong signal commands too much of the frontal cortex's resources, and there isn't enough left for the cortex to put together the ideas in the speech, or perhaps even to coordinate the motor commands for the mouth and vocal chords. Such speakers are "struck dumb." The resulting message to the brain that sounds are not coming from the mouth of course amplifies the anxiety even more. Game over. Take a seat.

Some people who are prone to high anxiety and yet must do things such as give speeches get help from drugs called beta-blockers, such as propranolol and nadolol. These block adrenaline—the adrenal hormone—from pumping up in the large muscles, which causes them to become tense when preparing for fight or flight. They also act to lower blood pressure and pulse, breaking the spiraling feedback between body and brain.

Ironically, a little bit of performance anxiety can be a good thing. Many actors, for example, say that they actually welcome a touch of stage fright because it "puts them on edge" and gives their performance more passion and energy. The heightened attention and emotion systems engaged in a degree of survival response drive them to a superior performance by activating keener attention in the frontal cortex. Indeed, a mild amount of stress and activation is what constitutes stimulation.

EXPRESSING EMOTIONS

UNLESS WE FIND OURSELVES in a fight-or-flight situation, we take the milliseconds required to "think" before we react to incoming stimuli. Much of that thought involves the emotion system.

Most incoming sensory information is sent first to the thalamus, which then relays it to the sensory and frontal lobes for detailed analysis and response. But when emotionally charged information comes in, the thalamus sends it on a more rapid pathway to the amygdala, bypassing the upper brain's input since there is no time to think about how to respond. Based on the limited sensory information it has received, the amygdala uses primitive, general categorizations—primary emotions—to activate an immediate aggressive or defensive response. Specialized cortical networks in the right hemisphere and frontal lobes are responsible for secondary emotions and for modulating the more primal emotional responses of the amygdala and the limbic system.

Movement is a physical expression of e-motion. For example, before we feel sad, a thought is triggered either from memory, a current situation, or an imagined future. As this is happening, an array of chemicals and hormones are produced that act as internal messengers throughout our bodies. The physiology that is created changes how we "feel" internally. Our internal feelings, or emotions, can also lead to physical changes, such as the production of tears by the tear ducts. Bodily states such as breathing, blood pressure, pulse, and heart rate may change as well.

We and members of other species communicate our emotions primarily through facial gestures. These and other expressions of emotion such as crying and laughing are controlled by the amygdala and brainstem. The best evidence for this comes from a disorder known as pseudobulbar palsy, which is caused by damage to the outputs of the motor cortex, on the second floor. Individuals with this disorder cannot make voluntary movements of facial muscles; however, external events can still elicit laughter, tears, and facial gestures of emotion. They are responding with the first floor, which is still intact.

What's even more fascinating is that the motor control of movements related to emotion is not in the same location as the control for a voluntary movement of the same kind. For example, when a stroke destroys the motor cortex in the brain's left hemisphere, the patient experiences paralysis on the right side of the face. When asked to smile the patient cannot move the right side of his mouth. However, when the same patient is told a joke and laughs spontaneously, the smile is normal; both sides of the mouth move as they should. The cortex cannot exercise its usual control over the muscles, but the muscles still respond to the more automatic and implicitly learned responses that are located on the first floor—the basal ganglia.

MEMORY, THINKING, AND LEARNING

AS WE HAVE SEEN, the brain's motor function affects so much more than just physical motion. It is crucial to all other brain functions—perception, attention, emotion—and so affects the highest cognitive processes of memory, thinking, and learning. To help illustrate the intimate connections of these systems, imagine the following scenario:

You're late for work and you can't find your keys. You scratch your head as you try to remember where you put them last: "When was the last time I had them?" You may get frustrated (emotion) as you extend your fingers one by one (movement) to recount (thought) all the places that you could have left them. You may talk yourself out loud through this exercise. Perhaps you begin to visualize yourself as you returned home yesterday; after all, you needed the keys to drive. Your facial muscles, posture, and breathing rhythm will have changed to reflect your emotion of frustration. Suddenly you find the keys. You smile as you think to yourself, "Now I can get going."

In each step of this scenario, movement or action ran parallel with thought, memory, and emotion. Even though we would generally associate the thinking process with purely cognitive activity, could it be that the motor function is the home of cognition and that "thinking" is all about evaluating, ordering, and deciding on action plans? How does the motor system integrate into the cognitive process?

Motor activity takes place in three stages. First we analyze the incoming external and internal data. Next we formulate and monitor a response plan. Then, we execute the plan. The second stage of formulating and monitoring a response plan is the step that involves thought-processing.

Thinking is indeed a process, a biological function performed by the brain. Thought-processing is the act of receiving, perceiving, comprehending, storing, manipulating, monitoring, controlling, and responding to the steady stream of data. The ability to link information from motor, sensory, and memory association areas is crucial for thought-processing and the ability to contemplate and plan future actions.

The cerebellum is very much involved with the integration of information and the timeliness with which the information gets processed—all crucial to thinking, learning, and memory. Timing provides that critical sense of how long things should take, such as gauging the oncoming traffic to judge when you can cross the street safely. We rely on our internal clocks and memories of how long these actions seemed to take last time. Our motor programs continually reorganize into sequences of motor movements that reflect what we learn each time, to lead to well-thought-out and successful performance. We are always modifying and learning through movement.

Loss of a sense of timing often accompanies movement disorders such as Parkinson's disease and Huntington's disease. The basal ganglia are heavily involved and are now being shown to have significant influence on thought and memory. They are believed to be the gatekeepers, or controllers, of sensory influences on cognition as well as motor control. When they are not working properly, the structures of the basal ganglia fail to appropriately shut down certain movements and thoughts, which leads to irregular movements, tics, and obsessive-compulsive behavior. One study of Parkinson's patients examined the process of learning and repeating new procedures. The results showed that the patients' degree of motor disability correlated with their impairment in recalling tasks.

Recent research in Gainesville, Florida, provided unexpected and astonishing findings about the relationship between movement and memory. While performing a new surgical technique to alleviate symptoms of Parkinson's disease, physicians were surprised to find that the procedure also improved memory. The procedure, known as pallidotomy, involves precise burning of a small spot in one of the basal ganglia, the globus pallidus, which provides relief from some of the tremors and muscle rigidity associated with Parkinson's and other central nervous system disorders. The researchers did not expect the procedure to help memory because the surgery does not touch an area of the brain that affects cognition. Or so they thought!

Sequencing is a motor activity that involves maintaining and organizing the serial order of information and integrating this information with previously learned data. The ability to handle sequential information, maintain it in accurate order for a finite period, and reorganize it for subsequent processing is obviously important. Without this ability we would not be able to remember, learn, or even think.

Motor function even affects the highest order of mental function: self-awareness. This level of thinking is the ability of the human cognitive process to monitor itself and reflect not only on immediate responses but on past and future potentials as well. Sequencing, evaluating, judging, and thinking of ideas and possible responses are essential to this process. Self-awareness requires mental rehearsal, imagery, thinking, decision-making, and voluntary actions. We must have the ability to monitor and evaluate the self in a variety of mental settings based on prior experiences and the ability to project future outcomes. The neural basis for self-awareness includes cognitive action—the human capacity for forming and manipulating imagined constructs.

LEARNING AND MEMORY OF THE FUTURE

MOVEMENT, MEMORY, AND LEARNING are so closely interrelated that it is difficult to talk about one without referring to the other. Learning, like thinking, requires the ability to sequence and manipulate information and memories in order to perform a new task. Learning can be thought of as an act that is eventually carried out as a memory of already formulated plans. In order to have created this "memory of the future," one must have rehearsed a variety of mental actions. Getting up to have a snack, reading a bit before going to bed, and stopping for groceries on the way home demand planning future acts. The actions are thought out, even rehearsed mentally, so that when the time comes to actually perform them they are at least partially performed from memory of the future. Executing the plan requires maintaining the current goal through working memory.

Learning, memory, and thought require the manipulation of knowledge about previously acquired knowledge. To think, learn, and remember, we need to interpret or convert this knowledge into action. For example, the manipulation of knowledge occurs in the interpretation of the proverb "One shouldn't cry over spilled milk." In order to interpret this statement, one must go through an analysis. "Spilled milk" becomes an action. "Shouldn't cry" represents the futility of reliving or remembering irremediable events from the past.

Our physical movements can directly influence our ability to learn, think, and remember. It has been shown that certain physical activities that have a strong mental component, such as soccer or tennis, enhance social, behavioral, and academic abilities. Although the reasons are not completely understood, many reports indicate that this is so. Evidence is mounting that each person's capacity to master new and remember old information is improved by biological changes in the brain brought on by physical activity. Certain kinds of exercise can produce chemical alterations that give us stronger, healthier, and happier brains. A better brain is better equipped to think, remember, and learn. Practicing an activity such as modern dance or figure skating requires the mastering and coordination of many moves. After taking up these kinds of activities, people report an increase in academic ability, memory retrieval, and cognitive abilities. During these physical activities, we not only exercise our muscles, we also exercise our brains, particularly our ability to sequence motor actions and information as well as access memory.

Learning requires us to practice, rehearse, and step through the process to develop new skills, thoughts, and ideas. Using motor adjuncts with speech, such as mouthing the words, helps us learn to read. Our physical movements call upon some of the same neurons used for reading. We approach the task from different modalities, using shared neurons, increasing our chances of cementing the learning.

No one really understands how we learn to tie our shoes or play tennis. A team of neuroscientists at Johns Hopkins University has found that within the first 5 or 6 hours of practicing a new motor skill, the brain shifts the new instructions from short-term memory to the areas responsible for permanent motor skills. As subjects initially learned a task, the prefrontal cortex—involved in short-term memory and many kinds of learning—was relatively active. When the subjects returned 5½ hours later, they had no trouble retracing the movements. But at that point, the premotor cortex, the posterior parietal cortex, and the cerebellum—regions that help control movement—had taken over.

During the intermission, it seems, the neural links that form the brain's internal model of the task had shifted from the prefrontal region to the motor control region. Even without practice, after 5 or 6 hours the formula for the task was virtually hard-wired into the brain. This suggests that a newly learned skill could be impaired, confused, or even lost if a person tried to learn a different motor task during the critical 5-to-6-hour period, when the brain is trying to stabilize the neural representation and retention of the original task. The team at Johns Hopkins is now studying whether interference occurs during this window of vulnerability. If so, their findings could change the way skills are taught and training is conducted in sports and in educational and industrial settings.

Visiting a new place, seeing a new movie, singing a new song, or solving a new problem are all ways to stimulate the brain. In each instance, movement is a major player in learning. Motor development in infants, especially at the crawling stage, has been found to be crucial in the development of learning readiness. It greatly impacts reading and writing skills. Often children don't get the motor stimulation they need because of increased periods of time spent in front of the television or strapped in a car seat. Movement provides practice handling objects and interaction that is needed for visual development. Babies can learn to see, point, and say a word; however, none of these has much meaning until they have sufficient life experience to match the words and their experience, that is, movement.

Studies suggest that challenge and feedback are necessary to maximize learning. The brain is exquisitely designed to operate on feedback, both internal and external. The substantia nigra, part of the basal ganglia in the midbrain and the area of dysfunction in Parkinson's disease, is critically involved in this feedback process. This structure seems to initiate and strengthen connections between perceptual states and responses. During the feedback process, what is received at any one brain level depends on what else is happening at that level, and what is sent to the next level depends on what is already happening at that level. The brain is self-referencing, which allows our interactions to provide constant feedback, crucial to our ability to learn.

As we gain more knowledge of how the brain works, we are finding extensive linkages between movement and learning. According to Linda Acredolo and Susan Goodwyn, at the University of California at Davis, gesturing and pantomime speed up the process of learning to talk, stimulate intellectual development, enhance self-esteem, and strengthen the bond between parent and infant. Certainly, babies make certain gestures before they can say the corresponding words. A baby waves good-bye and shakes its head no prior to saying the words "good-bye" and "no." Acredolo and Goodwyn studied three groups of forty children to learn about the relationship between signing and vocal development. One group was encouraged to sign and the other two groups were not. The children were evaluated every six months until the age of four. The children who signed were found to be significantly ahead of those who didn't in acquisition of vocabulary as well as in cognitive and IQ tests. Signers showed high motivation to communicate and talked earlier than nonsigners. Parents of the children who signed felt that their children were less frustrated, more confident, and happier.

Dale and Beverly Ulrich, kinesiologists at Indiana University, studied the impact of early walking intervention on Down syndrome children. Walking impacts human development as it facilitates cognition, spatial relations, communication, and social ability. Normal babies begin walking between the ages of nine and seventeen months, while children with Down syndrome usually don't take their first step until thirteen months to four years of age. The Ulrichs helped Down syndrome babies practice stepping by placing them in a device that supported upright posture on a miniature motorized treadmill. They found at least an eight-to-twelve-month improvement in the time the babies began to walk. The movement exercise strengthened neural organization, increased strength in leg muscles, and helped develop postural control to balance weight from one leg to the other.

The motor activity "play" helps learning and social relationships. Play is a physical activity that helps children gain a sense of mastery and become adept at social interactions, including a sense of sharing and enjoying one another. Interactive competition for toys, disagreements with peers, and making friends are all beneficial in preparing children for adult socialization. Children also learn fundamental skills, concepts, and principles through play that lay the foundation for academic skills, especially when the children lead the way and their parents provide esteem-building emotional accompaniment to their achievements.

Highly intelligent social species such as wolves, bears, and dogs continue to play as adults as a way of cementing social bonds. Researchers are discovering that in a variety of species, motor play is almost as important as food and sleep. Playful maneuvers help animals learn to interpret the signals and actions of others and respond appropriately—all necessary for successful socialization.

5

MEMORY

REMEMBER THAT TIME when you were a kid and you were with your mother in the grocery store? You were shuffling alongside the shopping cart down one of the aisles when all of a sudden you saw a five-dollar bill on the floor. Money! You grabbed it, beaming at your luck.

Then your mother said, "You can't keep that. It's not yours."

"But I found it," you protested.

"Yes, but it belongs to someone else who probably lost it," she said. "Let's go tell the store manager."

All you could think was, "This stinks."

Sound familiar? Did it really happen to you? It may have. And it may not have. But weeks from now, when you're talking about old times with a friend, you'll be more likely to "remember" this actually happening to you now that you've imagined yourself experiencing it.

What am I talking about? False memories. We all have them. Despite our great certainty about what we have and have not experienced, the fact is that given a few bogus details and a little prodding, about a quarter of adults can be convinced that they remember childhood adventures they never had. Our memories are much more malleable and fallible than we like to think.

While this may be disconcerting, it's also necessary. Memory must be stable so that we can learn by building up experiences, but it also has to be flexible enough to adapt to our changing environment. Otherwise, a person with poor eyesight would wake up each day searching for his old glasses even though he's since purchased contact lenses. Memory's very ability to adapt, however, means it may occasionally make mistakes.

The study of false memory was pioneered by Elizabeth Loftus at the University of Washington in Seattle in the 1970s. Her findings continue to be intriguing. In recent experiments, she asked aging parents to list some incidents that occurred in the early lives of their children, who had since become adults. She then told their adult children that she wanted to compare their memories with those of their parents.

She walked the adult children through a list of the real incidents, but slipped in a fake one—for example, as youngsters they had been lost in a shopping mall, were frightened, and cried until an older person found them and reunited them with their parents. About a quarter of the adults agreed this had happened to them, even though it had not happened to any of them. Some even went on to provide further details about the experience, and many refused to believe the memory was fake when they were told that they had been fooled.

In another recent experiment, Loftus gave adults a written list of forty possible childhood events and asked them to indicate the likelihood that they had experienced them. They rated each event on a scale ranging from "definitely did not happen" to "definitely did happen."

Two weeks later the adults were asked to imagine experiencing some of the events. For each adult Loftus suggested an event that the person had earlier identified as "definitely did not happen"—things such as being pulled out of the water by a lifeguard. In each case, she led the adult through a minute-long exercise in which he imagined the event happening to him, for example, being in the water, starting to gasp for air, and seeing the lifeguard dive into the water to come and get him.

Afterward, Loftus gave the adults the original list of forty events and again asked them to check off the events that had actually occurred in their childhoods. Again, about a quarter of the adults indicated that they had more confidence that the event they had imagined had actually happened to them.

For a few of the adults, the act of imagining the event might have reminded them that it actually had happened. But it is more likely, Loftus says, that the act of imagining made the event more familiar, and that this familiarity was later mistakenly "remembered" as a childhood experience.

Loftus learned for herself how realistic false memories can seem when she had an upsetting experience several years ago. She was shocked when, at a family gathering, an uncle informed her that thirty years earlier, when her mother had drowned in a pool, she had been the one who discovered the body. Loftus, who was fourteen when the drowning occurred, always believed that she had never seen her mother's dead body. Indeed, she remembered little about the death itself. She recounts what happened next in her book *The Myth of Repressed Memory.* Almost immediately after her uncle's revelation, "the memories began to drift back, like the crisp, piney smoke from evening camp fires. My mother, dressed in her nightgown, was floating face down. . . . I started screaming. I remembered the police cars, their lights flashing."

A few days later, she writes, "my brother called to tell me that my uncle had made a mistake. Now he remembered (and other relatives confirmed) that Aunt Pearl had found my mother's body." This shocked Loftus even more than her uncle's false revelation. If someone so specially trained as she is to recognize fallible memories could suddenly believe her own false memory, just think how readily the average person can be fooled.

Other studies replicate Loftus's findings, indicating that memory may well succumb to the power of suggestion. This troubles Loftus greatly. In the 1990s there has been an explosion of incidents involving dramatic and horrible recovered memories. A whole family remember engaging in Satanic rituals that never happened. Young women all over the country suddenly remember being abused by family members at ages as early as six months. The stories have touched off scandals in the clergy and triggered lawsuits, suicides, murders, and court trials of adults who are accused of long-ago sexual assaults against children. Many cases are legitimate. But in some others, the people who have "recovered" supposedly suppressed memories have been helped along by the often unconscious suggestions of others. Loftus contends that the techniques some psychologists use to "unblock" memories are very similar to those she uses in her experiments with suggestion, which prompt false memories.

The main point to these observations is that people for too long have taken their view on memory issues as truth. Our improving understanding of memory's constancy yet flexibility is opening up fascinating windows into how we learn, why we behave the way we do, and how we evolve in our lives. The results offer clues to choices we can make about diet, drugs, stress, and environment to build and maintain strong memories in ourselves and our children, and even to postpone the disheartening effect that aging has on memory. A better understanding of memory is also helping doctors develop effective therapies for people who suffer from stroke, brain injuries, trauma, mental illness, and Alzheimer's disease and other disorders that can beset memory itself.

MAKING A MEMORY

WE CAN KNOW ourselves only because we can remember. Memory is the centripetal force that pulls together learning, understanding, and consciousness. In the past, we believed that in the brain, one neuron equaled one memory, and that each section of the brain performed its particular operation in isolation. Today this notion seems as ludicrous as Franz Joseph Gall's early-nineteenth-century pursuit of phrenology, the study of bumps on the skull, which he claimed reflected personality traits hidden inside. Modern instruments such as PET scanners show us that the brain is more like an active ecosystem than a static, preprogrammed computer. There is no single center for vision, language, emotion, social behavior, consciousness . . . or memory.

Science has always wanted to know where memories are "stored." Is it in the perception neurons, where we saw or heard something for the first time? The hippocampus, which pulls memories together? The frontal lobe, which triggers recall? None of these, and all of these. The even more fundamental question is: What is a memory? Endel Tulving at the University of Toronto has been searching for memory for more than forty years and he still does not know what to call memory. Is it the storage space or the act and strategy of the retrieval? Is memory the act of searching for the memory or the energy devoted to forming the memory in the first place? A memory is only made when it is called upon. In its quiescent state it is not detectable. Therefore we cannot separate the act of retrieving and the memory itself. Indeed, bits and pieces of a single memory are stored in different networks of neurons all around the brain. We bring the pieces together when it is time to recall that memory. We pull the pants, shirt, and shoes out of different parts of the closet to re-create the single image of how we looked last Monday, when we thought we looked so good. When the day is over we put the pieces back, and even if they're not in exactly the same places we still know where to find them and how to put them together again.

The formation and recall of each memory are influenced by mood, surroundings, and gestalt at the time the memory is formed or retrieved. That's why the same event can be remembered differently by different people. One person isn't necessarily "right" and the other "wrong." Memory also changes as we change over time. New experiences change our attitudes, and thus how and what we remember.

Memories—from two minutes, two years, and two decades ago—come and go every waking hour. Each one arises from a vast network of interconnected pieces. The pieces are units of language, emotions, beliefs, and actions, and here, right away, comes the first surprising conclusion: because our daily experiences constantly alter these connections, a memory is a tiny bit different each time we remember it.

Consider the effect of mood, for example. The frontal cortex is the part of the brain that neatly organizes the bits and pieces into a temporal, logical, and "meaningful" story. However, it must be set in motion by the amygdala, which provides an emotional tag to a memory, a "meaning" that helps cement the pieces. Given this, one's emotional state at a given instant affects how the amygdala processes the emotional tag of a memory, perhaps changing ever so slightly how that memory is reconstructed. An individual who is depressed is predisposed to see a certain memory in a negative light—so it's a different kind of memory than it would have been had the person been generally happy.

For example, say a woman is telling her husband about a brief meeting she had with her boss concerning a memo she had written. She tells her husband rather enthusiastically how the boss had taken the time to go through her memo personally, and encouraged her to further develop the interesting points. This was a welcome impression, since the company had not been doing that well and several people had lost their jobs. She was excited about the boss's interest and the challenge of refining her points. A week later the woman finds herself feeling depressed after hearing that several more co-workers have been laid off. She tells a friend over lunch about the same meeting, focusing on the boss's seriousness and solemn tone of voice. She decodes this as his seeing the memo as inadequate and his suggestions that she go further as criticism of not having gone far enough. Her worry about losing her own job colors her memory of the event. Anxiety mounts over whether she might be the next to go, and she becomes overalert, recalling every nuance of the boss's demeanor and seeing only the signs of potential displeasure. It is through this lens that the memory becomes distorted. The string of events in the memory may be the same, but the tone and hence the meaning are very different.

Even if memories are not catalogued in a central repository and must be reconstructed each time we recall them, another question still remains: Where does the memory reside once it's assembled? Neurologist Antonio Damasio in *Descartes's Error* proposes that the elements come together at "convergence zones" near the sensory neurons that first registered the event. He and his wife Hanna used MRI scans to locate convergence zones that oversee the recollection of the names of objects and animals and other zones that unite sensory information about people, perception, and emotion.

Convergence zones also enable us to automatically conceive of objects, ideas, or interactions as a whole, if the pieces have been put together enough times. We are not forced to think of a screwdriver as a piece of equipment with a wooden handle and a metal head, which past experience has shown can be used to screw in a metal fastener. We think quite simply "screwdriver." Each time a screwdriver is encountered, we recognize it by igniting the connections that we have already made with the object's many features, and it is precisely when we have acquired not only a name but a holistic concept of the item that true understanding can emerge. Once a toddler has been sternly warned enough times about staying away from the oven because it is "Hot!" and will "Hurt!" the very mention of "oven" is enough to elicit the full memory of needing to stay away from it.

The Damasios also propose that there is a hierarchy of convergence zones. "Lower" convergence zones link the cues that allow us to understand the general concept of "face" while "higher" convergence zones allow us to recognize specific faces. Linking the two are intermediate convergence zones that differentiate details in individual faces—nose line, pallor, eye shape.

The beauty of the Damasios' system of memory is that it illustrates the brain's wonderful efficiency. Instead of storing an infinite succession of daily movies, the brain reconstructs them from a manageable number of reusable elements of experience. The sense of feeling "cold" is one puzzle piece that is available to help complete many different puzzles: a winter wind, a cave, ice cream. There is also beauty in the fact that there is no need for any logical ordering or classification of memories, which would be a true burden on the brain.

While the plasticity of the brain's neural networks allows us to store and bring together the pieces, research at Yale University by Patricia Goldman-Rakic has shown that some neurons may be specialized for different types of pieces of memories—features, patterns, location, direction.

If memories consist of pieces, it would seem likely that something in the brain must be responsible for divvying up an event into bits and later bringing them back together to form a memory of that event. We don't know what process, or brain region, might be responsible, although mounting evidence suggests that the hippocampus might serve as the master regulator, the hub at the center of the wheel. The hippocampus is in both the right and left hemispheres, and we do know this much: without it we learn and remember nothing.

The hippocampus does not store memories. It has been likened to an intelligent collating machine, which filters new associations, decides what is important and what to ignore or compress, sorts the results, and then sends various packets of information to other parts of the brain. It is a way station that hands out the pieces.

Where the pieces are dispersed and how they are reconnected are processes that are not well understood. Sleep may play a role. In an ingenious experiment, researchers planted electrodes in different cells in the hippocampi of rats and watched each cell fire as the animals explored different parts of a box. After being returned to their cages, the rats slept, and during sleep the very same cells fired. As I noted earlier, recent experiments also show that sleep, specifically the sleep associated with dreaming, is important to human memory. In Israel, researchers Avi Karni and Dov Sagi at the Weizmann Institute found that interrupting REM sleep sixty times in a night completely blocked learning, but interrupting non-REM sleep just as often did not. These findings and others suggest that REM sleep is crucial for organizing pieces and the associations between them needed for forming lasting memories.

What about memories that are not sensory in origin, such as reflections, beliefs, or emotions? When the brain forms concepts, it constructs maps of its own activities. The maps categorize, discriminate, and recombine the various brain activities needed to form ideas and emotions. The bits and pieces are of a different nature, but they are dispersed and pulled together in the same way.

Not only are memories more widely distributed than we once believed, but the plastic brain seems to be able to change how it distributes certain types of memory if there is an extreme need to do so. Consider the case of Martha Curtis.

By the time she was in kindergarten, Martha could play the violin. She also began having strange convulsions. Within a few years her condition deteriorated. Doctors said she had epilepsy and put her on drugs in an attempt to control the seizures, but they got worse, sometimes rendering her unconscious. Still, by age eleven, Martha was sitting with the junior orchestra at the Interlochen Center for the Arts in Michigan. But the promising young musician faced a major problem as she began to have seizures on stage.

Martha tried to hide her disability, but the seizures were alarming. Through her teen years they became frequent. She could only play with local orchestras because no others would risk having her appear in public. By her twenties the seizures were relentless and horrifying, and something had to be done. In 1990 she went to see Hans Luders, a neurologist at the Cleveland Clinic. Luders stopped the drugs she had been taking and did a series of MRI scans of her brain during a seizure. The episode began as a local electrical disturbance in the right temporal lobe, a large region, then spread, eventually taking over her entire brain in a global thunderstorm. Luders told Martha that surgery to remove part of her brain was the only option, because the amount and strength of the drugs needed to control her seizures were reaching toxic levels.

Various kinds of surgeries are used to treat severe epilepsy. Most often the approach is to remove the offending region that triggers the electrical disturbances that ignite the thunderstorm, which in Martha's case was part of the right temporal lobe. Surgery was scheduled for January 1991. Martha feared the worst: that she'd never play the violin again, because the right temporal lobe is the portion of the brain associated with music memory.

As soon as she got out of intensive care, Martha picked up her violin, and tried to play a piece by Bach, what she called the hardest music to play by memory. She played it beautifully.

But the surgeons had not cut away enough. The seizures returned. Martha underwent a second operation. The seizures continued. Martha wanted a third surgery, but her doctors resisted; removing so much of her right temporal lobe could result in paralysis, and the trauma to the brain could possibly even be fatal. Martha pushed. Reluctantly, the surgeons agreed. They carefully cut away as much of the right temporal lobe as they dared—a full 20 percent of it.

The seizures stopped. But could Martha play? Better than ever. In fact, she was able to memorize pieces she had been unable to memorize before. Her doctors concluded that her brain had been damaged early in life, perhaps as the result of measles at age three. But because she had been practicing the violin since she was very young, her brain had somehow rewired its music memory, recruiting other brain regions so that the problematic right temporal lobe had no role in it.

Now forty-one, Martha Curtis is a thriving soloist. She also plays for surgeons, psychologists, and epileptics to show how amazingly adaptable the brain can be in recording and decoding memory.

Martha's case shows that if we need and use certain types of memory more than others, our brains can extend the regions responsible for specific functions, even recruit new regions to help. Her triumph also holds a lesson for all of us: By exercising our brains we can strengthen our memories, just as weight-training strengthens our muscles.

LONG-TERM POTENTIATION

LEARNING AND MEMORY PROCESSES exist in a circular relationship that we take for granted. Learning enables information to cross over the lines of perception into memory, but once stored these memories affect future learning. To get a handle on how memory and learning work together, researchers have become good friends with *Aplysia,* a sea slug that Eric Kandel of Columbia University made famous. *Aplysia* was the first creature in which researchers were able to show that the one-neuron, one-memory scenario was wrong. It was also the first creature used to shed light on what is now the most recent, and most powerful, explanation for how memories become encoded: a process called long-term potentiation (LTP).

Each and every new experience causes the neuronal firing across some synapses to strengthen and others to weaken. The pattern of change represents an initial memory of the experience. However, the pattern soon disappears unless it is made more permanent by LTP, which is the cellular mechanism that causes synapses to strengthen their connection to one another, coding an event, stimulus, or idea as a series of connections. When a stimulus is received, LTP blazes a new trail along a series of neurons, making it easier for subsequent messages to fire along the same path. The more the path is refired, the more permanent the message—the new learning—becomes.

As neurons in the chain strengthen their bonds with one another, they then begin to recruit neighboring neurons to join the effort. Each time the activity is repeated, the bonds become a little stronger, and more neurons become involved, so that eventually an entire network develops that remembers the skill, the word, the episode, or the color. At this stage, the subject becomes encoded as memory.

This process, however, is not standardized. Motivation can affect how encoded a memory becomes. Michael Merzenich did much of the early work in showing that when there is a reward, the pieces of a memory are more strongly bonded. He placed a slowly spinning wheel beside monkeys' cages, which the monkeys could touch with their fingertips, and monitored the region of their brains responsible for the fingers. The cells responsible for feeling the wheel and remembering the sensation were mapped. Merzenich then added a learning task; when the monkeys could recognize a designated pattern of spinning and press a buzzer, they were given a food reward. They soon became experts at recognizing the right pattern, and literally within hours the nerve cells responsible for the task multiplied as the monkeys' discriminatory powers increased. Neighboring neurons were recruited to help perceive and then remember the perception. The adding of a reward led to having many more neurons code the memories. The monkeys were motivated to remember the event. The adage that reward is part of learning is backed up by real neuronal proof.

The monkeys learned in part because the process was repeated often. Each time an experience is recalled or repeated, the neurons can practice their chemical volleys and strengthen their connections. If the fledgling network is not reinforced, the connections will disband. Once memory connections become firmly bonded they tend to last, but over many years they can fade, as we all have experienced. If a memory unit is not occasionally reused or reinforced, the connection may weaken, disband, or die.

LTP may also explain why many people can remember where they were, even what they were wearing, when they heard that John F. Kennedy had been shot or that the space shuttle *Challenger* had blown up. Such shocking events send messages to every nook and cranny in the brain, triggering a kind of super-LTP that recruits neurons from all over the brain, cementing the event immediately in memory. Because the process is happening everywhere in the brain, insignificant everyday details, such as where you are or what you are wearing, are swept up into the potentiation and also become hardened into long-term memory. Researchers sometimes call this flashbulb memory, as if every detail of a single, sudden moment had been captured in a photograph. New flashbulb memory studies are now being done with people's recollection of the news of Princess Diana's death.

The model of memory as a set of distributed pieces that are pulled together on demand, the need to repeat the firing patterns to etch them into long-term memories, and the role of LTP in making this happen are all supported by modern sleep research. The culmination of decades of work by researchers such as Allan Hobson at the Massachusetts Mental Health Center shows that brain wave activity in the hippocampus during dreaming actually rehearses memory patterns, either to harden newer experiences into long-term memories or to keep fading connections alive. As noted earlier, this is also what Karni and Sagi found in their experiments with rats. The mechanism most cited is the cortex's processing sensory information during a new experience and sending it to the hippocampus, which initiates replay and consolidation of the experience into long-term memory during sleep.

This theory is substantiated by PET and MRI scans. PET scans show that during REM sleep there is communication between the amygdala, the anterior cingulate gyrus, and the occipital lobes, structures that have long been linked to attaching emotional significance to memories and dreams.

More evidence comes from the evolutionary ladder. In the one mammal that does not experience REM sleep, the spiny anteater, the prefrontal cortex—the major center of learning and behavior—is so disproportionately large relative to the animal's body mass that memories are encoded at the moment an event is first experienced. Higher mammals, lacking this massive reservoir, were perhaps forced to develop and reserve REM sleep as a time for solidifying memories; recall the study showing that the exact neuronal firing patterns present when rats explored a maze were repeated precisely when the rats were in REM sleep.

The possibility that LTP is the mechanism behind memory storage has several enormous physiological implications. One is that each particular memory is built in and stored and retrieved from a specific neuronal network. Another is that a given neuron may participate in many memories at once. Could a neuron become overcommitted? Not really. There are an astonishing number of connections within the layers of the brain—many more than there are neurons. If one neuron is getting close to its "capacity" to participate in various memories, a neighboring one will simply step up to complete the needed connections. The serendipitous result is that our brains are capable of constantly recognizing and reorganizing relationships in everyday experience while simultaneously comparing our current flows of experience to past memories.

This juxtaposition of the past and the present is an important aspect of LTP, for although it is rapidly induced, it is also easily disrupted by new stimuli, shifts in attention, high brain temperature owing to illness, and neuronal electrical disturbances such as seizures and electroconvulsive shock. As in so many areas of brain research, there are still many unanswered questions about how long-term memories are permanently stored, and lost.

Another implication of LTP is that learning "exercises" the brain, giving it the stimulants it craves. A well-toned brain often has more blood capillaries and glial cells, which, together, cater to the furious metabolic and nutritional needs of the brain's neurons. If neurons in the sequence are allowed to weaken, the memory weakens.

SHORT-TERM VERSUS LONG-TERM MEMORY

AS WE ALL know, there is a difference between short-term and long-term memory. The former lasts for minutes or hours, and the latter for longer than a day. For certain memories, the time in between is used to consolidate them from less stable to more permanent forms.

Short-term memory is also referred to as "working memory" because it allows us to carry out the hundreds of tasks we busy ourselves with each day. Working memory gives continuity to what we're aware of from one moment to the next. As noted in Chapter 3, it enables us to remember a phone number from the *Yellow Pages* long enough to dial it, and to keep track of the conversation we are having once it starts. It allows us to recall where we cut-and-pasted paragraph C while we are editing paragraph B in our draft book on the brain. Here I will permit an analogy to the computer—which otherwise is a dangerous exercise because a computer is far too simplistic a device when compared to the dynamic living organ inside one's head. Short-term memory acts like a computer's RAM: it holds the data we are working with at the moment, but loses them once the machine is turned off. Long-term memory acts like the computer's hard disk: information is only put there when we hit "Save," but once it's put there it stays there so that we can access it again and again.

Short-term and long-term memory are easily distinguishable. The complex question is how short-term memories make the transition to long-term memories. Initial consolidation of a short-term memory occurs in only a few hours. But conversion to a long-term memory does not happen until the information has been sent by the cortex to the hippocampus. Research suggests that there is a special window in time during which the transition to long-term memory is possible. This window is essentially the time needed for neurons to synthesize the necessary proteins for LTP. An initial stimulation triggers a communication across the synapse between two nerve cells in the brain. Further stimulation then causes the cells to produce key proteins that bind to the synapse, cementing the memory in place. If LTP—and hence a memory—is to last for more than a few hours, proteins produced in the first neuron must find their way to specific synapses and bind to them, an event that changes the structure of the synapses and increases their sensitivity to an incoming signal. This may explain why we must repeat a list of words over and over in order to memorize them. It may also validate the role of REM sleep as a process for reliving new and old experiences so they become more permanently etched as long-term memories.

Very recent research with mice, flies, and *Aplysia* indicates that remembering something in the short term uses proteins that are already present in synapses. But to shift the memory into the long term, new proteins that reconfigure synapses are needed. The synthesis of these proteins is controlled by a protein known as CREB, which is thought to act like a switch that triggers the production of new proteins.

CREB's importance may extend beyond memory. It is turning up in other situations that involve a long-term change in brain procedure, such as resetting the body's internal clock after crossing time zones and developing a tolerance to drugs, from medically prescribed antidepressants to illegal substances. CREB probably works with other proteins in these processes of learning, but it seems to be the one crucial for the process to occur.

We should point out that the many neuronal processes involved in learning and memory occur simultaneously each waking moment. Because there is such a staggering number of neuronal connections in the brain, the amount of parallel processing occurring at any instant is awesome. Even in the simple *Aplysia,* scientists can find a wealth of complex behaviors, all of which would be impossible without parallel processing. Parallel processing is essential to our ever-changing interconnected network of neurons. The activation of one particular firing pattern can inhibit or excite other firing patterns, which accounts for the existence of complex mental phenomena such as perception, thought, and impulse.

WORKING MEMORY

IN ADDITION TO helping us function from one moment to the next, working memory is also crucial to recall. It registers our current activity while retrieving information from long-term memory and dispersing or holding it at the ready. It is the mental glue that holds multiple connections together as we think a thought or enact an act, from beginning to end.

Working memory is also what makes us most human. It gives us "memory of the future," the ability to predict where we will be and what may happen when we get there. It allows us to consider our actions now in relation to what we're going to be like in the future, which gives us the uniquely human ability to make judgments, anticipate consequences, and take or shirk responsibility. As I said to my daughter as she went off to college: "Remember who you are." I was reminding her to bring up into working memory the image of herself, her beliefs, her goals, so that she could be true to herself.

Working memory is part of the executive function of the prefrontal cortex. Again, like the CEO of a company, it must always remember the goals and plans and see the needs and benefits of projects end to end. Ironically, one of the major functions of working memory is also to prevent information from being coded. Inhibition is a crucial part of learning because it prevents noise from distracting our focus on what matters. Although forgetting can be frustrating, it is a necessity! Otherwise trivial memories would clog our minds. We would succumb to the fate of the young man Funes, the subject of the memorable short story "Funes the Memorious" by Jorge Luis Borges. Cursed with total recall, Funes remembers every wrinkle in every face, the shape of every cloud, the sound of every raindrop. Overloaded with memory, he drowns in his own thoughts, sinks into despair, and becomes confined to a chair in his house, a hopeless cripple shut off from even a single additional stimulus.

There is an actual case in psychology of this very problem, involving S. V. Shereshevski, a famous synesthete mnemonist, who seemed to have lost his ability to forget. He was constantly troubled by an enormous and ever-increasing amount of trivial detail that constantly claimed his attention. The noise sapped his ability to make any sense of what he was experiencing. If he read a story or listened to someone talk, he could recall endless minute details of the content, but actually understood very little of the point or concepts conveyed.

Some people are born with difficulties in working memory. Often this can lead to behavior that is simplistically and mistakenly labeled as "antisocial"—individuals who have trouble tracking a conversation or activity feel helpless and so find it safer to not take part. They want to participate but can't manage it. It is essential that mental health pro-

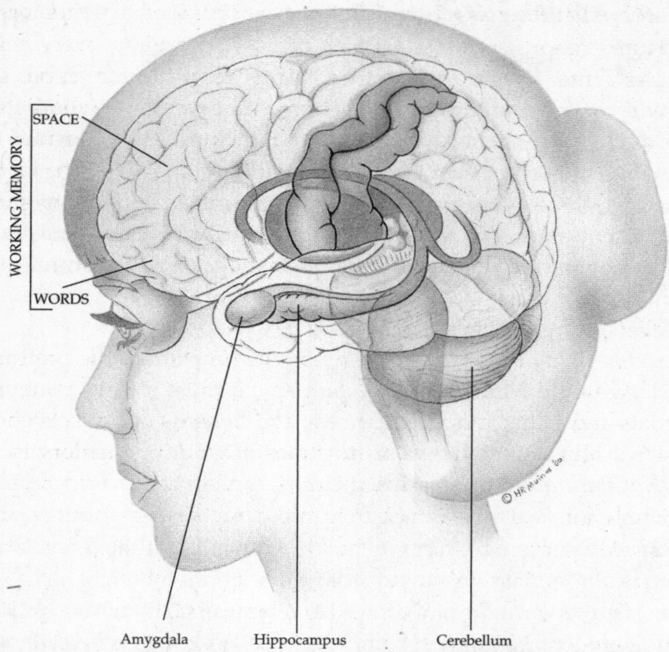

THE MEMORY SYSTEM Key parts of the memory system are shown here. After the information is perceived by the posterior cortex, the prefrontal cortex captures and stores it. The frontal lobe does this by holding the information in working memory for a while. After a few minutes, the frontal lobe washes its hands of the memory, and the hippocampus has to be recruited in order to retrieve it. When a memory is recalled with the help of the hippocampus, it is placed back in working memory in the frontal lobe. I call this our RAM, crucial for holding and manipulating words and spatial representations. These functions fit roughly into the two components proposed by Allan Baddely, the phonological loop and the visuospatial sketch pad. A few years after the memory is first acquired, the frontal lobe can access it directly without help from the hippocampus. Another type of memory, procedural memory, having to do with learning motor skills like riding a bike, is handled through the striatum and the cerebellum.

fessionals learn more about working memory because it underlies so many problems in coherent thinking—everything from poor evaluation and anticipation of consequences to attention and action disorders. The rest of us should also keep this in mind before we criticize too quickly. Problems with working memory are crucial to the many symptoms of ADHD. Those of us blessed with proper working memory can predict the consequences of our actions: we have memory of the future. People with ADHD lack this gift. Planning overwhelms them, and they "forget to remember" and "forget to remember that they will exist in the future" and so on until everything falls into an unproductive infinite regress. Further, they lack an ability to screen out extraneous stimuli. When these deficiencies come together in one unfortunate brain, the person suffers from a triad of poor memory, poor concentration, and poor planning. The executive system is overwhelmed by the noise, which creates an inability to pay attention.

SUBJECTIVE MEMORY

WORKING MEMORY ALLOWS us to function from one instant to the next. But the nature of working memory can change over time, even over short intervals. New research into how we move information from short-term to long-term memory is unveiling quite a revelation: we can never be "certain" about anything we remember.

The first evidence for this comes from James Haxby of the National Institute of Mental Health in Bethesda, Maryland, who is investigating facial memory. As time goes on, he says, a person's working memory seems to rely more on an analytic understanding of the face than on the image itself. In one experiment, his subjects were asked to remember an image of a given face for 21 seconds while PET scanners imaged the brain. The scans revealed that at first an area in the right visual cortex was activated. This soon faded, however, and as it did, activity in the left prefrontal cortex intensified. Haxby concluded that while the right visual cortex stores the working memory, the left prefrontal cortex encodes thoughts, impressions, and memory-related connections about the currently viewed face. This shows that the visual cortex was no longer needed; the picture itself wasn't needed for the identification, just the analyzing part of the brain. Studies of human amnesia have also shown that working memory can transfer information to long-term memory within 60 seconds of encoding; the memory is quickly reorganized to minimize dependence on the fleeting short-term memory function, and it is the subjective, interpreted information that is later retrieved for use.

The staggering implication of this research is that we can never describe in exact detail what was actually presented to us. For as soon as we experience something, we immediately interpret it and rewire it.

Daniel Schacter and his colleagues at Harvard University recently broke new ground on the true memory/false memory problem. Schacter had subjects listen to lists of twenty words each, read aloud. Ten minutes later, the subjects were given printed lists that consisted of either the word that was actually spoken, a different but related word, or a completely unrelated word. For example, if the original word was "cake," "sugar," or "candy" the false target word would be "sweet." The subjects then attempted to remember which words were on the original lists while a PET scanner recorded their brain activity. The recollection of true and false memories correlated with differences in blood-glucose metabolism in different areas of the brain.

True and false memories of the original words both stimulated the left medial temporal lobe, the area closer to the hippocampus. True memories, however, also stimulated the left temporoparietal cortex, the region where sensory information first enters the cortex and is encoded. A possible explanation may be that subjects who recalled true memories also remembered the actual sound of the word when it was spoken. While these results might encourage your imagination to run wild with visions of false-memory detectors in criminal trials, Schacter asserts that such PET scan results are reliable only for events that have occurred just a few minutes in the past. Still, they show there is a deep, strong connection between sensory perception and memory. A true memory of, say, learning the meaning of an unfamiliar sentence in a foreign language by speaking and then writing it would involve not only the conceptual meaning of the words but also the sounds of the words, how they appeared on the page, the movement of the arm and hand, and perhaps even what the page felt like under your pen.

EXPLICIT VERSUS IMPLICIT MEMORY

AS SCHACTER'S WORD experiments show, human memory is distributed throughout the brain, yet certain memory functions are still dependent on certain areas. The brain's development encourages this.

Procedural memory or skill learning is the first memory function to develop in the brain's early stages of growth; a baby learns how to extend its arm. Then, its perceptual representational system solidifies, and the baby recognizes objects. Next it acquires semantic or "factual" memory, and the baby calls the round object "ball." Lastly, it develops the capacity for episodic memory, which allows it to consciously recall past experiences, and the baby remembers the last time it threw the ball and the dog chased it, and throws it again because watching the pup pounce is just so much fun. This hierarchy and layering of memory systems allows for many complex memory functions.

The last stage of the brain's development creates the specialization of the hemispheres. This is when certain memory functions get localized, though most still have some basis in different regions, and is what makes explicit and implicit memory possible.

Explicit memory encodes factual knowledge—names, faces, events, things. It depends on an initial dialogue between the hippocampus and the temporal lobe. Explicit memories are directly accessible to our conscious awareness. They are flexible, rapidly retrieved, and occasionally unreliable. Implicit memory is responsible for the laying down of skills and habits that, once learned, do not have to be consciously thought about, such as eating, talking, walking, riding a bike, and the way to go about making friends. They are inflexible, slow, but extremely reliable, and involve the basal ganglia and cerebellum.

Research is helping to delineate further which regions of the brain are involved in explicit and implicit memories. For example, patients suffering from disorders of the basal ganglia, such as Parkinson's disease and Huntington's disease, seem to have no problem remembering facts and events. However, they are no longer able to properly perform an appropriate sequence of habituated movement, such as walking. They also cannot recognize the next item in a mental sequence that has been explained to them over and over again; they can retrieve memories that are stored, but can't store new ones.

A great deal of our everyday functioning and learning is the result of turning explicit memories into implicit ones. For example, explicit procedural memories can become implicit when we are able to complete a task without referring to how we have done it in the past. When we are learning how to ride a bike, each time we get on that seat we have to remember what we did right and wrong the time before. But as we master the task, we no longer have to actively remember what to do. We just do it. The procedural memory, at first explicit, has become implicit. Larry Squire at the University of California at San Diego has used EEG and PET scans to show that the location of the memory in the brain changes as it becomes implicit.

One important example of implicit memory is metamemory—the ability to have knowledge of one's own memory capability. It is a "feeling of knowing." New research indicates that the frontal lobes may oversee metamemory, because individuals who have had portions of the frontal lobe removed lack it. In everyday life, they have to function with no intuition about what they actually do and do not know. Metamemory operates when a word is on the tip of our tongue and we know that we know it but we just can't jog it to the front of our heads. People who have lost significant parts of their frontal cortex do not have the notion of what they have forgotten.

EPISODIC VERSUS SEMANTIC MEMORY

ALTHOUGH FOR SIMPLICITY we are describing each type of memory function as fairly straightforward, it is important to know that memories are anything but. This is evident in episodic and semantic memories, which are different yet interrelated.

Episodic memory is the capacity to place facts and events in time and to refer to them freely. It involves looking into the past to remember the birth of our first child, as well as looking to the future to envision an anticipated vacation next month. Episodic memory also constitutes the storytelling shaman in all of us. In tribal societies with richly developed oral traditions, the stories and knowledge of the elderly command a special respect because they are viewed as the source of important cultural memories that guide tribal life. One Seneca Indian elder tells the story of "the remembering," the moral of which is that people who spiritually incorporate the stories of the past become greatly gifted, while people who disdain the past are doomed to repeat its mistakes. Perhaps this moral is deeply relevant for all of us because it shows that our memories are not dead tokens of the past but the powerful forces behind what we believe in the present and imagine about the future.

Semantic memory is detached from personal experience. It is cognitive rather than autobiographical, the impersonal basis of one's repertoire of knowledge. Semantic memory allows for the retention of facts and everyday functions, including categories of events, objects, spatial knowledge, and symbolic description. Semantic facts differ from episodic facts only in that they are removed from a specific moment and place. Once the big bully Bill is removed from the episodic memory of your grade-school playground, he is reduced to the semantic, generic category of "aggressive, fear-inducing child."

Episodic memory is by necessity far more plastic than semantic memory, but it is also far less reliable and can be distorted by all sorts of distractions, including fear, anxiety, and stress. Semantic memories are often acquired by rote, aided by our ability to generalize and categorize. Episodic memory cannot by its nature be acquired this way. In fact, the original Greek meaning for episodic memory is "the spinning of a tale"—a creation made of thoughts, beliefs, interpretation, and emotion. Language depends largely on semantic memory. In order to have a universal system of symbolic representation, we need a system of recall for impersonal knowledge—knowledge such as the meaning of words, grammar rules, and syntax.

Like explicit and implicit memories, episodic and semantic memories also have a circular relationship. Episodic memory is embedded in semantic memory, yet semantic memory is dependent on episodic memory. Indeed, meaningful episodes seem to enter the consciousness in the form of scripts that are later converted into autobiographical memory. People who have a form of amnesia in which they are unable to recall many specific events that have happened to them in the past may recall the sequence of events that is required to turn off a computer, and will remember the general personality traits they may have had prior to their amnesia, but cannot recall a single event illustrating any trait. Endel Tulving and Dan Schacter reported on a patient, Gene, who had suffered damage to his frontal and temporal cortex, including his left hippocampus. He was unable to recall any specific event of his past and could not learn anything new. However, he had semantic knowledge; he was able to remember the route he took to get to school and the details of changing a tire, even though he could not remember himself at school, or any time he might have changed a tire.

Though episodic and semantic memory are related, a recent study of three British children suggests that the hippocampus is critical only for episodic memory. London neuropsychologist Faraneh Vargha-Khadem reported that children with severe amnesia due to damage to the hippocampus can still have surprisingly good semantic memory. The three children, Beth, Jon, and Kate, are now, respectively, fourteen, nineteen, and twenty-two years old. Each had suffered brain damage from oxygen deprivation—Beth at birth, Jon at birth or during convulsions at age four, and Kate owing to respiratory arrest at age nine. The children baffled doctors: They couldn't remember what day it was or what TV program they had just watched, and they routinely got lost in familiar surroundings. Yet they somehow learned to read, write, and spell as well as their classmates. They got average grades in mainstream schools and rattled off facts and definitions. Yet they would forget conversations they'd just had and even what day it was. The tragedy is that, despite their academic smarts, their amnesia is so severe that they have to live under strict supervision, and will never be able to lead independent lives.

SENSORY MEMORY

WE'VE SEEN THE ways in which memory operates. Now let's quickly look at the basic types of memory: sensory, motor, visuospatial, and language.

Each of our senses gives us a part of the world—sound, sight, taste, smell, touch—and so our memories can be recalled from any number of sensory cues. A famous clinical example of the melding of the senses and memory involves, once again, the famous mnemonist S. V. Shereshevski, who in the early 1960s astonished crowds with his infallible memory for meaningless detail. Shereshevski perceived with criss-crossed senses. He would see sounds, hear colors, feel tastes, and taste shapes. In response to hearing a tone at 2,000 cycles per second, for example, Shereshevski said, "It looks something like fireworks tinged with a pink-red hue." He would describe perceptions in ways like, "The strip of color feels rough and unpleasant, and it has an ugly taste—rather like a briny pickle. You could hurt your hand on this."

Today Shereshevski's "talent" would have been recognized as a rare condition known as synesthesia, in which the sense that is triggered in response to a normally unrelated stimulus—the seeing of smells, for example—is perceived not in the mind's eye but as an actual external event. As a result, most synesthetes have astounding memories. This arises from an ability to eidetically reexperience the synesthetic sense tied to the memory itself. For example, Shereshevski used the common memory technique of envisioning a town. As someone in the audience gave him a fact to remember, he would take the synesthetic experience tied to his perception of being given this fact, and place it somewhere in the town. For example, hearing a fact might cause him to see a white square. Another fact might evoke a black sphere. He would then place these sensory objects in his imaginary town.

To recall the facts, Shereshevski had only to "stroll" through his town. As he saw each object—the white square or the black sphere—the fact tied to it would emerge easily. This ability could be compared to photographic memory, for Shereshevski remembered his town as vividly as if he were walking through a real town.

Synesthesia provides an example of how important sensory experiences are for creating memories. Most of us do not realize the extent to which our sensory experiences are our memories. Think of your last birthday; it's likely that what first comes to mind is a visual image of a cake, or of people who were there. Think of the last meal you had in a restaurant; it's likely that your first recollection is a smell, or the noise in the room. This is a fun game to play when you have time to kill, say, when you're driving the car. Try remembering some event or experience, and be aware of how much of that memory is made up of sensory information. You'll be surprised.

MOTOR MEMORY

MOVEMENT IS A FUNDAMENTAL BASIS of learning, because it is a major aspect of experience every second of every day. Without the fine motor control we have over our vocal chords, for example, speech would be impossible. Motor memory is important for purely "mental" tasks, too, from how to do long division to sequencing the steps in solving a management problem at the office.

Recent research has shown that the cerebellum plays a leading role in motor memory. Patients with lesions in the cerebellum have difficulty judging the velocity of an object, tapping a foot with a regular beat, and distinguishing different time intervals. Learning a motor skill requires the development and modification of increasingly more accurate motor programs, which coordinate not only precise movements but also the cognition needed to perform them.

Motor memory is achieved with a sophisticated feedback system that detects errors made as the movement is learned. The feedback system uses these errors as a basis from which to generate a new, more accurate sequence of commands, eventually leading to a successful performance. We modify and learn through movement every second of our waking day whether we are active or inactive.

Learning a motor skill is another story. Recent evidence suggests that skill-learning occurs not in a loop but in two stages. During the first stage, activity focuses on that population of cells that best represents the stimulus or movement. In the case of learning to play the piano, the main motor areas are brain regions that control the eyes for reading music, the ears for listening to sounds, and the fingers for manipulating the keys. In the second stage, additional neurons are slowly recruited to refine the critical neural firing patterns. That is why practice makes perfect.

For motor memories, the frontal cortex plans and organizes events while the basal ganglia and hippocampus act together to store the memories for the long term. It must be remembered that the hippocampus bridges the transition between short-term and long-term memory, and so motor memory, like sensory memory, has a crucial window in time of vulnerability.

Motor memory and skill-learning are intimately interrelated. Interrupting either one of the systems interferes drastically with the other, and like any other higher cognitive skill (such as language or emotion) motor memory is a global enterprise. Studies have shown that people who learn to sing or play a musical instrument benefit from greater communication between the hemispheres. Playing the piano exercises the entire brain. As a result, other cognitive signals fly faster and are read more accurately. This has a significant impact on a person's mental acuity, because the communication between their hemispheres becomes better than that of the average person.

Creative and artistic individuals do indeed possess higher levels of interhemispheric communication. The creative meanderings and patternings of the right hemisphere are not enough for creativity; they must be joined with action or language (motor functions) coordinated by the left hemisphere to be demonstrated to the world. (The opposite is true for lefties.) A beautiful sonnet or painting in someone's head must be expressed through an understandable medium, which requires fine-motor movements. Other studies show that creative people also have a higher degree of cortical arousal. So it is conceivable that a child's systematic daily practice of a musical instrument or a dance step leads to an increase in cognitive capacity. A raft of studies that hit the popular media in 1996 showed that training in the arts, which is in large part rehearsal of movements, extended to good learning in other areas.

The goal in playing the piano, for example, is to achieve expression. Along with the parallel processing that must occur to evoke sounds from the instrument, the musician is constantly adjusting decisions on tempo, tone, style, rhythm, phrasing, and feeling—training the brain to become incredibly good at organizing and conducting numerous activities at once. Dedicated practice of this orchestration can have a great payoff for lifelong attentional skills, intelligence, and an ability for self-knowledge and expression.

The famous example of this phenomenon is Albert Einstein, who played the violin regularly. At times he would suddenly stop playing, jump up from his chair, and scribble down an idea or part of an equation. People who hum or whistle a tune while they are contemplating something—or walk the Stairmaster exercise machine, as I do—are using motor programs in the brain to help them wander along in thought in search of neuronal connections. One reason motor function and memory are so closely linked, as noted in the previous chapter, is that they are both coordinated by the frontal lobe, home of the brain's executive function.

VISUOSPATIAL MEMORY

YOU'RE AT HOME, it's late at night, a thunderstorm rolls in, the lightning flashes, and the lights go out. You can't see, but you manage to find your way to the junk drawer in the kitchen without stumbling, and fish around for the candle and matches.

How were you able to do this in the dark? Because you rely on vision, even when you can't see! Visuospatial memory enables us to remember the visual appearance of objects or scenes, including the three-dimensional placement and shape of furniture in a familiar but dark room.

Visuospatial memory is not limited to sensory data. Imagine you are seated in an empty room and someone has arranged to have an unfamiliar desk suspended in the air above you. Your task is to identify this object while looking up at it from the ground. While you have probably never seen a desk from this strange angle, chances are you will identify the desk as a desk. We can instantly recognize objects even when we see them from unfamiliar or constantly changing angles.

Visual processing begins with the establishment of a neural representation in the visual cortex. Later this information moves on to the temporal lobe for additional ordering. The resulting internal representation remains stable over time even though we are presented daily with innumerable varying perspectives. Our visuospatial brains constantly compensate for a lack of information, mostly by making up what is needed to fill in the gaps.

Painters make regular use of this ability of the viewer. One of the most intriguing things about any painting is what's not actually painted. It isn't necessary to paint anything but the front brim of a boy's hat to convey that he is indeed wearing a hat. Our visuospatial memory sees the rounded brim line and fills in the rest.

Imagery and perception have a unique interrelationship. Alzheimer's patients, for example, have two forms of impairment of visuospatial working memory. While lesions in the left hemisphere impair the perception of small details, lesions in the right hemisphere impair global representation. If such patients were hypothetically at the opening ceremonies of the Olympics watching the marching band move into the formation of an apple, they would see either just the apple or just the people.

LANGUAGE AND VERBAL MEMORY

LANGUAGE IS FUNDAMENTAL to the development of sophisticated memories. In millennia gone by, our capacity for naming became possible only when the brain developed connections between higher-order motor and sensory areas such as the Broca and Wernicke areas, respectively, and the CEO in the frontal cortex lying in front of Broca's area. In this way we evolved a working memory system for naming that later led to the miracle of language.

Verbal memory dysfunctions distort not only communication but also one's representation of reality. This is the cause of confabulations—false statements made without any intent to deceive, and with no root in any underlying psychopathology.

In order to have an adequate conception of reality, we must be able to discriminate among the possible origins of our memories. People who confabulate are constantly lost in a kind of fourth dimension. In those of their daily activities that draw on memory, they must deal with the insecurity of not knowing where their own memories are rooted. What is true? Where did this story come from? Am I talking about a real experience or a fantasy? Confabulators often cannot tell.

Many people with brain damage to the frontal lobes often confabulate. The statements themselves can be complex, ranging from subtle falsehoods to elaborately bizarre tales. One patient who had been in the hospital for days claimed he had spent the previous evening in a club playing cards with his doctor and the head nurse.

Confabulation should not be mistaken for amnesia. Although confabulatory individuals have obvious memory difficulties, the confabulations are not due, consciously or subconsciously, to a desire to "fill in the gaps." Many confabulators readily admit to the gaps in their memories and do not feel compelled to fill them. Some patients can even correct their confabulations without prompting. One patient said, "I have been having tea with my wife—oh, I haven't really. She's not been here today."

Vilayanur Ramachandran from the University of California at San Diego explains these behaviors as a miscoordination between the brain's hemispheres. The left hemisphere is charged with creating a model or story that makes sense. It organizes information, beliefs, and impressions we take to be generally true and makes them accessible for daily functioning. The right hemisphere detects and interprets anomalies of experience. Ramachandran hypothesizes that if anomalous information reaches a certain quantity and intensity, the right hemisphere forces the left to either revise its story or start over. Stroke patients confabulate because their damaged right hemispheres can no longer detect anomalies in their lives, the most obvious being their left-arm paralysis. With the left hemisphere free of intervention from the right, the patient's mind runs wild with uncontrollable self-deceptions of a fully functional body.

To test his theory, Ramachandran filled a syringe with ice-cold water and gently squirted it into a confabulating patient's left ear canal. This immediately induced rapid eye movement and dramatically cleared away the confabulations. The patient suddenly recognized her left-arm paralysis, which she had previously denied. Half an hour later, however, when the ice water had warmed, the patient again confabulated a denial of her left-arm paralysis. Ramachandran believes that the sensory shock of the cold water either stimulated the right hemisphere or triggered rapid eye movement, which can evoke the retrieval of remote memories, as sometimes happens during dreaming. In REM sleep, people often discover disturbing facts about themselves; perhaps in dreams the right hemisphere takes over, causing us to be more aware of oddities in our lives.

TRAUMA

IN GENERAL, EVENTS that cause us great joy or pain are easier to recall than other memories. In fact, the encoding and retrieval mechanisms of emotionally charged memories are structurally different from those of other forms of memory.

There is a dramatic difference between the way people recall traumatic memories and their actual sensations of the experience. Bessel van der Kolk at Boston University, a leading researcher of traumatic memories, discusses the issues in his book *Traumatic Stress*. In one study, he and his colleagues found that of all their subjects who had been traumatized only one came to develop a narrative of the trauma over the passage of time. Five of them who had claimed to have been abused as children could not, as adults, even tell a complete story of what had happened to them. They had only the intuition that they had been abused. Some had fragmentary memories that were supported by other people's accounts of actual incidents. Essentially, these people had dissociated themselves from their experiences.

Memories of traumatic or other highly emotional events remain unusually stable over time. This is remarkable given how easily other memories deteriorate. Since the work in 1889 of Pierre Janet, a colleague of Sigmund Freud, it had been accepted that conscious memory is colored by the totality of a person's experience—distorted by both one's experiences in the past and one's emotional state at the moment of recall.

According to van der Kolk, however, the important distinction is that sensory elements of *normal, everyday experiences* are easily integrated into an ongoing personal narrative. Traumatic experiences are exceptional because these intensely emotional events are not encoded into the ongoing narrative states. A traumatized individual is often unable to formulate a unified conception of the harrowing experience, yet continues to be haunted by the powerful emotions of the experience in the form of fragmented sensory perceptions and emotional states.

Perhaps traumatic memories cannot be explicitly recalled because they are established too heavily in long-term memory. When a subject tries to recall the event, the actual sensory experience interferes by flooding back in a mass of vivid and painful but seemingly irrelevant details. The amygdala overreacts while Broca's area, crucial for language and speech, shuts down. As a result, the subject is "struck dumb" with each attempt to recall the traumatic episode and is unable to express the experience in words. This is important because the formation of words often acts as a delaying function, giving the brain time to sort out the information needed to remember. But if no words are being formed, the brain is overcome with a confusion of incoming data.

Therapeutically, helping such a person find words can begin a process of dealing with the terror, moving it from a sensation to a concrete experience that can be tackled. When the emotional response can be tolerated, then words can come. When the problem is known and can be named, it becomes less terrifying. The same is true with a psychiatric problem; once a patient has a name and form for it, it is less threatening.

A recent study by Antonio Damasio has pinned down the brain regions involved in this mismatch. The study looked at individuals with brain lesions on the hippocampus only, on the amygdala only, and on both. All three groups were shown slides of different colors. When the blue slide was shown, a loud, jarring noise was sounded. People in all three groups showed a startle or fear-related skin-conductance response to the loud noise.

After several cycles, the blue slide was shown a few times without the noise. The control group of normal individuals with no brain atrophy showed a classical conditioned response; they still reacted with a strong skin conductance. The patients with hippocampal lesions did too, although they could not remember the training involved. The patients with the amygdala lesions, however, showed no emotional conditioning to the blue slide alone, even though they could recall what had occurred during the training sessions. The people with both hippocampal and amygdala lesions neither recalled the training nor showed evidence of emotional conditioning. Damasio concluded that emotional conditioning is dependent on the amygdala and that emotions are processed independently of the events with which they are associated.

Once the amygdala attaches emotional significance to sensory information, its emotional evaluation is passed on to the hippocampus, which organizes the information and integrates it with previous memories of similar sensory details. The greater the emotional significance assigned by the amygdala, the more intently the memory is permanently recorded by the hippocampus.

There is a limit, however. If the emotional arousal is too high, the hippocampus is hindered in making a proper categorization and evaluation of the traumatic event. What happens, chemically, is that a traumatic experience or painful memory—or a high incidence of stress, for that matter—causes the level of cortisol in the brain to rise. Cortisol, the stress hormone, works by binding to receptor sites in the hippocampus, but when emotion gets too high too much cortisol binds to each neuron. The onslaught increases the metabolism of the cells so much that they essentially overheat and die. As a result, the hippocampus can't organize the components of the traumatic experience into a unified whole. The person can only reexperience the painful sensory fragments, not the event. The traumatic experience is recorded as separate and dissociated from other life experiences, and takes on a timeless and alien quality.

CONTROVERSY OVER TRAUMATIC AMNESIA

YEARS AGO, PIERRE JANET described a case in which a woman "froze in terror" every time she passed a certain door in the house leading to the outside. Although the woman could not explain what frightened her about the spot, Janet later discovered that several men had once decided as a joke to tell her that her drunken husband, whom they had placed on her doorstep, was dead. Ever after, this woman associated that doorway with intense fear. Much more recently, researchers B. A. Tobias, John Kihlstrom, and Daniel Schacter described a woman who had a phobia of running water. She had no memory of how this response had been acquired. She eventually was visited by an aunt who elicited a childhood memory in which the woman had strayed away from a picnic and become trapped under a waterfall.

Traumatic amnesia has been documented since the late 1800s, and in the decades since, the breadth of cases has never ceased to be remarkable. For example, van der Kolk describes his contemporary work with one traumatized woman who lost all explicit memory of being in the Boston Cocoanut Grove nightclub fire, and yet continued to reenact her experience every year on its anniversary. Another patient, who was a Vietnam veteran, tricked the police into re-creating a shootout with him on the anniversary of an army buddy's death. He, too, had no conscious recollection of his traumatic memory.

Research has shown that the younger a person is at the time of the trauma and the more prolonged the trauma, the greater the likelihood of significant traumatic amnesia. What is devastating is that even though the actual account of the experience is hidden, the emotional and sensory components of the memory stay with the person for life; and any sensation related to the traumatic experience, from intimacy to fear to sexual arousal, becomes a powerful cue for negative emotions. Together, these symptoms form a condition now known as post-traumatic stress disorder (PTSD).

The most controversial aspect of traumatic amnesia is the validity of delayed recall. Although the controversy is a complex affair that touches on issues of incest, family, social mores, and even religious beliefs, it is fundamentally a debate about accuracy, distortion, and suggestibility in memory. This is why scientists such as Schacter feel professionally obligated to try to uncover the truth about recovered memories, though as he points out, "Searching for the truth in this charged atmosphere is not easy."

Certainly, some traumatic events can be temporarily forgotten and subsequently remembered. For example, we know that child abuse is a major problem in our society. There's no reason to question the memories of people who have always remembered their abuse or who have spontaneously recalled it on their own. However, as Elizabeth Loftus's tests about false childhood memories proved, there is reason to seriously question whether memories newly found in suggestive therapy were indeed there all along or were invented under the powers of suggestion. This issue comes down to distinguishing between dissociation and repression.

Proponents such as Ernest Hildegard and John Kihlstrom, who argue that dissociation is the cause of traumatic amnesia, base their theory on the assumption that it is possible for our thought, feeling, and conscious memory systems to lose communication. Trauma or stress may break the links between these systems, which results in the fragmentation of past events, so the traumatic experiences can never be explicitly reconstructed.

Repression is an important Freudian defense mechanism believed to hinder conscious recall of terribly difficult experiences. But contemporary psychiatrists such as David Speigel at Stanford University claim that repression works only for an isolated traumatic experience; it is not powerful enough to explain the total amnesia of entire periods of a person's past. For Speigel, traumatic amnesia can only be explained through dissociation.

One problem with the dissociation theory, however, is that for normal subjects repeated experiences enhance memory. So if abuse, say, occurred regularly it should be easier to remember than not. In order to achieve a profound long-term amnesia, the repression mechanism would have to overcome the strong reinforcing action of repetition. Meanwhile, studies also show that the recollection of a single traumatic incident is very high. Schacter cites examples such as the Chowchilla kidnappings, a sniper killing at an elementary school, and the collapse of skywalks at a Kansas City hotel. Some forgetting and distortion do occur in the people who experienced these events, but not one person involved in them has complete amnesia regarding them.

In response, Lenore Terr at the University of California at San Francisco suggests that repeated traumatic events defy the natural tendency people have for reinforcement. The repeatedly abused child, for example, becomes more adept at using repression to dissociate the experience from his or her conscious awareness. Schacter goes further to say that the general knowledge of abuse is retained while the event-specific details may become blurred, which might help explain why the memories of sexual-abuse survivors are often sketchy on details. Episodic memory can fade out and semantic memory will take over.

The research seems to support the possibility of forgetting details of repeated trauma, but does not support full amnesia of the experience. In a study by Linda Meyer Williams at the University of New Hampshire involving women with documented histories of abuse, 16 percent of them said there were times when they did not remember that they had been abused. But in most cases, the women said that they began to forget only years after the abusive incidents. The memories faded like any others that were not called up for long periods of time. They did not instantly disappear, which is what amnesia is all about.

There is good reason to believe that different brain mechanisms are at work for temporarily forgotten traumas versus amnesia of extended periods of time. For example, the flashbacks of war veterans have often been cited as evidence for the accuracy of recovered memories. However, these memories often reflect a mixture of fantasy and reality and are heavily influenced by expectations, beliefs, and fears. The content of a flashback may say more about what a person believes or fears about the past than about what actually happened. So while single incidents may be repressed, dissociation offers the best explanation for traumatic amnesia. In his book *Searching for Memory,* Schacter describes one particularly compelling case:

> In the mid-1980s a collaborator of mine, Dr. Nissen, a cognitive psychologist, came across a remarkable patient: a middle-aged woman who apparently harbored multiple personalities. Nissen said she appeared to have 22 personalities. One was of a 5-year old girl. Another, 39-year-old Alice, was studying to be a counselor, spent a good deal of time reading the Bible, and enjoyed painting religious subjects. Bonnie, 36, was interested mainly in the theater. Charles was an abrasive 45-year-old man who drank heavily, liked to watch televised wrestling matches, and painted wild animals. Gloria, 32, was one of several left-handed personalities; she also painted, but more abstractly, and she adopted a different last name from the others so that she could obtain her own social security number. Each personality came forward to deal with the external world at a different moment in the patient's life.

Some of the personalities knew about each other, but many had no memory for the others' experiences and were unaware that any other personalities existed.

Dr. Nissen could find no motive for the patient to fake the disorder, nor any evidence that she was doing so. Besides, the woman had a relatively low IQ and would have been incapable of the enormous mental effort required to keep 22 feigned personalities straight. The patient had numerous gaps in her memory and often failed to remember where she had been or what she had done. And her history contained signs and symptoms of a disturbed identity that dated back to childhood; at the age of five or six, she had displayed bursts of aggressive, violent behavior, and family members noted that she referred to herself by different names during the outbursts. Her attendance at school was irregular and her behavior was erratic.

The woman's problems were typical of children with dissociative disorders. According to Schacter, these children have severe behavior problems, are frequently in trouble, and are often referred to as pathological liars or persistent daydreamers. A person with a true dissociative disorder leaves behind a trail of serious pathology, and the trail in Dr. Nissen's patient was easy to follow.

Indeed, Schacter strongly warns against claims made by therapists who suddenly recover a "suppressed" traumatic memory in a patient who has no such track record. The warning goes double if the therapist uses suggestive techniques such as hypnosis. Schacter cites a recent survey of twenty women who retracted their recovered memories of sexual abuse. There were some striking similarities among these women: nineteen of them recovered their memories during therapy and all of them stated that their therapist clearly influenced the development of their memories. Notably, 90 percent of the retractors also reported that some type of trance induction was used in therapy to "recover" their memories, primarily hypnosis.

While there can never be indisputable evidence indicating that a memory of trauma has been falsely implanted, anecdotes support this as a distinct possibility. Way back in a 1982 paper, Loftus quoted two psychotherapists who admitted to using a controversial treatment for obese patients that involved implanting entirely false histories into their autobiographical memories. The patients had been obese all their lives, but the therapists successfully induced false childhood memories of having grown up thin. The therapists' goal was to use false memories to inspire the obese patients to become thin again, as they ostensibly had been in their past. The therapists even commented that they could "very easily install memories in you that related to real-world experiences that never occurred," and that this "happens a lot in therapy."

Many researchers have aligned themselves on opposing sides of the false-memory issue, and see each other as evil opponents. Those who doubt the validity of recovered memories say there is no basis for the amnesia of trauma. Those convinced that PTSD is real say painful repressed experiences need to be fully remembered, dealt with, and gotten rid of.

The battle is intensifying even more rapidly as claims for or against amnesia of trauma begin to come up in legal cases that are reported, usually sensationally, in the media. Stories began surfacing in the mid-1980s about adults who, in the course of psychotherapy, remembered incidents of abuse, usually sexual, during childhood. They would then publicly accuse the perpetrator, usually the father, and sometime sue for damages. In 1992, the parents of Jennifer Freyd, a psychology professor who accused her father of such abuse, started the False Memory Syndrome Foundation to combat what they saw as unfair smearing of innocent people by unscrupulous therapists. Those convinced about trauma amnesia see such efforts as attempts by abusers to suppress the truth.

Despite the passion in both camps, everyone is quickly forgetting (!) that the scientific evidence about genuine or false memories is painfully inconclusive. In 1995 the American Psychological Association created a task force to review the research on memory and repression. Its conclusion was that "absent other corroborating evidence, it is impossible to tell whether a memory is a real memory, partially real, or false, and that because of this point there is no way for science to determine the differences there." The task force found that "recovered memories are extremely rare but that they can happen. The bottom-line recommendation is caution and education." Until better research is done, the APA takes the position that the courts are not the best place to resolve questions of childhood abuse based on such memories.

More unbiased diagnostic research will help resolve this burning issue. A start may come from a recent discovery by a team led by

Schacter, using PET scans, which shows that local patterns of cerebral blood flow are different for real and false recollections. Wary of the volatile nature of the subject, Schacter was quick to note his team's worry that their very initial findings would be prematurely used by zealots on either side and prompt media hype for the possibility of a biological "lie detector."

FORGETTING IN OLD AGE

PRIMING PEOPLE WITH suggestions can be useful in certain cases. For older folks, it can help them recover real memories. So many elderly people seem unable to "put their finger on" a past experience. But often this is not because the memory has been erased; it's just that the person can't initiate the process of retrieving it. Give such people a beginning—some fact to organize around—and they can then pull all the pieces together. They can remember the word, the name, and the action, and then feel very much relieved. Aging is the most common factor that compromises the memory of us all, and its effects are being studied intensively.

Cognitive changes assumed to accompany aging are seriously misunderstood. Many people, for example, confuse normal age-associated memory changes with the severe clinical condition of Alzheimer's disease, a form of senile dementia. Statistics show that no more than 10 to 15 percent of people from age sixty-five to one hundred show symptoms of clinically diagnosed senile dementia, yet thanks to prompting—or perhaps priming—by the popular press, a great majority of the aging population would swear to having the disease.

Some memory loss is common as people get older, and it differs significantly from dementia. In normal aging, individuals may have a "tip-of-the-tongue" memory loss for words that haven't been used in a while. That's why Grandma, having a "senior moment," may confuse her grandchildren's names when they first come to visit after having not seen them for several months. With Alzheimer's, people lose the names for common objects they run into every day, like glasses or ovens.

The age at which a normal person's memory begins to decline significantly is highly individual, though it usually doesn't happen much before age seventy. Researchers don't yet know conclusively why almost everyone experiences some memory loss with aging. Neurons might become less effective owing to shrinkage, cell death, degeneration from lifelong exposure to stress-released chemicals such as cortisol, or just a drop in efficiency, as we see in muscle cells. There may also be a decrease in neurotransmitters (notably acetylcholine), neurotransmitter receptors, or the dopamine that keeps the receptors receiving.

Not all memory functions are affected equally. Long-term memory seems to suffer little, while working memory is very much affected. It is also interesting to note that the minds of elderly people work quite well as long as they are not pushed too hard. One experiment showed that older people performed almost as well as younger adults on memory and cognitive tests when given ample time and comfortable conditions, but when they were pushed under stressful conditions their performances dropped much more sharply than those of the younger adults.

The effects of clinical dementia are vastly different. What marks Alzheimer's is a sudden decline in cognition (an abrupt drop in scores on dementia tests from one year to the next) rather than the gradual decline of normal aging. The consequences can be severe, and include the deterioration of memory, language, and perceptual abilities.

To illustrate this point, Robert Ornstein and Richard Thompson, in their book *The Amazing Brain,* relate an anecdote about Donald Hebb, a pioneer in explaining the brain's plasticity and memory. When Hebb was forty-seven, he published an article entitled "On Watching Myself Act Old," about his own memory changes. In it he described the first occasion on which he detected signs of memory lapse. One evening while he was reading a research paper he turned back a page in order to pencil in a note. Much to his surprise, he saw that his intended note was already there, penciled in in his own handwriting. He had completely forgotten that at some point in the past he had already read the article! The experience was a terrible shock to him, and he decided to slow down his busy life by ceasing to work in the evenings.

As we age, most of us tend to accept more and more responsibilities, and despite our amazing brains, we do have physical limits. When Hebb experienced his memory lapse, he was doing extensive research, teaching, writing, directing a new laboratory, and acting as chairman of McGill University's Psychology Department. Can anyone doubt that his memory capacity was being fully challenged? Yet most older people are all too ready to attribute their memory changes to an inevitable onset of dementia.

When Hebb reached seventy-four, he noticed further changes in cognition. His vision was poorer, his balance was less steady, and his forgetfulness had increased. He also thought that his vocabulary was declining and that he tended to repeat his thought patterns. In another article, he referred to all of this as a "slow inevitable loss of cognitive capacity." These "losses," however, seemed rather imperceptible to others. The editor of the magazine that published the article quipped, "If Dr. Hebb's faculties continue to deteriorate in the manner he suggests, by the end of the next decade he may only be twice as lucid and eloquent as the rest of us."

Even researchers have misunderstood aging, believing that age-related memory changes result from a widespread loss of cortical neurons. Recent developments in brain-scanning techniques refute this idea, showing that age-related neuronal loss is insignificant. However, it has been shown that parts of the hippocampus atrophy as we age, and that this correlates closely with problems with explicit memory (that of facts and figures, faces, and things).

One reason for this may be that while few neurons are lost overall in the brain, the basal forebrain, which provides the hippocampus with acetylcholine, suffers markedly. Without acetylcholine, the hippocampus's synaptic plasticity hardens, though this conclusion is quite controversial.

Another study showed that both the young and the elderly have an increase in hippocampal blood flow when they recollect a recently studied word, but use a region of the prefrontal cortex when trying to retrieve the word later. For effective memory, the frontal lobes must work just as well as the hippocampus. The frontal lobes are also strongly affected by aging. Changes there include neuron atrophy as well as a reduction in blood flow and glucose metabolism. As the frontal lobes are the center of the executive function, which logically sequences memory organization, it is not surprising that poor frontal lobe functioning leads to a breakdown of temporal order and recall. Aging people often have difficulty remembering the order and timing of events.

Dopamine may also play a role in the loss of plasticity, and thus of memory. The weakening of synaptic connections, sometimes referred to as long-term depression (versus long-term potentiation), occurs when receptors on neurons at either end of the synapse begin to close down—to stop receiving messages. The presence of dopamine, which seems to act as a chemical reward and indicator for a neuron that continues to receive, keeps them open and receptive. Thus, a decline in dopamine might lead to the degradation of synapses and memory.

As we age, our general worldly knowledge and verbal ability do not change at all, but the speed at which we store new information slows, mainly after the age of sixty-five. The memories most affected by this are those of tasks that require manipulation or a transformation of information and tasks that require the establishment of new routines. R. C. Mohs, Director of Psychology at the Mount Sinai School of Medicine in New York City and a longtime researcher of interventions for memory loss in older people, reminds us that although the rate at which we store new information decreases with age, information registered deliberately and consciously can remain securely retrievable. And good retrieval methods can be learned.

For example, Mohs's 1993 study of people ages seventy to seventy-nine found that those with more education had more efficient memories and experienced less memory change with the passage of time. Years of schooling trains people to learn the best ways to encode and recall memory. Effective learners, for example, tend to look for patterns, group information according to category, and use a greater, more effective variety of memory strategies. Mohs suggests that such mental exercising keeps memory strong by reinforcing synaptic connections in the brain . . . as we saw in Chapter 1 with the nuns of Mankato.

POPCORN!

MEMORY IS AFFECTED by so many things, even the wacky and wonderful. Recall, for example, that implicit memory (popcorn) works in the background. It is not readily made available (hungry) to our conscious minds. Yet it can be affected (now) by stimuli we may not even be aware of. For example, A. G. Greenwald and S. C. Draine of the University of Washington recently demonstrated—for the first time through reproducible methods—what psychologists have long believed: that subliminal messages can and do influence behavior.

The nation's attention suddenly became riveted on subliminal messages in 1957, when James Vicary's tomfoolery with the owners of a New Jersey drive-in were leaked to the public. Vicary told the owners he could influence movie patrons to purchase more popcorn and Coca-Cola by flashing short commands such as "eat popcorn" and "drink Coca-Cola" extremely rapidly on the movie screen. Just enough movie frames of each wording were spliced into the movie so that the words would register in the language areas of the brain, but not remain on the screen long enough to register in the visual cortex, so they'd be understood but not consciously. No one tested the results, and when the scandal broke, the experiment stopped.

Greenwald and his colleagues used reliable scientific methods to show that subliminal messages are in fact influential. However, they also showed that the influence lasts for as little as one-tenth of a second. Furthermore, "messages" that can be registered and incorporated into implicit memory can only be extremely simplistic. Even fragments like "eat popcorn" are too complex for the subliminal mind.

The researchers asked more than 300 subjects to identify nearly 500 target words as either male or female, or as pleasant or unpleasant. The words were flashed briefly on a computer monitor. The subjects responded by pressing keyboard keys. Just before seeing certain target words, however, the subjects were also exposed to a subliminal string of letters containing a priming word such as "girl." On some trials the priming word agreed with the target word, but on others it did not. To test the brain's ability to receive the subliminal message, Greenwald and his colleagues required subjects to make progressively faster choices, flashing target words at intervals of six-, five-, and four-tenths of a second. When subjects were time-pressured, the rate of error increased dramatically, and the priming word altered the pattern of the error.

The subliminal messages have an effect because they are priming the brain to pursue a memory. This is the same mechanism that caused Elizabeth Loftus's patients to falsely remember childhood events. Whether the primer is the words "eat popcorn" or a picture of a buttery bagful, it generates activity in the area of the brain where related representations are stored. That area then starts to send out messages that arouse the brain's smell and taste functions. These turn on the amygdala and hypothalamus, which say, "Go get some popcorn. Gotta get it. Hungry. Gotta survive."

6

EMOTION

JASMINE WAS A thirty-year-old dancer who had become quite well known in the United States as well as in Europe for her renditions in modern dance and jazz. She had studied widely and was entirely in tune with her body. This helped her tremendously in controlling her movements, but it also made her anxious, and she would overrespond to the slightest sign of change in her body. The resulting panic disorder plagued her for years, until she learned how to counter it by engaging in yoga, deep breathing, and vigorous exercise.

Jasmine had recently married, and the life change, though joyful, had also led to a recurrence of severe panic episodes. Her body would tense up. She would feel that she was losing control. The alarms in her amygdala would go off and juice up her entire emotional system. Her heart rate would shoot up. She would enter a state of chronic startle—inordinate fear. The recurring episodes of intense feelings that overwhelmed Jasmine soon led her to become depressed, paranoid, even semidelusional at times. She became distrustful and would fly into a rage at almost anything when she was in a "state." She could not calm herself or look at anything in a logical way.

Having reached this degree of difficulty, Jasmine decided to take my suggestion and try beta-blockers, drugs that, among other actions, block the epinephrine receptors in the muscle spindles, which set the resting muscular tension and the uptake of adrenaline (the alarm hormone) in the brain. They had a profoundly calming effect on her behavior.

Why was it that yoga, deep breathing, and exercise—purely physical acts—helped calm Jasmine's historic anxiety, an emotion? And how could beta-blockers, which interfere with cellular action in the muscles as well as in the brain, free her from her gripping panic? The answer is that her emotions were largely due to her exaggerated interpretation of signals that she was getting from her body. The success of her control techniques clearly indicates that both the brain and the body contribute to emotions and do so in a complex, interdependent way.

Exciting research has recently challenged several long-standing assumptions about emotion. For years psychologists have maintained that emotions are purely mental activities, some of which, such as fear, elicit a physical response by the body. But while a few unique emotions, such as altruism, are dominated by mental processes, the rest are equally due to the body.

What the public, at the hands of some scientists, clinicians, and popular movements in psychology, has missed for so long is the fact that emotion wells up from the brain and the body acting together. The role of the body in emotion has been discounted, especially since the psychopharmacological revolution, with Thorazine, Valium, lithium, and Prozac. But we are now bringing the body back into the analysis of emotion.

The other leading historical idea is that emotions are all localized in one system in the brain, commonly referred to as the limbic system. However, we are learning that emotions are the result of multiple brain and body systems that are distributed over the whole person. We cannot separate emotion from cognition or cognition from the body. It has always been our need as humans to divide and conquer, to separate out two kingdoms as heaven and hell, but separating the body and the brain is rapidly coming to be seen as ridiculous.

The new view shows that emotion is not the conveniently isolated brain function that we once were taught. Emotion is messy, complicated, primitive, and undefined because it's all over the place, intertwined with cognition and physiology. Despite this distributed complexity, science is beginning to be able to look at one emotional pathway and then another, and to figure out how these bits of brain are interacting. The hope is to use this information to learn how to better help people with emotional difficulties and to help all of us manage our everyday emotions.

BRAIN OR BODY?

HISTORICALLY, QUESTIONS SUCH AS "How do emotions work?" and "What is the role of emotion in people's lives?" have been answered on philosophical grounds. The modern study of emotion began when William James described the internal path between an emotion-provoking stimulus and an emotional response. Conventional wisdom says that human beings cognitively assess a situation, which creates an emotion, which then directs a response expressed by the body. We lose our fortune, are sorry, and weep; we meet a bear, are frightened, and run; we are insulted by a rival, are angry, and strike. James, however, maintained that cognitive assessment could only take place after physiological changes in our bodies signaled us as to the nature of our emotions; we feel sorry because we cry, angry because we strike, and afraid because we run.

The James-Lange theory, as it came to be called, has been refuted for many different reasons as scientists have learned more about the nervous system and the brain. However, James must be credited with bringing the importance of the physical experience of emotions into the equation. In the early part of this century, researchers Walter Cannon and Phillip Bard, also of Harvard, maintained that there are just general states of arousal and that these states are subjected to cognitive assessment and then are assigned an emotion. In some cases, physiological changes such as butterflies in the stomach or a racing heart are too slow to be the determinants of emotion; when we see a bear we begin to run away before we even realize that we are afraid. In other situations, we not only have time to think before physiological changes take place, we sometimes have time to act; we fear being confronted by a bear if we wander into the woods, and so we stay on the beaten path. According to the Cannon-Bard theory of emotion, information about an emotional stimulus enters the brain through the thalamus and from there follows two pathways: to the cerebral cortex, where cognitive assessment is made, or to the amygdala and hypothalamus, which direct body reactions.

In the 1960s Stanley Schacter and Jerome Singer moved the debate to the next level. They agreed with the general arousal theory, but thought that the brain's perception of this arousal interacted with the reality of the social environment to create emotion. They conducted research that consisted of injecting experimental subjects with adrenaline. These individuals then spent time in a room with either a euphoric or an angry lab assistant who was acting as another subject. The genuine subjects reported feeling happy or angry depending on the environment created by the lab assistant with whom they were placed. The Schacter-Singer theory of emotion suggests that when the brain receives feedback that the body is physiologically aroused, it then looks out at the world to evaluate and decide what the emotion may be.

Meanwhile, debate also advanced over how emotions work within the brain. Is emotion generated from one "center" in the brain or is it the result of interactions of different parts of an emotion "system" that creates giddy joy or paralyzing fear? Cannon and Bard thought of the thalamus as an emotional center. In 1937, Heinrich Kluver and Paul Bucy did some interesting research that implicated the amygdala as an emotional center. The credit for the idea of an emotion system goes to James Papez, who proposed in 1937 that the thalamus divides information into two "streams"—one that provides cognitive assessment and the other that creates physiological arousal and physical reaction to a stimulus. In the 1950s, Paul MacLean at the National Institute of Mental Health named this visceral brain the limbic system, and it is still generally assumed to be the network in the brain that senses and generates emotions.

Even though different brain structures have roles to play in emotion, a few stand out. Andrew Young and his colleagues in Cambridge, England, have recently had the opportunity of working with a female patient who had a rare disorder that required removal of her amygdala in both hemispheres. While not cognitively impaired at all, the woman has some deficits in recognizing emotions of all kinds and a complete lack of recognition of the emotions of fear and anger in people's voices. She understands what fear and anger are and when and how they might be expressed, but she cannot comprehend fear or anger as they are manifested in real life. Imagine how your life would be if you could not understand that someone was angry with you or that you were angry with them. Imagine the danger if you could not understand the urgency in a command like "Look out for the bus!"

The debates about body versus brain and centers versus systems are still with us. Joseph LeDoux at New York University is combining the centers-versus-systems paradigms, postulating that there are different pathways for different emotions, rather than one region or system that underlies them all. Contemporary researchers do not all agree on which specific feelings make up the human emotional palette. Most agree, though, that there are four basic emotions—fear, anger, sadness, and joy—and that the other emotions are created from combinations of these four, just as all colors are made up from combinations of the three primary colors. For example, worry, anxiety, and stress all derive mostly from fear, with a little anger or sadness thrown in. However, some researchers claim that surprise, disgust, and guilt are their own unique emotions. Research with brain surgery patients has also shown that emotions are much more subtle and complex than we may realize. The patients report that stimulation to certain brain areas results in complicated feelings that often include the presence and reaction of other people; for example, they might feel as though they are standing at a cocktail party and have made a social faux pas. These types of emotions reveal the importance of the social realm to our psychological functioning.

Emotions are not nearly as distinct as we would like to think they are. There are also individual differences in the way people experience emotion, and thus mood. Throughout the ages people have found different emotions and different levels of emotion more or less adaptive. Innate temperaments for these emotions become genetically determined and are subjected to success or failure in the survival of the fittest.

One interesting new theory, developed by Edward and Carol Diener at the University of Illinois in Urbana, involves the notion of a "set point." According to this theory, people have an inborn set point for mood, similar to the set point for weight. The set point is your basic level of happiness or sadness, which is subject to the ups and downs of life but will inevitably return to some kind of base line, even in people who experience dramatic changes in their life circumstances. Research with lottery winners and victims of spinal-cord injury has shown that despite these life-changing events, people's moods return to their set point, often within a few months. Christopher Reeve, who became paralyzed from the neck down after falling off a horse, is as incredibly enthusiastic in his new life as a high-profile supporter of research into cures for spinal-cord injuries as he was in his formerly high-profile life as an actor. In some people, however, set points decline with age.

THE CRUCIAL ROLE OF MOVEMENT

THE TERM "EMOTION" is derived from the Latin *movere*—to move. It is important to realize that emotion is a movement outward, a way of communicating our most important internal states and needs.

The brain mechanisms that evolved to display emotion are the same as for all of our sensory and motor input. The difference is in the intermediate state of processing information. Input from a person's face that will lead to identification is channeled via different pathways from the information about the emotional expression on the person's face. The emotional information goes directly to the amygdala and the insula, which then send directions to act to our motor systems in the brain. So there is a splitting of the information, and you can identify a face and have no emotional confirmation about it and claim that the person is an imposter, which happens in Capgras's syndrome.

The motor and emotional systems probably evolved concurrently in primates. Geographically they are right beside each other and intertwined, and whole-body postures that signal aggression or mating in invertebrates evolved into behavior patterns and facial expressions in mammals and primates. The limbic system comprises the amygdala, hippocampus, medial thalamus, nucleus accumbens, and basal forebrain, all of which connect to the anterior cingulate gyrus, which is the major gateway to the frontal cortex. This system is the launching point of emotions and the emotional connector to the cognitive prefrontal cortex. Yet all of it is wrapped around the system for movement.

Emotions are played out physically in the body through internal motor activity, such as a more rapid heartbeat, and externally in such movements as a smile or a frown or a change in body posture—whether jumping for joy or sitting slumped in sadness. All of the outward behavior that results from emotion is composed of movement.

Bodily expressions, especially of the face, are the first means of emotional communication between a baby and its mother. Movements of expression—a kiss or a slap—also give emotional meaning to words. Some emotions are expressed bodily more than verbally; you can tell someone is angry, as Darwin noted, by "the body being held erect." Social relationships depend greatly on proper body language.

Darwin believed that the muscular movements of facial expressions were inherited behavioral patterns. Facial expressions of emotion and other behaviors such as crying and laughing are implemented by other neural circuits in the brain. These responses are hard-wired into the brain. They are present or appear soon after birth without any training.

From an evolutionary perspective, emotion is the result of behavior that has been repeated over and over through the generations, such as escaping from danger, finding food, and mating. Primitive movements such as goosebumps, snarling, erection of body hair, flashing of feathers, and biting are all intimately connected with emotion.

A good illustration is the emotion of fear and the movement involved in the fight-or-flight response. During threatening situations, many interacting parallel pathways that include neuronal, chemical, and hormonal activity are activated. The autonomic nervous system creates internal visceral movements, while the voluntary motor pathways orient us and begin activating for external movements to fight or flee.

The upper cortex and the lower limbic structures are in continuous communication with each other. There are many more connections from the small emotional limbic center into the large logical and rational cortical centers than the reverse, which may be the reason that emotions are more dominant in determining behavior and why we sometimes react or speak before we think. Activation, whether by fear or arousal, causes an outpouring of activity toward the motor cortex to initiate and guide a movement response. At the same time, there are messages from the aroused limbic area to other areas of the cortex to evaluate the incoming data. After a decision is made, guidance is sent from the cortex back to the amygdala to tell it to act, to cool off, or that it is not advisable to act. For instance, the brain responds to the sound of a shutter banging by preparing to ward off burglars, until the frontal cortex intercedes and takes into account the wind as opposed to the likelihood of burglars. The amygdala is cooled off and the brakes are applied.

The hypothalamus activates the amygdala, the anterior cingulate, and the brainstem. Stimulation of the amygdala produces anger, rage, or threatening behavior. The amygdala uses primitive general categorizations of the limited sensory information that it receives in order to activate an immediate aggressive or defensive motor response. Certain key characteristics of objects, people, or situations are enough to produce a reaction. The shadowy figure on a sidewalk standing next to a building at night leads to an immediate response as we walk down the street.

The amygdala, in turn, activates the anterior cingulate and the hypothalamus, which then switch on the autonomic nervous system, the motor system, and the endocrine system, which causes body organs to adjust to the demands of the situation. There is an increase in heart rate and stronger heart-muscle contractions, constriction of blood vessels and increased blood pressure, an opening up of airways in the lungs, decreased movement of digestive organs, and increased blood flow to the skeletal muscles. Meanwhile, chemical neurotransmitters are sending messages throughout the body and activated hormones are significantly influencing nervous-system reactions and organ systems throughout the body.

The physiological reactions of the fight-or-flight response are recognized by the individual as fear. This primitive, hard-wired emotional response prepares us for the strenuous motor efforts required for fighting or running. It also provides clear evidence of the intimate link between emotion and movement.

WHERE EMOTIONS LIE

BASIC EMOTIONS SUCH as happiness and sadness are separate functions, and they represent opposite patterns of activity in the hemispheres of the brain. Increased activity on the right side of the brain often signals depression, while activity on the left side often indicates happiness, euphoria, and even mania. Richard Davidson at the University of Wisconsin has done studies that indicate that people with more general activity in the left hemisphere have a more positive mood, while people with more activity in the right hemisphere have a more negative one. Remarkably, research has also shown that infants are born with an innate predisposition toward a more active left or right brain, meaning a happier or sadder temperament.

The neural processes that underlie "worry" may reside in the right hemisphere. Researchers at Johns Hopkins monitored subjects' brains with PET scans as the subjects listened to tapes of themselves describing family crises, work-related stress, financial problems, and other concerns. There was significantly more activity in the right frontal lobe, a central structure in planning and decision-making. By contrast, when the subjects listened to tapes of themselves describing neutral everyday events, activation in the right frontal lobe was significantly decreased.

Scientists are not completely in agreement about how the brain's hemispheres interact to create the feeling and function of emotion. In 1996, M. K. Mandal of Banaras Hindu University found that hospital patients with right-hemisphere brain damage were significantly less accurate in identifying the emotional expression in a photograph than patients with left-hemisphere brain damage or than general medical patients. Other studies show that the right hemisphere has more to do than the left with the final processing of emotions, once they've made their way up the emotional pathway from the limbic system to the prefrontal cortex. If this is the case, the right hemisphere may play a leading role in the comprehension and production of emotion, just as the left hemisphere plays the primary role in language.

Both the right and the left frontal lobes are very important for the regulation of emotion, needed for making decisions in the social and personal realm. It may be that this area connects the limbic system and the motor cortex, establishing the link between areas that plan and those that carry out the actions. There may even be an "upper" path between the limbic system and the cortex through the cingulate gyrus that deals with pleasure and sociability and a "lower" path involving the amygdala and ventromedial prefrontal cortex that deals with issues of self-preservation.

Studies of specific patients support this model. Once a patient of Antonio Damasio's who had suffered damage to his ventromedial prefrontal cortex drove to Damasio's office on a winter morning when ice had made driving extremely hazardous. He told Damasio in a matter-of-fact manner about a number of people who had gotten into a wreck in front of him because they overresponded to a tailspin, which is the usual reaction. Unlike them, he did not brake when he hit the huge patch of ice, but gently pulled away from the tailspin. His dulled response to emotional information saved him; it allowed his cognitive appraisals to "coolly" direct his actions.

I once had a patient who ran the blackjack club at a local university. He told me that the best players that they sent out to the casinos to apply their system of card counting and weighted betting were the ones that could sit there all night and follow the plan without being tempted to up their bets prematurely. These were normally the ones who drove everyone else to distraction. For instance, asked whether he would take a ride or drive his own car to a casino, the star player endlessly debated the wisdom of each choice and exhausted all concerned. These people's behavior is characterized by not being able to decide what is good for themselves in the future. They can list option after option, but can't make a decision. They can follow a logical system perfectly and thus win at blackjack, but it is as though they had little in the way of preference because they had few feelings to direct them. They are tiresome and make others angry with their talking on and on. They act like patients with impaired frontal lobes.

The case of Phineas Gage in 1848 is famous in the annals of psychiatry for demonstrating the crucial role that the frontal lobes play in emotion. Gage was a railroad foreman who survived an accident in which a 3-foot-long iron rod 1.25 inches in diameter was driven through his skull, and returned to work. Soon afterward, however, Gage's temperament changed so drastically that he lost his job. He had been known as a remarkably responsible and even-tempered individual, but after the accident he was often compared to a wild animal with no moral sense. He cursed in front of women—unheard of in his day—and fought irresponsibly.

Gage often said he felt that he had lost his ability to feel emotion. That loss of feelings may have been directly responsible for his objectionable behavior, because without emotions moral judgments and socialization become difficult if not impossible. Hanna Damasio and Thomas Grabowski analyzed photos of Gage's skull and utilized computer technology to re-create a three-dimensional image of his brain.

They found that the areas most likely to have been damaged by the iron rod were the left anterior prefrontal cortices and the ventromedial prefrontal cortex—areas crucial to decision-making.

How emotions work—and how we might help ourselves and others control them—will become clearer with a closer look at the four primary emotions: fear, anger, sadness, and joy.

FEAR

FEAR IS A UNIVERSAL EMOTION that includes everything from the decision to fight or flee to the insidious mounting of stress. It can also cause us to "freeze," which is not an indicator of indecision in the face of fear, but stems rather from an ancestral skill used to respond to a stalker or predator. A fearful stimulus primes the body with adrenaline and prompts the fastest physical reaction possible. When the brain is triggered in fear, the autonomic system and stress hormones are activated. The amygdala gets immediate input from the thalamus and acts to start up the internal readiness and reaction system. This bypasses the cortex and any consideration of the context and such—it is just responding. In fact, the feared stimulus and the programmed response to it are indelibly etched into the amygdala, as its job is to alert the animal to dangerous, novel, and interesting situations and to direct its response.

The physical and mental responses to fear were so important to the survival of primitive man that they remain very powerful and long-lasting. Unfortunately, this adaptive response is not always appropriate in today's world. Our civilization has evolved away from the need to overrespond, but we still do. Regularly overresponding to life's minor troubles can lead to high blood pressure, heart disease, migraines, and ulcers. (As Robert Sapolsky's book title informs us, zebras don't get ulcers.) Other malfunctions of the fear system are shown in disorders such as panic and phobias. Once we learn to be afraid of something, our brains become programmed to remember that stimulus in the same way, so that it's hard to get rid of our conditioned fears.

The startle response is a good example of an adaptive fear mechanism that can sometimes get out of control. A loud, sudden noise will elicit a startle response from most people. As this type of noise is often associated with danger, it is important to be immediately alert and have the adrenaline pumping. However, if a stimulus like a loud noise is repeatedly paired with a dangerous situation, some people will develop an overactive startle response. This is often the case in PTSD. People with this disorder—war veterans or victims of abuse—startle easily and often. They suffer from physical ailments more frequently than the general population, and have an increased incidence of cancer, which is associated with a lowered immune response and raised levels of cortisol. Many of the physical and psychological symptoms associated with PTSD can be traced to the frequent, sometimes constant state of startle and hyperalertness that afflicts these individuals.

Jackie, a victim of early child abuse, was afraid of everything, from new situations to her own shadow. She was not agoraphobic (afraid of going out into the world) but nevertheless stayed at home. She always overresponded to new situations, seeing them essentially as threats, bringing with them the possibility of her being hit again.

The most poignant example of PTSD is seen in women who have been raped and cannot allow themselves to enjoy sex again. Such a woman is often indelibly programmed to be vigilant and fearful. She may consciously want to engage in sex with her mate but has an inner resistance to it. Clearly, this can cause real trouble in her most valued relationship.

The amygdala is the area of the brain most involved in fear. Stimuli have a direct pathway through the sensory filter of the thalamus to the amygdala, which can then mobilize the body through its brainstem connections. If you see a snake, or anything that looks like a snake, in the corner of a shadowy garage, the amygdala is immediately triggered and you react before cognizing the image. The image triggers the optic nerve to send a signal into the brain. On its way to the cortex, the signal takes a short route to the amygdala, which shouts "Emergency!" to the rest of your body, triggering a cascade of reactions: your heart rate soars, your blood pressure increases, and your senses become heightened as your body prepares to take action.

With enough time or experience, reason can stop the action. There is another, slower pathway for fear, where the information about a fearful stimulus goes from the thalamus to the frontal cortex and then to the amygdala. This occurs when you realize that the "snake" is really an old coiled garage-door spring. The response to the second pathway overrules the indication of the first. Now all systems reverse. Your blood pressure comes down and your heart rate returns to normal. The lower brain, the amygdala and the rest of the limbic system, is inhibited by the upper brain. You then begin to "think" about what just happened rather than just respond.

The two pathways can be seen as the low road and high road of fearful responses to danger. The path straight through the thalamic projections to the amygdala (the low road) is rough and crude but fast. The pathway using the cortex (the high road) gives a more accurate assessment and can be expected to lead to a more considered response, but it takes longer.

Fear responses to sudden, potentially life-threatening stimuli such as explosive noises or the attack of an animal are automatic in most people. But many other fear responses are learned. Most of us have marveled, for example, at how young children seem to have no fear of heights. We also have to teach them to look both ways before crossing the street, for fear that a car might hit them. New MRI studies also show that teenage brains may not have fully developed the reasoning pathways to adequately assess fear, which may contribute to teens' difficulty in dealing with emotions. Neuropsychologist Deborah Yurgelun-Todd of McLean Hospital flashed forty faces showing expressions of fear to sixteen adolescents age eleven to seventeen. The younger teens reacted with heightened activity in the amygdala but only a modicum of activity in the frontal lobe. The older teens had greater activation in the frontal lobe. In previous tests, adults showed greater activity in the frontal lobe and less in the amygdala than adolescents. Apparently, there is a gradual shift of emotional and cognitive processing from the instinctive to the cognitive regions as the adolescent brain learns and grows. While this growth of wisdom or activation of the frontal cortex can help teens learn how to stay calm in stressful situations, it can also cause them to learn from parents or friends fears they didn't have, or need to have, such as an undue fear of heights or of social situations.

As the snake example shows, fear involves contextual conditioning—those other stimuli that are present. A garage corner is dark, cool, and dirty, making it much more likely to be the place to find a snake than a corner of the living room. Context is a collection of many stimuli and is dependent on accurate memory of situations. The hippocampus is the brain area responsible for assessing this function. It receives processed information from the cortex that has already been associated with the context of the situation and the fearful stimulus, bringing the whole picture into perspective.

Contextual conditioning can be used in reverse to treat panic disorders and phobias such as fear of snakes, dogs, or heights. The technique, which is called "flooding," involves a step-by-step process of gradually experiencing more and more of the feared stimulus so that the patient can learn that snakes or dogs or heights are not invariably dangerous. First the patient is asked to visualize the least fearful aspect of the experience—the snake's interesting skin design, the dog's cuteness, the great view from the bridge—and then to practice relaxation or meditation, which gradually lessens the anxious firing of the brain's neurons and relaxes the tense muscles of the stomach and legs, relieving the fear input from the body. Note that both the brain and the body symptoms must be dealt with, again supporting the theory that emotions are sustained by varied systems throughout the body. Eventually the patient works up to actually experiencing the feared stimulus: holding the snake, petting the dog, standing on the bridge.

Hans Sieburg, a psychiatrist at the University of California at San Diego, has developed a virtual-reality treatment for acrophobia that he calls City Project. Patients wear high-tech goggles that provide a realistic, three-dimensional image of what it looks like to be standing on the top of a skyscraper, and while wearing them are calmed with music and reassurance. With practice in repeat sessions, they realize that they are not going to fall. Their bodies stop swaying at the sight of the ground far below. The off-balance feeling subsides. They learn with their bodies that they are not going to fall, and thus they conquer the irrational fear. They train their cortex to re-evaluate the situation and quickly respond to inhibit their amygdala.

The flooding process is straight cognitive behavioral training; it is rearranging the circuits in the brain, reducing all the neural connections that have long supported the thesis that height equals falling while strengthening the circuits that convey "safe." By gradually rewiring, the patient begins to refocus on the fact that he's not going to fall off the building. Separating the low (bodily) and high (cognitive) roads in this way seems to be the key to successful treatment.

The lesser cousins of fear are worry and anxiety. Chronic worry can be an uncomfortable emotion. Anxiety disorders plague a significant portion of the population and reflect some of our most human concerns, ranging from social position to acceptance by God. Recent research is starting to show how anxiety works in the brain. National Institute of Mental Health researcher Dennis Murphy and several colleagues have identified an "anxiety gene." In a study of 500 people, they found a difference in a gene that affects the level of serotonin in the brain. Serotonin is the brain's brake and policeman; it prevents the brain from getting out of control from fear or worry. It has a calming effect that helps us to assure ourselves that we are going to survive and elevates mood and self-esteem. Some 30 percent of the subjects who had the longer form of the gene, which promoted more serotonin in the brain, had lower levels of anxiety, while 70 percent who had the shorter form of the gene were found to have higher levels of anxiety. The fact that the majority of people are more anxious may mean that they are more uncomfortable on a daily basis, but are also more ready to respond to the environment. A little healthy anxiety leads to a greater ability to survive in our constantly changing world.

ANGER

THE SECOND UNIVERSAL emotion is anger. Everyone experiences anger at one time or another and it is easy to recognize in the faces of others. Learning to control anger is a natural and important developmental step for toddlers, and yet one out of five people experience attacks of rage that they report they cannot control.

Aggression is an important part of the natural world. Violent combat between males before mating upholds the rule of survival of the fittest and ensures the strength of the gene pool. Mothers also engage in aggression to protect their children from predators. Human anger is closely connected to the fierce defense of territory, mate, and self that many animals display. Anger evolved as a unique set of feelings and behaviors that has its own value in changing other people's behavior.

As in any evolutionary analysis, one must consider the costs and benefits of a behavior. It is important that all social animals be able to control their anger and aggression. While it might be advantageous to win out over other members of the species, it would hardly be helpful to indiscriminately kill or hurt them. The sociopath is an example of this brake on aggression gone wrong. In everyday situations, anger toward others can be costly, because it undermines future positive interactions. The benefit of changing someone else's behavior to one's own advantage can be enormous, however. We must walk a fine line to get a decent benefit without an outrageous cost. Therefore, the most important thing to learn about anger is when and how to use and control it.

The popular psychology of the 1960s—"let it all hang out"—has translated for some people into letting go with their anger and not trying to control temper tantrums. Research has shown, however, that there is little health benefit from this kind of behavior. Anger can get out of control and create both mental and physical health risks. Unreasonable anger is a symptom of many disorders. It has been identified as the primary cause of heart attacks in Type-A personalities, and, psychologically, we know that anger can create an unhealthy environment for anyone in contact with the angry person.

Biologically, researchers are still learning more about the pathways and expressions of anger. Aggressive people often have underactive frontal lobes, the areas of the brain that restrain impulsive action and that supply wisdom, and if these are not working correctly or actively enough, feelings of rage will not be inhibited. Partial evidence for this conclusion is provided by findings of low frontal lobe activity in people with antisocial personality disorder, who are characterized by their angry, destructive behavior. It is as if the amygdala is saying to the cortex, "Be still. Let the automatic pilot work," even though it would be better for the cortex's reasoning to interfere and stop the inappropriate behavior.

People in the general population also experience episodes in which they can't control their anger even if they want to. These occurrences may also be caused by a lower level of activity in the frontal cortex. In a heated confrontation, a person may feel that his brain is going too fast, considering all the aspects of the anger-provoking situation and maybe even events from the past that add fuel to the fire. With no inhibition from the frontal cortex, the thoughts are free to get out of control and the person quickly becomes overstimulated. This "noise" is very difficult to overcome in a rational way. The prefrontal cortex is less active than it should be; the underactive executive function is not as alert, gets overwhelmed, and subsequently has a hard time putting on the brakes. This imbalance can be caused by such things as ADHD, brain trauma, or the toxic effects of alcohol or drug abuse.

The problem can be exacerbated by an inability to express one's thoughts and emotions. Verbalizing aggressive thoughts and feelings is the best antidote to violence. A popular therapy for perpetrators of domestic violence is getting them to learn how to "talk it out." Breaking the cycle of low inhibition and overstimulation, however, is made more difficult when a person learns that acting on aggressive impulses will bring a kind of relief. Addiction to aggression as a way to solve problems and relieve frustration can make it very difficult for the angry person to change.

This was the case for Deborah, who as a girl grew up in a household that was chaotic and permissive to a fault. The home soon became ruled by the children. Deborah was affable, a perfectionist, and very smart, but she had a hard time with frustration. When she was four she began to tantrum periodically, though it was nothing compared with her two older sisters. They were outwardly bellicose and were hellions in school. Deborah's behavior was tolerated because she was a good student who got all As.

In sixth grade, though, Deborah hit a subject she could not master: diagramming. She just couldn't do it. In response, she disrupted the whole class and was suspended for a few days. This experience taught her that if she created a scene she could get more attention than her sisters. As she became an older teenager and encountered more frustrating challenges, she used her tantrums more often. They were manipulative. They got people to listen. They immediately resolved her frustration and disappointment. They were a high for her and she got addicted to them. She would even throw a tantrum simply because she was bored. This tantrum behavior ended up dominating her life, and she eventually spent most of her time in and out of psychiatric hospitals.

The chemistry of aggression is not well understood, but researchers do know that very low or very high levels of serotonin in the brain can contribute to aggression. Some clinicians have successfully treated anger and aggression with SSRIs (selective serotonin reuptake inhibitors) such as Prozac that make more serotonin available in the brains of people whose natural levels are low. Other research has shown that high levels of testosterone can increase aggression.

SADNESS

ALTHOUGH SADNESS MAY APPEAR to be a much more subdued primary emotion than fear or anger, it ranges just as widely, from mild melancholy to uncontrollable crying. Sadness probably evolved to emphasize and underscore losses of all kinds; it takes us off-line so that we can regroup and reevaluate. It may even cause us enough "pain" that we are motivated to change. In the brain, sadness seems to be related to an increase in activity in the left amygdala and the right frontal cortex and a decrease in activity in the right amygdala and the left frontal cortex.

Prolonged sadness can cause sustained overactivity in the amygdala and frontal lobe. Some speculate this can cause neuronal "burnout" in these areas, either by depleting their stores of neurotransmitters or crippling the ability of these chemicals to transmit messages. When this occurs, sadness can slip into depression, which is characterized by emotional numbness rather than intense feeling. This is distinct from depression that is accompanied by anxiety, which can cause a person to become feverishly active, even suicidal. In classical depression, typified by a person sitting motionless in a chair with no intention of getting out, the numbness may be adaptive, granting relief in the case of a terrible loss or giving a person some "down time" to prepare for the next stage in life or to incorporate a major change.

That's what happened to a patient we'll call Bobby Jack. B.J. was a happy-go-lucky guy who generally responded positively to most everything. He tended to see the world as full of answers, structures, and stories that were positive and had good endings. His left hemisphere was working all the time, fitting everything into what he would have predicted. The ongoing story of life generated by his left hemisphere was accurate and upbeat. Events had closure.

But one September, after his company had conducted its annual job reviews, B.J. was told he wasn't going to get a promotion he was expecting. He hadn't failed to work hard; he just wasn't right for the job. This upset him. He was sad. It was one of the first times real life didn't square with the story he had running in his mind. And no promotion meant no raise, which meant that he couldn't leave his apartment and finally buy a house, which he had wanted for so long.

B.J. tried to put matters back in order in his mind, but he couldn't because there was no resolution. He didn't get the job, period, for no reason that he could change. The spin he put on it was, "They made a mistake. They are fools. They will promote me later." He tried to act differently, plan differently, do something to fit the situation back into his internal story line. In doing so, his left hemisphere may have started to overactivate, trying to deal with the new reality, trying to impose logic on the illogical set of circumstances, trying to get back to happiness.

Meanwhile his stomach had begun to act up. He got diarrhea and became physically exhausted. He began to worry that his physical condition was deteriorating. This steady anxiety commanded even more attention from his left hemisphere, which tried to find a way to explain his physical maladies as well. But there was no story that gave him control. He got sadder, more tired. He started to feel hopeless.

In all the discomfort B.J. had also stopped exercising. He gained a few pounds, which hurt his self-image. He became preoccupied with his failings, convinced that he was no longer fun to be with. He stopped talking to his friends and spent more time alone. He withdrew into himself, running his left hemisphere nonstop in trying to solve what were now all these huge problems. His left hemisphere could no longer take it. It got worn down, then burned out. There were no solutions. B.J. became clinically depressed. Now that his left hemisphere was completely inactive, his harsher, more realistic right hemisphere was free to take over. He chased people away. He didn't have words anymore. He sank into a real blue funk.

Finally, B.J. went to a therapist. The counselor got him talking again. He had been stuck trying to explain events to himself, and it was extraordinarily helpful to thresh it out with someone else. As he did, he was able to put his experience into a bigger story line that could include a future. He saw that he hadn't made egregious mistakes in the past and that he wasn't a bad worker, but that someone else was in fact more suited to this particular job and that the choice was indeed his boss's. His future was still in front of him. He could still work effectively and make a few changes that would make life more interesting and more fun.

B.J. put the loss into perspective by activating the talking brain, then recruited even more of the talking brain to plan and get into thinking about the future again. This prompted him to start taking walks, and to talk a bit to one of his siblings, which reactivated his left hemisphere. But it wasn't quite enough, so the therapist prescribed an antidepressant, imipramine, which helped change the chemistry of his limbic system. He began to accomplish new tasks at work, to exercise, and to look forward to attaining new goals again. Bobby Jack was back.

Depression may be characterized by feelings of despair, guilt, helplessness, and hopelessness. People with depression may have symptoms such as less ability to concentrate, impaired memory, weight loss or gain, fatigue, sleep disturbances, and loss of interest in everyday activities. The onset of this disorder is typically in early adulthood, although it can occur with anyone at one time or another, especially if someone experiences a significant life trauma. Depression affects 3 to 5 percent of the population at any given time, and about 20 percent of people will experience major depression in their lifetimes. Even children only five or six years old can experience symptoms clinically similar to adult depression.

Depression is less genetically based than any other mental illness, and is the one most dependent on environmental factors. Life events can affect brain biology in even the most naturally cheerful people, like Bobby Jack. Mark George, a psychiatrist and neuroscientist at the University of South Carolina Medical School, has done PET scans of the brains of clinically depressed individuals and maintains that scanning technology can open up new treatment options by identifying subtypes of depression and differences in responses to medication. Helen Mayburg and her colleagues at the University of Texas recently used PET scans to locate an area of the brain that is different in depressed people who do not react positively to antidepressant drug therapy. In these people the front tip of the cingulate gyrus has below-normal glucose metabolism. Being able to separate out a subgroup of patients for whom current medications are not effective is an important step toward finding the right treatment for the group.

The traditional approach to treating depression—talk therapy—should still be pursued. It is helpful because it opens a straightforward connection to another person. Talking helped Bobby Jack create a palatable story of the past and a new story for the future. It connected him to his therapist; he felt understood and was encouraged. This allowed him to break free from the loop of self-hate and recrimination. At the same time, it helped break the lock in this pattern in his brain.

Also, the physical act of talking itself may have been helpful, forcing the language centers in the left hemisphere to work more, reactivating other structures.

For years, the last resort for people who did not respond to talk therapy or antidepressant drugs has been electroconvulsive therapy (ECT)—shock treatments. Electrodes are placed on the scalp and a strong electric current is sent through the skull to the brain. To be effective, the current must be so strong that it triggers a seizure. Because ECT succeeds in a majority of cases, some 50,000 people a year turn to it.

Like antidepressant drugs, ECT works by changing the chemistry in the brain, elevating mood. However, the side effects are significant. The typical regimen is three shocks a week for several weeks. To prevent pain and injury during each seizure, patients are put under general anesthesia. By the end of the cycle patients can suffer confusion and memory loss, some of which may be irreversible, and their mood may improve for only 3 to 6 months.

A new technique that has been found useful for treating severely depressed people is transcranial magnetic stimulation (TMS), which appears to have many of the advantages of ECT without the nasty side effects. A coil of magnets placed against the patient's scalp sets up a magnetic field inside the brain, which excites neurons, also inducing heightened levels of a number of neurotransmitters. No anesthesia is needed, and there seems to be no loss of memory or disturbance of other brain functions. Unlike ECT, this technique can target a specific region of the brain, notably the left prefrontal cortex, where activity is often lower than normal in depressed people.

TMS is still experimental, but early results are encouraging. In one study patients showed a 50 percent improvement on a commonly used depression rating scale—better than that seen in most antidepressant drug or ECT treatments. TMS may also be useful in treating PTSD and OCD, as well as Parkinson's disease.

JOY

HOW DO WE EXPERIENCE JOY? It is both the physiological experience of warmth and satisfaction and the cognitive assessment that this is the way things should be. Joy, happiness, pleasure, is its own incentive; it is what makes survival and propagation of the species worthwhile.

Some of the earliest research on happiness was serendipitous. In the 1950s James Olds and Peter Milner, hoping to influence learning, placed an electrode into the hypothalamus of a rat. When the rat pressed a bar connected to the electrode, the hypothalamus was stimulated. The researchers concluded that the rat perceived this sensation as pleasurable because it proceeded to press the bar up to 4,000 times an hour, and would allow itself to starve rather than stop. They had hit upon the pleasure center of the brain. Since then, research with humans has shown that the hypothalamus is just one of several pleasure centers of the brain, among them the septum and the nucleus accumbens.

Neurotransmitters and endorphins play an important role in the perception of pleasure. But dopamine is a key factor—and the one currently getting the most attention. Each of the pleasure centers uses dopamine as a transmitter. If a rat is trained to push a lever for internal stimulation to a pleasure center and is then given a drug such as pimozide or haloperidol that blocks the action of dopamine, the rat will stop pushing the lever. Pleasure is often muted in people who are taking conventional antipsychotic drugs, which block the dopamine receptors. The drugs are used to stop hallucinations and delusions, but often produce a state of joylessness and a lack of motivation and drive. As this can complicate treatment, newer drugs that have less of this effect are being developed to treat psychoses. Meanwhile, drugs such as cocaine and amphetamines work in the brain by increasing dopamine levels. But if dopamine—or any of its artificial substitutes—reaches levels that are too high, hypomania or even mania can result.

It is hard to imagine a disorder arising from too much happiness, but there are several that can result from not having enough happiness or enough internal reinforcement and feelings of pleasure. Reward deficiency syndrome, a concept coined by Ken Blum at the University of Texas, is helpful in understanding the complexities of addiction and compulsive behavior. This idea states that a lack of internal rewards leads a person to self-medicate with substances or with behavior that is rewarding. We see evidence for this in the statistics for conduct disorder and ADD; children who have either of these illnesses are 5.5 times more likely to be addicted to drugs or alcohol as adults compared with the general population.

While different neurotransmitter systems cascade upon one another in the reward mechanism of the human brain, perhaps the most important interaction is that of dopamine in the nucleus accumbens, a group of neurons that have a special relationship to reward and motivation. It is located just beneath the front of the striatum, a part of the basal ganglia which is involved in movement and cognition. If the nucleus accumbens is lesioned in lab rats that normally push a lever to receive addictive drugs such as cocaine, the rats will stop pushing.

Recent research at the University of Cagliari has shown that within the nucleus accumbens there is a further division of function. In a study that supports conclusions about the addictive effect of nicotine, rats were injected with nicotine directly into the brain and scientists observed corresponding increases of dopamine and activity in the nucleus accumbens. This area of the brain behaves similarly when cocaine, amphetamine, or morphine is administered. An important finding in this study is that a difference was identified between the action of the outer shell of the nucleus accumbens and its inner core. The outer shell seems to be most involved in emotion, motivation, and addiction. This area has direct connections to the limbic system and is part of the extended amygdala, which serves as a link between the midbrain and the forebrain.

This area is important for learning, in part because it tags information with a signal of intensity that tells the rest of the brain to pay attention. Stimulating this area with an electrode helps rats to learn more quickly and use more extensive areas of the cortex during learning. The extended amygdala's emotional coloring of learning affects our notions about the rewards and dangers of different stimuli.

There is still much to be learned from research on addiction. A research group at Yale University is examining the different roles that dopamine receptor subtypes play in creating and maintaining addiction. Understanding the processes of addiction and motivation more thoroughly could possibly remove the stigma and improve treatment for disorders ranging from alcoholism and drug abuse to gambling and sex and food addictions.

One of the most intriguing emotions in the spectrum of joy is love. While most of us wax poetic about it, some researchers are breaking it down in typically rigorous laboratory fashion. According to Helen Fisher, an anthropology professor at Rutgers University, there are three distinctly different physiological and emotional categories of love—lust, attraction, and attachment—and biologically, at least, they all relate to the ancient drive to mate. Fisher says that each behavior evolved with a different purpose. Lust evolved to get you out looking; attraction evolved to make you focus and expend your energy on one specific individual; and attachment evolved so you would stay with that individual and raise offspring once mating was accomplished.

Using MRI scans that show chemical activity in the brain, Fisher has found that lust is associated primarily with estrogen and androgens. Attraction, however, is associated with elation and a craving for emotional union, which may be linked to the monoamines such as serotonin. The neurotransmitters associated with long-term attachment—a behavior evidenced by close body contact, separation anxiety, and a sense of calm, security, and peace with a partner—have been harder to find.

Fisher hopes that the results of her ongoing work will show that the stages of love are based at least as much in brain chemistry and physiology as they are in psychology, further evidence backing up the thesis that emotion is not one system in the brain but multiple systems that tie together workings of the brain and the body. We all know this is true: early in our relationships with our current lovers or spouses, our hearts raced when we suddenly heard their voices on the phone; we had butterflies in our stomachs when we prepared to meet them.

These kinds of physical sensations are linked to increased quantities of neurotransmitters such as dopamine, serotonin, and norepinephrine in the brain's pleasure centers, as well as other chemicals such as oxytocin, endorphins, and phenylethylamine (PEA), known as the "love drug." These brain chemicals are also the ones long associated with various states of euphoria and in particular with the ecstasy caused by drugs such as cocaine and amphetamines, as well as with the high that long-distance runners report experiencing. The chemical compounds in chocolate act like nicotine, causing the release of dopamine in the pleasure centers.

Of course, not all joy is brought on by physiological stimuli. We are happy when we receive praise, find a dollar, or finish a puzzle. Whenever I show a picture of my basset hounds to someone, the person inevitably smiles. These stimuli start the pleasure ball rolling by eliciting a small squirt of dopamine, serotonin, and oxytocin in the pleasure centers.

One of the most joyful of emotions is laughter, but the neurochemistry of it is hard to explain. We laugh when something strikes us as funny. But we also laugh when we are nervous and sometimes just because someone else is laughing. Laughter derives from the primary emotion of joy, but it's a bit confounding because of the many and varied circumstances that trigger it.

Robert Provine, a behavioral neurobiologist at the University of Maryland, studied students on college campuses to find out exactly what made people laugh; 1,200 "laugh episodes" later, he was convinced that most laughter has little to do with jokes or funny stories. Clearly, social context is important; people laugh as noted when they're nervous as well as when they are amused, and they may laugh cynically when disappointed. While laughter is evoked by a punch line, indicating that the brain's conscious, cognitive regions must decide that the circumstances are right for laughter, most people cannot will themselves to laugh on command or to suppress an unwanted case of the giggles. Laughter arises from our conscious minds and from a primitive, precognitive part of our brains, something that's very deep in our animal nature.

Recent study also indicates that laughter may be primarily a function of the left hemisphere. In 1998 doctors at the University of California of Los Angeles reported that they were able to make a sixteen-year-old girl laugh by stimulating a tiny region in the left frontal lobe, the supplemental motor cortex. They were testing her to try to find the source of her epileptic seizures. When they stimulated that particular region with an electric current, the girl burst out laughing. She was asked to perform various tasks, such as naming objects, reading, counting, and extending her forearms, but regardless of the activity, she consistently laughed when that area was stimulated. Even more interesting was the fact that although the girl's laughter was being triggered electrically, each time she laughed she had a different explanation for it, attributing it to whatever object was in front of her or whatever action she was engaged in at the time. She saw a picture of a horse as hilarious, giggled over a book she was reading, and once told the researchers, "You guys are just so funny."

Provine says that, in part, laughter functions as a kind of social signal—just like a smile or a scowl. Indeed, studies have shown that people are thirty times more likely to laugh in social settings than when they are alone. Even nitrous oxide—laughing gas—loses much of its potency if taken in solitude. Laughter occurs when people are comfortable with one another, and the more laughter, the more bonding within the group, lending credence to the old saw that laughter is "contagious." If there is a feedback loop of bonding–laughter–more bonding, it may explain one of the most bizarre incidents of contagious laughter ever recorded: in 1962, an epidemic of laughter among schoolgirls in Tanganyika lasted for six months and forced officials to close schools to break up the group and end the marathon.

MOTIVATION

WE CANNOT EXPLORE the topic of emotion without understanding motivation. Motivation is not an emotion per se, but a process that ties emotion to action. Motivation is the director of emotions. It determines how much energy and attention the brain and the body assign to a given stimulus—whether it's a thought coming in or a situation that confronts one. Motivation is essential for survival.

The term "motivation" also comes from a Latin root related to movement, in this case the word *movere*—to set in motion. Motivation creates and guides the goal-directed behavior that fulfills our basic needs. One theory of motivation is that it arises from internal instincts or drives for food, sex, and other basic needs, but this hypothesis is difficult to defend in view of the vast array of different and novel behaviors that human beings are capable of. In the 1930s, Walter Cannon suggested that individuals seek to maintain a state of equilibrium that he called homeostasis. If this is true, and the body has a low level of water, say, the person will feel thirsty and thus be motivated to drink.

Cannon's theory doesn't explain all the human behavior that doesn't seem to satisfy a need or even be good for us. The incentive theory of motivation tries to explain these phenomena. Incentives are external stimuli that pull the individual toward them in some way—everything from a ripe apple on a tree to the prospect of a raise in pay. Abraham Maslow gave a structure to this idea when he proposed his theory of a "hierarchy of needs." In this pyramidal organization of motivating factors, basic biological needs form the broad base, and the psychological need for "self-actualization" is the narrow pinnacle.

Whatever theory is used to explain it, motivation is the pressure to act. Because motivation is at the heart of all goal-directed behavior, many levels of the brain are involved. The brain must perceive and assess both internal and external stimuli—internal physiological cues such as hunger and external environmental cues such as a plate full of steaming spaghetti. The brain structures for memory are also involved in motivation, so that present stimuli can be assessed and compared with previous similar or different stimuli.

Another important function is the ability to emotionally label certain stimuli or situations, which is really the center of motivation. Weighing our feelings for or against something determines our movement toward or away from it. The brain structure responsible for this ability is the extended amygdala, the primary pleasure center. Because motivation is so closely tied to physical behavior, the structures that produce and maintain it are closely related to those that regulate motor function and movement.

The cingulate gyrus is the main link between motivation and emotion. It has the appropriate sensory inputs to receive processed visual, auditory, and olfactory information and also receives inputs that reflect the internal states of the body. After it gets all this information, it must transmit the total message to parts of the brain that can enact a behavioral response. The cingulate has outputs to the basal ganglia for motor reaction and to the brainstem for physiological arousal. It also has connections to the hippocampus, important for memory. Owing to all these connections, the cingulate is able to assess motivational aspects of the environment and compare them with memory in order to give incoming stimuli different motivational priorities. This system provides a person with the ability to judge what is worth pursuing.

Several subcircuits are also involved in motivation. Structures of the limbic system, thalamus, and basal ganglia interact to perform different parts of the whole task of perceiving, assessing, and communicating motivational influences in the environment. They hold the various motivations in working memory and compare conflicting goals. Ultimately, this leads to choice, inhibition, and the seeking of reward.

This complex science works rather simply in real life. A pro football coach gets his team psyched up to beat the opponent by portraying the other team as the enemy, causing an overreaction of the subcortex which turns on the motivation machinery that improves focus, energy, and desire.

We can see how motivation ties emotion to action by looking at a case of retaliation, perhaps not the most laudable of motivations but one with a clear impetus and a clear goal. Sara was mad at Jim because he insulted her at a cocktail party held to raise money for the local summer-stock theater, where they both volunteered. As soon as he had uttered his smart-aleck statement she was flushed with anger, but she suppressed her reaction because, unlike Jim, she was concerned about acting in a socially acceptable manner. She didn't dismiss her feelings, though. Instead, she decided she would get back at this guy. She turned her anger into complete motivation, which fixed her attention on a plan to retaliate. She decided that she would set him up for an embarrassing fall during the next volunteer meeting, drawing him out with a sequence of questions that would show he was a fool. She planned each word carefully; it was the most creative she'd been in years, and it captivated her attention for two weeks until the meeting was held.

An interesting version of Sara's motivation belonged to a professor of surgery I had in medical school. We nicknamed him the Velvet Harpoon. He would not suffer fools, but he would not immediately condemn them. He often asked us questions that we had trouble answering, and most of us would admit to not being sure of the right response. But once in a while a student who knew he didn't know the answer would try to gloss over his response by quoting some expert or manufacturing what sounded like a plausible response. Rather than flatten him right there and then, the professor would play along with him, getting him further and further out on a limb by asking leading questions. Then, suddenly, he'd slip him the velvet harpoon—a question that would immediately show the absurdity of where his responses had led, and embarrass him totally.

The professor's behavior was motivated by anger. He was dismayed that the student did not know the answer. But he was angry that the student lied, trying to smooth over his ignorance by fudging the details. The subsequent embarrassment was awful, but there was a potent lesson: a surgeon can't gloss over a procedure, making up his moves as he goes along; if he does, he'll kill the patient. Our beloved Velvet Harpoon used retaliation in a constructive way.

A disorder of the motivation system is apathy, which can have a neurological basis or accompany another medical disorder. Apathy can be particularly difficult to treat because the behavior may be seen by the patient's family as moral weakness or be misinterpreted by a therapist as passive-aggressive behavior.

Apathy seems to be a specific malfunction of the motivation circuits in the brain, not just another manifestation of depression, and is best understood as a neurological disorder. There is no evidence that antidepressants are particularly helpful in treating apathy, but some drugs that increase dopamine, as well as some psychostimulants, may improve motivation. When Parkinson's patients are treated with L-dopa, their motor coordination improves, but so does their mood and motivation. High doses of dopamine are usually needed to help patients suffering from apathy. In one study, seven out of eight patients who became depressed following liver transplants and were given methylphenidate, or Ritalin, a drug that elevates dopamine levels, showed improved motivation in pursuing their rehabilitation regimens and less social withdrawal and apathy than they exhibited before taking the drug.

For apathy patients, drugs are not the final cure. They also need assistance in practicing techniques to help themselves. Education of the family is an important job for the therapist, too. Treatment of apathy raises some complex human rights issues, because these patients may be competent enough to make some decisions but not others. They are also prone to anxiety, which must be relieved so that they will consider options and get involved with life again.

Treating apathy is also important in depressed patients. Drugs can be used to bring about rapid improvement at an early stage. Since depressed patients are often convinced that nothing can be done for them, they may not adhere to the treatment plan, and become uncooperative and neglectful. Rapid countering of this resigned apathy is an essential aspect of treatment.

EMOTIONAL INTELLIGENCE

EMOTIONS ARE ESSENTIAL to our identity as human beings. We are also learning that emotions are essential to the trait that makes us most human, the ability to reason. In contrast to the steadfast belief that we must be calm, cool, and collected to make a proper decision, it is as likely that our gut feelings, impulses, and intuitions guide us to any particular decision. Emotions tap into areas of our brains that judge situations effectively without our having conscious access to them.

Understanding emotions is also the key to treating many mental disorders. From the nervous back-seat driver to the florid manic-depressive hospital patient, the overaction of emotion can create problems. The lack of motivation is equally troubling, and the problem of addiction has taken on epidemic proportions in our society. Fortunately, treatments are improving all the time. Drug companies are bringing new tools to the pharmaceutical marketplace at an astonishing rate. Home remedies, such as light boxes and herbal and homeopathic treatments, are also having some success, as is the emotional and social-skills training of elementary school children.

Understanding our own brains and our own reward systems is essential, but as we seek better techniques, we have to be careful not to let our own motivation for improvement lead us into quick fixes. The current movement to define each person's "emotional intelligence" is a good example of such a temptation.

Emotional intelligence was one of the hottest psychological ideas of the 1990s. The phrase was coined by psychologists Peter Salovey of Yale and John Mayer of the University of New Hampshire in the late 1980s, as a way to sum up human qualities such as empathy, self-awareness, and emotional control. Daniel Goleman, a writer with the *New York Times,* picked up on the term and made it the title of a best-seller, *Emotional Intelligence: Why It Can Matter More Than IQ.* Today the term is everywhere, and magazines provide tests that help you figure out your emotional IQ, or EQ.

Emotional intelligence is an attractive concept because it can provide a convenient scapegoat for today's epidemics of violent crime, marital strife, and teenage drug abuse. Conversely, we'd like to believe that if we improve levels of emotional intelligence in the young they will be better equipped for life's trials. Much of the how-to advice that is supposedly a reflection of emotional intelligence, however, is just plain common sense. Obviously, being able to control rage or develop empathy, say, will enable a person to have better future success in life than someone who doesn't have these abilities.

There are two central questions in the debate over emotional intelligence: Can it really be measured in a meaningful way? And can it be "taught" as a skill to children and adults? We don't know the answers yet, but you can be sure there will be plenty of research in coming years to find out.

7

LANGUAGE

SALLY WAS A sprightly girl of eight who was cheery most of the time. However, she would start to scream at her three-year-old brother, Joey, whenever he would barge in on her uninvited, whether she was alone or playing with a friend. She wouldn't stop yelling, would berate her inconsiderate little brother, and go on to criticize her mother for not disciplining Joey enough. Sally would wind up in a tearful rage.

Even after months of many short and long "time-outs" imposed by Sally's mother, the behavior persisted. Then her mother tried a wonderful technique to help Sally use her language to control herself. Even though Sally could not put into words what her behavior was, after the next outrage her mother sent her to her room to write a full account of the incident, with a focus on how it all started, what she was feeling, and the indignity of it all. She was to write down all the details as best she could, and then they would discuss it. This worked marvelously well. In time Sally even began to write about how Joey might feel, which was the insight her mother was hoping for.

The writing exercise allowed Sally to use words to help govern her behavior, which is probably a huge part of why we humans developed words in the first place. As we evolved and our social groups got bigger and more complex, we needed to delay and react more deliberately or chaos would have reigned. Language may have evolved as a delay mechanism.

The exact connections between thought, language, and action have been debated throughout history. Although the three processes can be separated, they are woven tightly together, making it difficult to delineate boundaries. The ability to use language not just to communicate but to plan and direct future action is at the core of humanity. Language improves and refines our thoughts, allowing us to remove ourselves from the present, to symbolically hold objects in our minds and manipulate them into different potential sequences before taking action. It is the moment of delay that is so crucial to planned action. Owing to language, we don't have to act immediately on emotional impulses determined by our immediate surroundings.

Language is acquired with so little effort that we fail to realize its full impact on our lives. Mapping thoughts to symbols lets us define ourselves, function in a social world, evaluate our emotions, and change our behavior. It is the foundation of self-governance and whatever autonomy humans can claim. The universal ability of humans to communicate links us together into a powerful community. From forty-four distinct, basic sounds (phonemes) that can be arranged into an infinite number of combinations, we have created today's complex society. Yet when people like Sally fail to make proper language connections, or to stop and consider what they are saying, they wind up not only with speaking, reading, or writing problems—which are bad enough—but with difficulty sustaining social relationships, making moral decisions, controlling anger, and even feeling emotions.

For decades scientists thought that they understood how the brain acquires and creates language. But new research is revising the long-standing model—so much so that we are in the throes of piecing together a completely new picture. We no longer see language as a highly localized function that exists in a neatly defined section of the brain. Indeed, language functions are distributed throughout the brain, and the locations can vary significantly from one person to the next. Some of the functions are incredibly specialized, to the point of being comical: there is, for example, a specific site in the brain responsible for naming vegetables. Yet we're also finding that because the functions are so distributed the brain has a remarkable ability to correct language problems, once we figure out how to retrain it.

THE POWER OF SELF-TALK

THE DIFFICULTIES FACED by Sally, and her turnaround, show us how fundamental a force language is in our lives. In 1781, Immanuel Kant defined thinking as "talking with oneself," and it is now clear that the silent articulation of speech—our conversations and instructions to ourselves—is what links thought, language, and action.

Recently P. K. McGuire of the Department of Psychological Medicine at London University used PET scans to show the brain region activated during self-talk. It turns out to be the left inferior frontal cortex, the same region that turns on when we rehearse a string of letters to ourselves and when we produce audible speech. Further studies have found that deaf people who communicate in sign language also use a form of self-talk in their thought and planning processes, and that it activates the same brain area. The region in front of this area is where we consider and plan, an important executive function of the prefrontal cortex.

The use of language to guide actions is most apparent in children. Many studies have shown that young children who talk out loud when instructing themselves and doing things have greater chances for success at a task and more self-controlled behavior during problem-solving. Over the course of development this "private talk" diminishes to whispers, and by early elementary school it becomes internalized and silent.

Learning-disabled children, however, particularly those with ADHD, tend to use more audible self-talk. While this may help them in the task at hand, it can also cause them to miss the crucial step of listening to themselves in order to stop and consider, to delay action, to plan. Jack, a brilliant four-year-old, was brought to my office one day by his very successful father, who had been searching for treatment for his son. The boy was a dynamo. For example, he knew the names of the dinosaurs, their body parts, the ages in which they flourished, and how they mated and the anatomical differences between them—all learned in a heartbeat and repeated to one and all.

Jack also asked questions about everything. After two minutes of shyness the questions and commentary began in a rush: Who was I?

Did I have children? Where did I go to school? Did I believe in God? Do we really inherit our brains? This was amazing. I was captivated—and exhausted after fifteen minutes. Jack was forever talking, forever asking questions. He had a tremendous vocabulary and thought about things all the time, but constantly skipped from topic to topic. He ran me over.

I would learn later that Jack routinely dominated his friends. He found it easier to be around adults, who were glad to answer his questions . . . at first. The problem was that he never stopped. He was a classic ADDer. His parents had found a way to keep him in bed until 7:30 each morning by allowing him to come into their bedroom at that hour with a list of five and only five questions. Jack's parents were giving people, but were overwhelmed and exasperated by the talking.

Part of Jack's difficulty was that he could not stop and consider any aspect of what the future might be. He could not self-talk. He could hardly reflect. He did subvocalize a lot; he could talk himself through a task by mouthing or whispering the instructions. But he did not seem to consider what he was saying. He also did not consider the consequences of his actions. His parents were concerned about his total lack of regard for the feelings of others. His playmates soon gave up on him and his bossiness. His incessant talking left everyone feeling frustrated and he was soon hated. He now found himself alone, a shame for a bright, perceptive child.

I only saw Jack once, as a consultant, but before he left I told his father that they should try to work on getting Jack to stop and consider what he was saying. At a minimum, he needed basic training in how to shut up and listen to himself. He also needed to slow down and begin to learn to reflect.

Self-talk is at the root of the empathy, understanding, cooperation, and rules that allow us to be successful social beings. Any sense of moral behavior requires thought before action. The reflection that comes with internalized speech is crucial to allowing us to weigh different courses and their consequences, and make the best decision before taking action. When impulses are reined, an individual can refer backward in time to consult similar experiences, put together new combinations of possible behaviors, and refer forward in time to probable consequences. In this way, memory and goals work with emotions to shape deliberate behavior.

It is also this sense of past and future that underlies morals. If one acts before allowing oneself time to think of the consequences, there is no will power or self-control. Values and goals are automatically ignored in the maelstrom of activity. A child who stumbles through life in this way also gets no practice in using his inner theater to expand on possibilities found by twisting and turning reality. Multistage planning for new situations—imagining several possibilities and their probable results—is at the core of human intelligence. It is the stuff that problem-solving is made of. If stopping to evaluate the consequences of one's actions is not done enough in early life, the groundwork needed for later consideration of ethics and morals cannot be laid.

Plan-ahead abilities begin to be developed through childhood narratives. These can come from many sources. Religious instruction for children inculcates ethics through Bible stories. Many fairy tales and fables achieve the same. I attended a Catholic school and the stories there were useful to me, particularly as I was an impulse-driven child who had ants in his pants. Learning about the lives of the saints provides heroes galore. Confession in the Catholic Church encourages children to reflect on past behavior and envision better future behavior—as it does for adults. Confession is a painfully shame- and guilt-producing tool used by the Church, but it has probably prevented many a violent act in grade school. Although children having access to guns is a growing contributor to school violence, the real problem is that some children lack the ability to reflect.

There are two main sources of interference with the ability to self-talk. The first is impulsivity, the lack of a sense of time and response inhibition, which is the primary deficit in ADHD. Like Jack, individuals who talk constantly about what is happening at the moment have no time for the secondary processing needed to detach from the immediate stimulation and pull together a deliberate plan. This leads to the behavior problems and uncontrollable anger that frequently accompany ADHD.

The second major source of interference with self-talk is an impairment that hinders the ability to use language precisely or easily enough to create a delay between stimulus and action. This may be the case in severe dyslexia or other language problems. If an individual is unable to translate anger into words, there is nothing to create the delay that gives him time to think more objectively, consider the perspective of others, or, most important, foresee the consequences of different actions such as giving in to an immediate desire to yell or strike out. The function of the "time-out" that is used by many parents and teachers in dealing with children and in many adult anger-management classes serves to slow everything down, giving an individual time to move to secondary processing in the form of self-talk before acting on emotion.

A RECENT ACCESSORY

BEFORE THERE WAS SELF-TALK there was just plain talk. Its origins are also debated. In the past few years the media have been filled with accounts of the language abilities of chimpanzees . . . and computers. The writers of these stories inevitably trumpet the results by suggesting that both chimps and computers are only a small step away from exhibiting the same communications skills as humans. Not so! A huge gap remains. Chimps communicate with each other in the wild using approximately thirty-six sounds. However, for chimps—and all other animals except humans—each of these sounds has one meaning and is not combined or linked with others to create a new message. It is the same thirty-six messages over and over again.

Homo sapiens uses phonemes (such as "da" and "ba") linked together in different combinations to form words. By the time we graduate high school we know 45,000 to 60,000 words, and all are combined and recombined into phrases, sentences, and paragraphs, each of which creates a precise new meaning. From different combinations of forty-four phonemes have come Jane Austen's novels, Shakespeare's plays, and chats over the backyard fence. Each phoneme has no inherent meaning; it is only in the vast number of combinations that we communicate complex ideas. The Swiss linguist Ferdinand de Saussure calls this "the arbitrariness of the sign." As there is no direct connection between sound and meaning, we can only communicate by agreeing to use a common code: specific combinations of sounds that build words, phrases, and sentences. It is the set of rules that we use to link the meaningless sounds—our grammar and syntax—that allows us to understand and express new ideas.

Although chimpanzees can learn to name objects in a manner and at a rate similar to babies, they cannot move beyond this stage even after years of intensive training. They cannot develop the rules that allow humans to combine signs, or to use prosody, the distinctly human ability to modulate the voice, adding emphasis and emotional tones that help convey meaning. As the noted psychologist Merlin Donald at Queens University, Ontario, says in *Origins of the Modern Mind:*

> Humans did not simply evolve a larger brain, an expanded memory, a lexicon, or a special speech apparatus; we evolved new systems for representing reality. During this process, our representational apparatus somehow perceived the utility of symbols and invented them from whole cloth; no symbolic environment preceded them.

As language is such a quantum leap forward from our evolutionary ancestors, we are driven to learn how it arose. Unfortunately, there is still no widely accepted answer. Linguistics, philosophy, anthropology, psychology, and neuroscience all examine the issue from different perspectives. We can look at brain structure, vocal tract position, the use of increasingly complex tools, religious ceremonies, or the abstract thought represented in cave paintings when we speculate about early language.

Three million years ago the anatomy of our animal ancestors' vocal tracts started to assume the modern form that gives us the motor movements necessary to form the sounds in our speech. Fossils reveal that the earliest *Homo sapiens* appeared 100,000 to 200,000 years ago, and there is a general consensus that the fast-paced symbolic language we use in our speech today has only been in continual use for some 50,000 years.

How could this complex ability evolve so quickly? There are two basic views. In the first, natural selection over thousands of years imposed the organization that was most successful in allowing symbolic thought, communication, and all the higher cognition this creative ability allows in our current brain structure. The universality of language in all human cultures and its consistency in structure and acquisition timeline in childhood support this claim. The second view is that language is a chance result of the evolution of an overall bigger cortex. According to this theory, early language was not a specific skill that helped our ancestors in their quest to survive and reproduce; however, other abilities that provided an advantage drove the evolution of neural mechanisms that happened to allow language to emerge.

Fossils of earlier hominids and comparative studies of primate brains provide some evidence for the second theory, because they reveal possible intermediate steps in the move toward a larger cortex, which allows for more adaptive functioning in the world. Clearly, primates must have some brain structures that enable them to associate objects with names, since they can be trained to name and to achieve the protolanguage of a two-year-old human.

The most remarkable case of a primate learning language occurred almost by accident. Duane Rumbaugh and Sue Savage-Rumbaugh were training four older chimps and found that by pointing to symbols, the chimps could show that they understood the commands given to them. One of the chimps, Matata, was raising Kanzi, one of the bonobos monkeys that have been the subject of much study recently, as a foster baby. Kanzi was a constant distraction, demanding attention and playing all around as the humans were trying to teach his mother how to communicate. But then he surprised everyone by showing that he had picked up language as his mother was being taught. Kanzi's language became the best developed, because he was introduced to it at an early age.

If language is a by-product of a brain that was reorganized for another cognitive ability, we naturally want to find out what that ability might have been. It may have begun as an ability to piece together individual items into larger combinations. Michael Corballis at the University of Auckland, New Zealand, calls this "generativity," and says it was the major breakthrough in human cognitive evolution. According to this theory, the generative ability evolved to help humans make tools, part of the popular theory that tool-making and the advantages it gave in hunting and survival drove the brain to expand and reorganize, allowing for the emergence of language. Several studies have indicated that the regions of the brain that control sequenced hand movement and speech rely on the same mechanisms. Corballis and others have developed the theory that this general area and ability to be generative were used to create the language capacity. Thus, from making the appropriate sequence for tool making and using, we were able to combine phonemes into words and then words into sentences.

As we commonly think of language as a mode of communication among individuals, it makes sense that as the size of social groups increased, bringing together families into bigger hunting groups and agrarian settlements, it became more important to devise ways to warn about danger, share knowledge and desires, and formalize rules to ensure peaceful coexistence. Representations of the concrete also allowed humans to pass on religious beliefs, traditions, legal systems, and scientific discoveries into a collective memory device—language.

With the advent of writing and reading 5,000 to 6,000 years ago, thoughts and memories became much easier to share, preserve, and build upon, which greatly increased the collective power of humans. Evolution has moved forward increasingly rapidly in the past few millennia because culture and language have co-evolved, leveraging each other into a faster pace of advancement. Modern culture is tied to memory devices such as books and television that rely on language. With all of the storage and manipulation that computers can do for us, we will likely develop new ways of thinking and explaining the world around us; already, the creators of the World Wide Web are proposing "metadata" software that can analyze the patterns of links made between Websites, perhaps revealing all sorts of connections between people and information we've never known existed.

Whether language was directly selected for in the course of evolution or was a by-product of another skill or an overall increase in intelligence, there is no doubt that its emergence greatly affected overall brain structure and behavior. Some neuroscientists even see it as triggering all of the differences between the right and left hemispheres. Others argue that there is a tremendous benefit in having both sides working in parallel, which demands a coherent language function. However, it is possible that neither of these views is correct; we must constantly be aware of our desire to conveniently assign brain functions to one side of the brain or the other. We like to think in terms of binary and complementary functions—good or bad, logic or faith, left or right—so that we can map and thus try to control our universe.

If language was at the root of our expanding capabilities, it can be viewed as the foundation of human consciousness and abstract thought. Linguist Derek Bickerton proposes that we needed the representations of language to move beyond direct sensory perception to the "off-line thinking" that allows for reflection and novelty. This is self-talk expanded. Burial rites for the dead and cave paintings illustrating myths can be viewed as telltale signs of its emergence. They would indicate that our ancestors had moved beyond communication tied to objects in the present to communication that seems to have no evolutionary advantage. Yet this is part of the co-evolution of our culture that defies the usual "selfish gene" concept, which states that all evolution must promote only those skills that give the bearer an advantage and thus selectively push the genes into the next generation. Cave paintings and such seem to have no direct advantage; this is where civilization took over more. And it is when civilization began to evolve much faster and truer than our gene-driven brains had caused. The wish to make life more pleasant, more meaningful, and richer, to read novels and the like, has a secondary and not a primary advantage. And it is simply what happens when you give a man a brain.

Extensive memory is required for success in using symbolic communication, and it seems likely that as language evolved the brain adapted to improve short- and long-term memory. The sequencing needed to understand long speeches and complex documents requires short-term memory to keep track of individual phrases and how they connect plus long-term memory to understand the meaning of each word, figure of speech, and metaphor. We also have to be able to make decisions between rules in the grammar system when there is ambiguity, so language may have pushed our brain's ability to analyze. It is the decision-making we so effortlessly use in language that is impossible for even the most advanced computers with vast memory systems.

BORN AND BRED

THE DEBATE OVER HOW and when our ancestors began to speak runs concurrently with the long-standing argument over whether language is wired in our brains at birth or learned. As in most nature-versus-nurture debates, the answer lies somewhere between the two views. Throughout history most people have assumed that infants learned language from their parents. But in 1959, linguist Noam Chomsky at MIT revolutionized the study of language by providing evidence that language acquisition is a biological process, not a learnable body of external knowledge. Because children raised under all sorts of conditions by all sorts of parents master complex rules of grammar at a very young age, Chomsky concluded that infants are born with a built-in propensity for handling the basic rules of language; every child has a "universal grammar" embedded in his or her brain, which specializes for the particular language by which the child is raised.

This theory has dominated research over the last four decades. In its absurd form, it argues that language is completely coded for in our genes; some sort of language would develop spontaneously without any interaction with others. But most researchers, including Chomsky, propose that the capability is a sort of "language acquisition device" present at birth—a genetic ability of the brain distinct from other cognitive functions—and that environmental input is necessary to trigger it so that we can subsequently learn the words and grammar of a specific language. This is why a child growing up in Spain gains fluency in Spanish, not Chinese.

Strong evidence for a genetic language ability comes from the observation that children who are not exposed to any speech, but are able to interact with each other, will invent their own language, which is complex in syntax and meaning. This has been seen in deaf children who were not exposed to sign language. Amazingly, as long as they had someone to interact with, they managed to communicate complex thoughts by inventing their own system of signing.

Rachel Mayberry at McGill University in Montreal has uncovered powerful evidence for a critical period of language acquisition by studying deaf children. The 10 percent of deaf children in Canada whose parents are also deaf and who thus grow up using sign language become as fluent in signing as a hearing child would become fluent in speaking. The children of hearing parents, not exposed to sign language until later, are less competent. Mayberry has also found that people who learn a language in infancy, are subsequently deafened in childhood, and then learn to sign do much better than children who are deaf from birth but only learn to sign at the same later age, having had no previous language acquisition (usually because their parents could not sign). The children who once heard speech do not gain perfect fluency, but are like good speakers of a second language. They even have accents—they hold their hands too far from their bodies, fail to make signals clearly, and so on—and make mistakes a lifelong signer finds funny but understandable. Late learners not exposed to speech or sign language, in contrast, lack fluency and make incomprehensible mistakes. It seems the laws of language acquisition are a universal part of the brain's makeup, applying equally to babies who speak and those who never hear a word.

The argument that there is a genetic underpinning for language acquisition also rests on the principle that since every sentence is a new combination, children could not learn all of the rules necessary for comprehension and production through observation alone. Followers of Chomsky contend that as we learn language without explicit teaching, our brains must be preset to learn the syntax that allows us to extract rules for word combination from the speech we hear.

Further supporting evidence comes from the fact that there are striking similarities among all of the languages of the world. They all have the same basic grammar. There is also remarkable consistency in the timeline by which babies acquire language and learn words and grammar.

Some linguists, however, emphasize humans' great capacity to learn. They theorize that there is a language acquisition process without inborn abilities. Perhaps children can detect patterns in speech by comparing many utterances, and use these to extract the grammatical rules. Or maybe our overall intelligence allows us to store all of our linguistic experiences in long-term memory and group new experiences with old ones.

Researchers Jenny Saffran and colleagues at the University of Rochester reveal a study showing that infants' learning ability may greatly exceed previous expectations. When eight-month-olds are exposed to just two minutes of nonsense syllables strung together (for example, "bidakupado"), they are able to use simple statistics to find the word boundaries in connected speech. They can recognize whether they have heard the phonemes linked together in that particular order before or whether the arrangement or "word" is new. Eight-month-old babies can also detect clear patterns in the sounds of language after hearing only a two-minute sample of connected speech. For example, in the four-syllable phrase "pretty baby," the first sound "pre" is followed by the sound "y," and "ba," in this exercise, is also followed by "y." The "y" sound often indicates the end of a word. The pattern "y-ba" is less common. The babies were able to use this statistical information alone to learn where one word ends and another begins. In real life, babies also use other cues to tell what is a word and what isn't, including pauses and changes in pitch, stress, and rhythm.

The researchers maintain that the babies can pick out words, surprisingly enough, by thinking like little statisticians, speedily detecting clear patterns in the sounds of language. Such vigorous "computational abilities," they say, can help explain how very young humans learn so much about their world so quickly. Two psychologists at the University of California at San Diego explain that the Rochester study is an important new argument for the "learning" side of the ledger in the long scholarly debate over whether language is innate or learned.

Richard Aslin of the University of Rochester emphasized the importance of these results: "For many years, language acquisition has been thought to involve a strong innate component in part because children learn language so rapidly. No one believed that young infants could be such rapid learners. Now we have evidence that months before infants begin to produce words, they can very rapidly learn which sounds are likely to go together to form words."

A study at Johns Hopkins University showed that babies remember words, even complex words, that occur frequently in speech, an important prerequisite for learning language. Audiotaped children's stories containing words like "peccaries" and "python" were played to eight-month-old infants once a day for 10 days. Two weeks later, thirty-six words that occurred frequently in the stories were played back to the babies in list form. The babies recognized the words, even though they sounded different in list form than they had in the stories.

Whether our brains are prewired for language or not, the timeline of language development in children is incredibly consistent across cultures, and this is the strongest evidence we have that there is some sort of language acquisition device, or an innate capacity to learn language, present in every human brain at birth.

The process actually starts before birth when neural connections are made from the speech a fetus hears while in the womb. Although the sound is muffled by the uterus and the mother's heartbeat, and only low frequencies are registered, enough auditory information is taken in that newborns prefer listening to speech in their own language: four-day-old Russian babies will suck harder when they hear Russian than when they hear Portuguese. Babies are also responsive to Mom's voice; almost immediately after birth, they orient toward it as opposed to other voices.

Indeed, humans may have evolved to ensure a child's learning of language, as well as emotional bonding between parent and child. Research shows that 80 percent of mothers cradle their babies on the left regardless of whether they are right- or left-handed. This may be an evolutionary trait that allows sound to enter the infant's left ear, which means it will then be processed by the right hemisphere of the brain, the center for the emotional part of language. The earliest communications may be pure emotion with no literal meaning—the right hemisphere, which develops first, is stronger at interpreting the melodic tones of baby talk that mothers and fathers use when holding newborns. If the baby is held on the left, then the mother's and baby's right hemispheres will be in contact. Thus the left cradling of the baby activates, stimulates, and nurtures his right hemisphere while ensuring emotional feedback to the mother's own right hemisphere. This right-to-right communication begins the emotional bond and ensures the importance of first nonverbal and then later verbal interchanges to the bonding process.

It is interesting to note that music is recognized by brain circuits that are similar to those that recognize language, and that music centers, like language centers, are distributed throughout the brain. However, studies have shown that while the left hemisphere usually contains most of the specialized language areas, most of the specialized music areas are in the right hemisphere. Some scientists postulate this is because much of what determines whether incoming sound is considered to be music relates to the emotional content of the sound.

The exact time frame in which a child advances from crying to rattling off elaborate tales of dragons and princesses will vary, but the progression is always in the same order and usually each step is taken close to a particular age. After only crying and grunting for the first two months of life, babies begin to coo and laugh in the third month, as the larynx descends in the throat, opening the cavity behind the tongue and giving it the forward and backward movements needed for the production of vowel sounds. From the fifth to the seventh month infants play with the sounds of consonants and vowels, and soon are babbling in syllables. Babbling helps them tune up their brains, directing them to produce the sounds that they need for speech and learn how to move the muscles, tongue, and voice apparatus effortlessly to make the sounds appropriately. (Babies that have mechanical problems with their voice apparatus are slowed in speech and language later in life.) By the time they are six months old they group phonemes in all sorts of combinations, and by ten months they group phonemes to form syllables that only correspond to the language of their environment.

At a year to a year and a half, babies begin using words and begin to form short phrases, but these don't make grammatical sense until age two or three. Typically, half the early words are names of objects such as body parts and the rest are for actions and modifiers such as "hot" or "dirty." Into their third year children progressively increase sentence length and complexity of syntax by adding word endings that represent past, present, future, singular, and plural.

During this process babies are understanding syntax even though they cannot speak correctly. By the time they begin to put words together they have already learned, through observation, the peculiarities of their native language, and thenceforth will have trouble learning the peculiarities of other languages. English kids learn the irregular verbs of their language quickly and effortlessly, while German and French kids who learn English later struggle. Conversely, German and French kids almost never have trouble with the constant switching of gender for nouns, while late learners of these languages always find it confusing.

Children learning the sign languages found in every community of deaf people in the world follow a similar progression. The time frame for language development in deaf children is approximately the same and is, again, consistent across cultures. As they are exposed to parents' sign language they will babble with their hands, learn words, and begin to put them together in the same way and on the same schedule as hearing children. The conclusion is that some inborn brain capacity exists for language that goes beyond that for speaking.

Children growing up in various environments even tend to make the same mistakes at the same developmental stages. In all languages, for example, children have trouble forming past-tense verbs or plural nouns at about the same age (in English they add "ed" to everything past, as in hold and holded, and "s" to make a plural, as in tooth and tooths).

IT'S ALL IN YOUR HEAD, BUT WHERE?

SINCE THE 1860s it's been accepted that normally acquired language relies on the region around a long groove in each hemisphere of the cortex called the Sylvian fissure, a deep cleft that separates the temporal from the frontal and parietal lobes of the brain, beginning from behind the ear. Different language functions, the theory goes, are based in different parts of these regions. In trying to pin these down, scientists developed a nice tidy picture in which Broca's area in the frontal lobe of the hemispheres housed language production, Wernicke's area in the left posterior temporal lobe housed language comprehension, and a bundle of connecting nerve fibers integrated the two.

However, now we're finding that this model is just not accurate. Neither nature nor the brain always fits into discrete boxes. For example, recent MRI and PET studies and highly specific clinical tests of language abilities and impairments show that the ability to move the face and tongue in the sequence necessary to produce speech sounds like "da" and "ta" and the ability to hear and decode the same sounds are in Broca's area of the brain. This indicates that speech production and comprehension are not independent systems. A slew of studies provide evidence for "mirror neurons" in the brain—neurons that fire when we throw a ball as well as when we catch it. The same neurons are used to speak and hear the same words.

Unfortunately, we haven't yet been able to craft a new brain model of language. What research does indicate, however, is that language functions are distributed more than previously believed, and that some of the functions are remarkably specialized. Researchers have pinpointed sites in the cortex that control aspects of language as narrow as the naming of living things, gemstones, or fabrics. There are even different centers for regular and irregular verbs. The frustrating part is that these areas are not the same in all of us, so we can't draw a general map of the brain that locates each language function, or conclude that if a specific region is damaged it will necessarily cause a certain deficiency. More and more, we are finding two patients with lesions in the same general areas who have different kinds of impairments and patients with the same impairment who have lesions in different areas.

If specific language functions are located in discrete areas, we want to know how they are integrated. It could be that the functions are grouped into language comprehension, speech, and writing. Or perhaps they are grouped according to parts of speech, meaning, and context. Psychologist Alfonso Caramazza of Harvard University proposes independent brain systems for grammar, meaning, and word form,

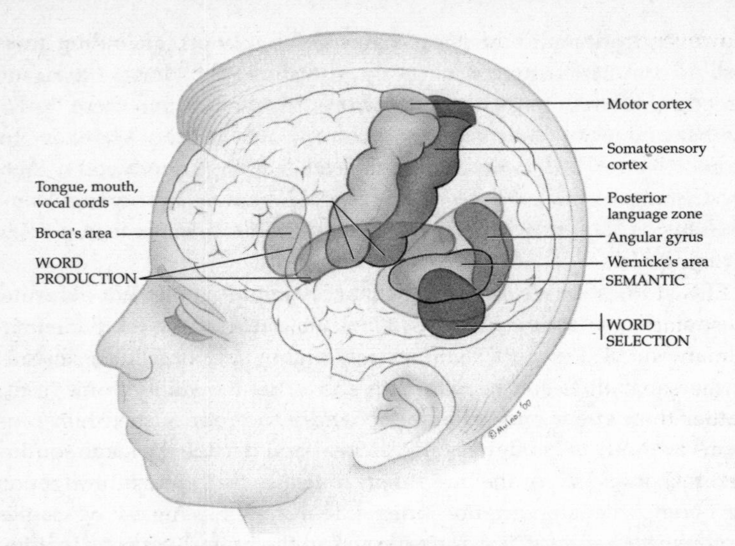

THE LANGUAGE BRAIN Language is the further elaboration of certain sounds used for communication, and mental images of sounds are used for thinking. The sounds are trapped in the back of the brain near the primary auditory cortex. The frontal speech area grabs these sounds and articulates them into spoken or silent speech.

each with its own neural mechanism and its own independent impairments. Within these three complex processing units, language function is divided into more specific tasks.

This basic organization is supported by observations of many patients with specific impairments. Certain brain-damaged individuals shown a picture of a snowman, for example, can accurately describe the concept (cold, man, children made it), but cannot say the actual word "snowman." Another person can understand and use the normal number of words, but can't string them together into meaningful sentences. If damage to the brain is restricted to a particular region, it may impact only one of the three systems (grammar, meaning, or word form).

The way in which language functions work together is also being questioned. Language functioning in the brain has long been viewed as a step-by-step process: for an object to be named, it must first be matched to an internal representation of the object (a memory), and then this representation must be recoded into a word to be expressed.

However, some individuals can name an object even after they have lost all comprehension of its meaning. People with certain types of agnosia (literally, "not knowing") can correctly name the object "doll" but have no idea what it is. This shows that a concept does not have to be decoded before traveling to a center for word formation and speech production. Instead, there must be a direct connection between perception of an object or printed word and oral production that skips any meaning system.

EEG studies reveal the neurons that actually become active during these stages of naming. When you look at a cat and are told to identify it, many small sites in the temporal lobe and in the frontal lobe activate at the same time and remain active until you say the word "cat." Rather than a step-by-step process, where one part of the brain performs a specific language task and then sends it to another area for the next task, there is parallel activation of many small areas throughout the cortex. Even thinking the word "cat" (silent naming) activates the motor speech areas. All these parts activate because, like visual representations in the brain, words are all split up and distributed, with the sound of the word in one area, the meaning in another.

It is important to remember that the brain's language-processing modules are not fixed in place. In fact, neuroscientists have found evidence suggesting that a person's facility or difficulty with language may arise, in part, from the combination of brain regions used to process language. What's more, the brain's language map may change as it becomes "smarter." By understanding such processes, neuroscientists one day may create new methods of learning that overcome language deficits or improve everyone's ability to use and understand language. The fact that the map is generally consistent among people, yet can vary, makes research difficult but the outcomes fascinating.

Nonetheless, it seems bizarre that language would be organized in a piecemeal way. Why would living things, gemstones, and fabrics be stored in distinct brain sites? One reason may be that our testing so far has been limited and we haven't found broader logical connections. It is also important to remember that naming difficulties have many possible underlying causes, such as damage to a system of semantic knowledge or error in word retrieval or speech production. Each of these tasks may be controlled by a separate region of the brain, in which case all the information related to the word "toaster" would not simply be held at one precise point. Each sense modality might store some data about the toaster—its size, its shiny silver color, the "ding" that it makes when the toast pops up. It may be that each word can be positioned anywhere in the left Sylvian region as long as it connects to parts of the brain that store shape, appearance, use, and syntax.

HELP FROM MOVEMENT AND EMOTION

OTHER REGIONS OF THE BRAIN apart from the many small areas clustered around the Sylvian fissure are also involved in language. Certain regions, for example, allow for the constant interaction of movements and emotions in everyday conversation, since the patterns of our muscle movements help us code certain words. We often use hand gestures when explaining an idea; when you stretch out your arm, point to a person, then curl your index finger back toward you, the other person knows you want him to approach you. We also mouth words to ourselves when learning to read or when we come across a particularly difficult passage of text. Dyslexic children are often taught to silently mouth words they are reading because the physical movement helps cement the learning. Gesturing and speech are closely bound; they are acquired together in childhood and break down together in aphasia.

Evidence of a link between speech and movement is clear in babies. Long before they start talking, babies become skilled at using eye contact, facial expressions, and nonverbal gestures to communicate. Children, even at the one-word stage of language development, spontaneously produce gestures along with their speech, just as adults do. Studies of aphasiacs show that verbal information received without visual context is perceived differently than the same information with visual context.

Gesture-speech combinations deliver a coherent message to the listener despite the fact that they consist of two different modalities of expression. This coherence is possible because before the communication unfolds, gesture and speech are part of a single idea. As expression proceeds the message is parsed, with most information channeled into speech but some information channeled into gesture. As with any two systems that work together, mismatches can occur. Observation of combinations in which gesture and speech convey different information may prove a useful clinical tool for distinguishing, at a relatively young age, children who will have problems mastering spoken language without intervention.

The crucial role of motor function in speech is seen most dramatically in deaf children. Researchers at the University of Chicago studied four unrelated deaf children in the United States and four others in Taiwan. None of the children had had training in a standard sign language, but each was able to develop his or her own form of communication with parents using gestures to form complex sentences. The researchers found that the children made use of over 10,000 individual gestures. Remarkably, the gesture systems, rather than resembling those associated with either English or Mandarin Chinese, were similar to each other.

The supplementary motor area, a region in the frontal lobe, is responsible for initiating and planning complex movements and is crucial to language. Clearly the complex sequences of movements of the face, tongue, and larynx require fine-motor selection. In fact, stimulation of some parts of this motor area block the ability to speak. At the University of Minnesota, fifteen different brain regions of six adult epilepsy patients were exposed to transcranial magnetic stimulation. When the magnetic fields impinged on the left frontal speech area that is part of the supplementary motor area, the patients suddenly could not speak.

The motor area may even help a person speak when speech areas are compromised. One trilingual patient—speaking Farsi, English, and French—who was being observed at Johns Hopkins prior to brain surgery for severe epilepsy, was unable to name objects in pictures correctly in any language when a small electric current was applied to the left lateral occipitotemporal area behind his left ear. However, he could repeat words that were said to him. The supplementary motor area is also activated when we need to recall a word without the external cue of an object or a picture. When we recall the months of the year, for example, motor areas become active.

Further evidence of the motor area's crucial role comes from MRI tests at Yale, which demonstrated that when dyslexic readers took phoneme tests, they showed activation in Broca's area, responsible for the motor output of spoken language, while normal readers did not. Sally Shaywitz, co-director of the Yale Center for the Study of Learning and Attention, maintains that the dyslexics are using the speech motor region to compensate for deficits in the speech-understanding region.

Several studies have linked language production with complex motor skills, indicating that the two functions share neural networks.

Doreen Kimura and Catherine Mateer at the University of Western Ontario found that aphasia is often accompanied by difficulties in complex movement sequences (apraxia); for instance, the patient is unable to demonstrate how to use a hammer to pound in a nail when asked to do it verbally, but can do it when presented with the actual hammer, indicating a disconnection between language and learned movements. It may be that the ability to sequence in relation to verbal commands—the true source of language and all of our unique intelligence—is what is actually damaged in these cases. Movement in and of itself isn't damaged in apraxia patients, as they can still perform each movement separately, and they also haven't lost their understanding of words. But the damage seems to impair their ability to sequence individual parts to produce more complex chains of movements. Many patients with apraxia can put a key in a lock, can twist, and can pull, but they can't put these three parts together to actually unlock a door. Also, there is a related condition called apraxia of speech. Nina Dronkers's research at Berkeley suggests that the insula, an area beneath the frontal and temporal lobes, could be the common sequencing site that binds together language and movement. She studied patients who had lesions in a discrete region of the left precentral gyrus of the insula; they could perceive verbal sounds but couldn't string together speech movements to form words.

One major implication of this avenue of study is that if a sequencing area is at the root of human language, we might be able to boost our language abilities by practicing other sequential activities, such as music, dancing, imagining multipart objects, and step-by-step logical reasoning. Recent studies question the efficacy of sensory-integration and perceptual-motor training in improving the learning of language skills. However, programs abound for improving motor skills that focus on groups who have language problems, and they are held in high esteem by the participants and their parents. In one study of first-grade children, an experimental group received Kodaly training, which uses folk songs and emphasizes melodic and rhythmic elements. The group exhibited significantly higher reading scores than a matched non-Kodaly-trained control group. Moreover, after an additional year of Kodaly training, the experimental group was still superior to the control group. The improvement appears to be mediated by the facilitation by music at the phonemic ("sounding out") stage of learning to read.

Although it's more of a stretch, the converse might be true, too: strengthening our language abilities with extensive reading and writing could actually help our overall ability to sequence concepts and perhaps even to master the physical motions of dance. As noted in Chapter 4, athletes who visualize their performance beforehand say the mental exercise helps them perform better physically.

The brain's emotion circuits may help language too. Human language lets us understand and convey not only the literal meanings of words but rich emotion, from a bitter retort to a gleeful exclamation. Emotional language seems to be an independent system outside the Sylvian region. Some researchers have linked emotion in language to a distinct brain area in the cingulate gyrus of the limbic system, right above the corpus callosum. Many victims of strokes in the left Sylvian region lose almost all language abilities, yet retain emotional speech. They may not be able to talk at all, but can shout swear words. One trauma patient in the Netherlands, who was in a persistent vegetative state, showed increased blood flow in the anterior cingulate and right middle temporal and right premotor cortices when doctors played a tape of a story being told by his mother. When the patient was presented with nonword sounds, there was no change in blood flow. The older emotion structures of the cingulate gyrus may be our main connection to primate vocalizations. Our crying, laughing, and shouting use the same brain structure, the cingulate, as a monkey's warning call when it spots a predator.

The emotional part of language, or emotional prosody, provides the melody of speech, variations in emphasis, pitch, and timing, as well as cues to the beginnings and ends of phrases. Research indicates that the ability to comprehend the emotional aspects of language or speak with feeling can be impaired without affecting the comprehension of the literal meaning of words.

LEFT IS RIGHT FOR MOST PEOPLE

ONE STRONG CLUE as to how the specialists work together can be found in the way that the two hemispheres in the brain are coordinated. Even though research clearly shows that the left hemisphere almost always "rules," language functions are found in both. We also know the corpus callosum is critical to linking the functions in the two hemispheres. However, the degree of asymmetry is not fully understood, nor is the role of the right hemisphere. Only humans have brains in which there are two hemispheres that differ significantly in function and structure, and language is our most lateralized function.

Language resides predominantly in the left hemisphere in 90 percent of the population. About 5 percent have their main language areas in the right hemisphere, and another 5 percent split language fairly evenly between hemispheres. This distribution indicates that there must be a genetic inclination for left-hemisphere dominance, but that the right hemisphere also has the neural mechanisms necessary to support normal language.

For nearly 150 years, since the Broca and Wernicke aphasias were first described, language has been attributed to the left hemisphere. Stroke victims have been the major source of evidence. After strokes that damage the right hemisphere, language usually remains intact, but strokes to the left hemisphere often cause language difficulties or loss. This is true for all forms of language, from sign language to oriental pictographs.

Further evidence comes from neurological studies in which one hemisphere is temporarily paralyzed. When sodium amytal is injected into the carotid artery on one side or the other, the hemisphere on that side goes to sleep. This is called the "Wada test." A person with a paralyzed right hemisphere can still talk, while one with a paralyzed left hemisphere usually can't.

If you want to verify the lateralization concept for yourself, take this test: Try to repeat a passage of poetry while you simultaneously tap a finger on a table. It is significantly more difficult to tap a finger on your right hand than your left, because the movement of the right finger is controlled by the left hemisphere and competes for neurons with the language areas there. The brain is not limitless. If you found the reverse, you may be one of the few people whose right hemisphere dominates language. If you can tap fingers on both hands equally well, you might be linguistically ambidextrous!

Despite the asymmetry of language, more connections between the hemispheres owing to the corpus callosum may result in better language functions. Women have more such connections than men and have higher verbal IQs. During phoneme tests, males activate only the left hemisphere, while females activate both. This is why after a left-hemisphere stroke, women are less likely to suffer a severe impairment in language skills and why they can compensate for dyslexia more often than men.

Language disabilities such as stuttering are more common in the minority of people whose language areas are more equally split between hemispheres, perhaps indicating difficulties in coordination. Dyslexics also show less difference in the size of language areas in the left and right hemispheres. The same is true for left-handers—a surprise, as most people think lefties have right-brain dominance; only about 20 percent show right-brain dominance. Concurrently, left-handers also have a higher incidence of language impairment, stuttering, and dyslexia.

Although one hemisphere usually tackles most language functions, the less dominant hemisphere still controls crucial abilities. In most cases, the right hemisphere controls functions that are just one step beyond the essential functions of relating the word symbol to meaning. The right hemisphere, the major hemisphere for most emotions, determines the emotional state of speakers from their tone, and also is responsible for allowing us to understand metaphor and humor. The right hemisphere appreciates the whole picture and thus can see when you switch sets of meaning, which is the basis for most humor. After a right-hemisphere stroke, patients can still communicate quite well, but they often lose prosody and can only interpret language literally. They have difficulty using and understanding figures of speech, tones of voice, humor, and expression of feelings. Recent studies indicate that the less dominant hemisphere may be more active than was thought in normal language functioning. For example, the right hemisphere activates not just in prosody, but also when a sentence is complex in structure, indicating that it may provide extra support in demanding situations. Comprehension of sign language increases activity in parts of both the left and right hemispheres.

These studies also show the extent to which the brain is specialized. Regions within the left half of the brain control facial expressions according to linguistic content: "You look great!" carries one meaning when the speaker smiles and another when the speaker frowns. The right half of the brain, in contrast, activates the very same facial muscles for the smile or frown that is your response to hearing "You look great!" Most remarkably, these different aspects of language processing fit together seamlessly to provide us with our ability to understand a message. The brain can coordinate the action of its many processing circuits to produce an integrated response to stimuli.

The neurobiology of sign language gives us greater insight into how language is organized in the brain. Individuals who rely on sign language are much more likely to be impaired from damage to the left hemisphere. This is a paradox; one would think that sign language, which is visual and spatial, should be processed in the right hemisphere, where visual-spatial analysis occurs, but instead the left hemisphere is dominant for sign language. This was shown in twenty-three brain-lesion patients at the Salk Institute for Biological Studies in La Jolla, California. These results support the view that the left hemisphere is specialized for language and does not just utilize a more general sensory or motor function.

PET studies on deaf signers show that inner speech or self-talk occurs in the left inferior cortex, just as in hearing individuals, which confirms the notion that the left hemisphere houses language, including sign language. It also suggests that inner signing is controlled by the same regions that control inner speech, even though it appears to involve internal representation of hand and arm movements, which are normally under the control of the visuospatial brain areas in the right hemisphere. However, comprehension of external signing increases activity in parts of both the left and right hemispheres, bringing the right hemisphere back into play to help assess visuospatial relations. These findings, plus studies of brain damage in signers, indicate that the left hemisphere's role derives from higher-order properties of language.

Even though the right hemisphere normally plays the supporting role in language function, it can take over if there are problems in the left hemisphere early in life. In a rare congenital malformation called Sturge-Weber syndrome, it is necessary to remove one hemisphere in infancy for the child to survive. Bruno Kohn and Maureen Dennis at New York University tested a group of these children years later, at age ten, to see how language had developed. Boys and girls whose left hemisphere had been removed had adequate language function, but showed more problems in using complex grammar and the future tense than the children without a right hemisphere; their performance IQ was lower. Otherwise, their IQs were pretty much the same. This offers evidence that in most people there is a genetic basis that prewires the language function in the left hemisphere, but that the right hemisphere makes a good pinch hitter if the brain is altered at an early age.

PLASTICITY

THE LIMITED LANGUAGE IMPACT of removing the left hemisphere in young children makes it unlikely that special neural tissue is necessary for language functioning. So does the retraining of aphasiacs; in some, the right hemisphere is activated during relearning and the language function stays there, while in others language shifts back to the left hemisphere after relearning takes place in the right. Clearly, the brain's plasticity allows neural connections to rewire after damage and provide for some recovery of function. It would therefore seem likely that plasticity also underlies the great variability among individuals in the location of language functions.

The brain's language areas need to be able to adapt to environmental input so that a child can master the phonemes and words in his or her own native tongue. The fetal brain's language areas can accept all phonemes, but by six months in the womb the fetus is already beginning to group together sounds in terms of phonemes it hears in its mother's speech. As noted earlier, a four-day-old newborn will suck harder when it hears its native language over others. As early as six months after birth, the brain loses its sensitivity to phonemes that are not part of the language the infant hears every day, as the brain, preprogrammed to accept all types of phonemes, specializes owing to environmental input.

As time goes on, adaptability decreases and the connections become permanent, which is why a massive left-hemisphere injury will affect people so differently depending on their ages. An adult who must have the left hemisphere removed will lose many language abilities. The brain can no longer reorganize these language functions because the critical periods of sensitivity to environmental input are over.

If brain damage occurs before the age of two, the brain can reorganize extensively, creating language areas in different regions. By four to six years of age, there is less language plasticity and a stroke or injury will cause a severe learning deficit, though most of the language already learned will remain. After the age of six or seven, the window of opportunity for forming language connections is largely gone and language loss can be permanent.

The fact that we lose flexibility in forming new language connections by age seven or so has enormous implications for education. Research shows that second languages learned after this age are stored within neural systems that are distinct from those for the native language. These systems tend to be less lateralized and display a high degree of variability among individuals. By contrast, people who grow up bilingual from birth store their native and second languages in the same area. This information may help educators develop new language-learning strategies, and it is useful to neurosurgeons who, during surgery to remove a brain tumor, must navigate their scalpels carefully around the brain's language-sensitive areas to avoid impairing a patient's ability to speak.

Language plasticity may be lost over time in part because ultimately permanent connections are needed to encode a specific language, which by definition limits flexibility. Before six months of age, infants of all nationalities can distinguish among the sounds used in all the languages of the world. But as phoneme categories are formed their brains select for the ones that they hear and those that are not heard are lost forever. Infants can hear subtle differences between sounds that adults perceive as identical. As young children mature they begin to hear each phoneme as belonging to a distinct category, though there is a range of incremental sounds between "pa" and "ba." Humans learn to deal with variations in how speakers pronounce sounds by creating broad categories, so "pa" comes to represent many acoustic wavelengths. The Japanese, for example, use a phoneme between the English "r" and "l," so Japanese children learn to ignore slight variations around this wavelength. As they get older they hear no difference between the English "r" and "l"; both sounds fall within one mental category.

As environmental input will have less and less effect on reshaping language connections after the age of six or so, a child must experience language before this time to ever acquire proper functioning. Abused children who have grown up in complete silence—like Genie, the girl we met in Chapter 1, who was locked in her room in Los Angeles until the age of thirteen—can never adjust. After extensive rehabilitation Genie learned many words, but she never achieved syntax. It seems that by age three this ability weakens and by age six the critical period for learning proper sounding and syntax—the ability to manipulate, connect, and create—is over.

Unfortunately, the fact that this window of opportunity comes so early creates a fairly common problem in the deaf population. If a child is not identified as deaf and exposed to sign language at an early age, he or she will have extreme difficulty achieving full sign-language mastery.

A classic example of the need for appropriate stimuli at an early age for language to develop fully was found in Nicaragua in 1985 when the new government there asked Judy Shepard-Kegl, a linguist at Rutgers University, for help in developing educational programs for deaf children. When she arrived in Nicaragua, she found that the country had no sign language of its own. Prior to the educational reforms of the Sandinista government, deaf children had remained isolated; indeed, there were few schools even for hearing children. Shepard-Kegl arrived at a vocational school in Managua that had a large deaf population. There was a group of teenage girls who had not been exposed to any sort of signing in their younger years and who attended classes taught by teachers who had no knowledge of sign language. The gestures the girls had developed were based on the physical characteristics of objects and people, rather than on any phonology, and formed a rather crude, or pidgin, kind of language. However, there was also a younger group of deaf children who had grown up together and had developed a much richer, more complex, and coherent sign language of their own.

READING AND WRITING

ORAL LANGUAGE HAS BEEN AROUND for tens of thousands of years. But the ability to represent sounds in written symbols—the writing and reading that enable us to preserve and pass on information among groups and across generations—has only been around for 5,000 years, and it is only in the last century that a sizable percentage of various populations has been able to read and write. It seems most likely that the neural mechanisms involved in reading and writing didn't evolve specifically for these purposes, but are being applied in a secondary manner as they fulfill their primary missions in the cortex. Evolution of the brain occurs in the most economical way possible; the brain co-opts one structure that might have evolved for an entirely different purpose and applies it to another function altogether.

All writing systems are made up of symbols that designate phonemes, syllables, and words. The invention of the alphabet allowed us to link written language to sounds, but reading and writing are not natural abilities prewired in the brain, as oral language seems to be. Extensive teaching and practice through years of schooling are necessary for mastery.

Most evidence indicates that reading and writing are separate systems that are controlled by different brain regions. Certain individuals who can identify letters, write, and have normal speech comprehension and production just cannot read no matter how they are taught. Reading deficits can be very specific, too; some stroke victims can read normally except for specific kinds of words, such as adjectives, nouns, verbs, or abstract words, while others lose only the ability to write. This is an indication that many small discrete areas are involved, as in oral language.

Electrical stimulation studies on the conscious brain have helped define the reading areas. They overlap with naming sites, but aren't always in the same area, and exact locations vary from one individual to another, as is the case for naming sites for oral language. Other tests show that people with high verbal IQs have reading sites in the superior temporal gyrus and naming sites in the middle temporal gyrus. Ironically, the reverse pattern is found in people with weaker verbal skills. William Calvin of the University of Washington proposes one explanation: We learn to name before we learn to read. If we are genetically less efficient at this task, the greater neural area in the superior temporal gyrus is necessary for success. So when we begin to learn the complex task of reading when we start school, the superior temporal gyrus is already dedicated to naming sites, so reading sites form in the less optimal middle temporal gyrus.

A basic model of the reading process can give us a better idea of how this function is organized in the brain. Words that we see on a page are processed as a visual representation of letters that are grouped into words. This process is distinct from the visual perception of everything else. We process words as visual units rather than as a series of single letters, and recognize whole words as fast as single letters. Reading is further sped up by regularity of the words and by our previous knowledge. Speed reading mainly makes use of context, the ability to guess at a word one sees rapidly in passing based on the words that one saw prior to that.

There is also ample evidence that we visually process words along parallel routes of sight and sound, each with its own separate neural system: there are people who read primarily by sight and those who read primarily by sound. The Japanese language has two separate scripts; one is syllabic and based on sound, and the other is ideographic and based on visual characters that represent words, groups of words, and concepts. Remarkably, people can lose the ability to use one script and not the other; the two systems seem to be based in different brain regions.

These two independent routes explain why some children learn to read better with phonics—sounding out words—while others learn better with whole-language techniques, where the whole visual word form is learned in context. Phonics teaches the mechanics of written language, stressing the letter sounds, combination sounds, and rules of combination. Children group the pieces together into words, sentences, and ideas, and are taught to sound out new words and to sound out spelling while writing rather than memorizing the whole word. Whole language, the most widely used system in U.S. schools today, uses reading immersion, which follows the idea that exposure to written language in the context of interesting stories will enable children to learn to read naturally. Most of us use both pathways simultaneously and learn to read by combining the two systems. Whether schools should teach reading by phonics versus whole language has become a hot, almost political debate, but brain research provides a simple answer: they should use both.

The whole-language trend assumes that reading is a natural, genetically programmed part of language development, and that children will pick it up as easily as speaking. However, as noted, since writing has only existed for 5,000 years and literacy has only been widespread for a few centuries, it is highly unlikely that the human brain has evolved structures specifically for reading and writing in this time. It is our ability to learn through experience that allows us to achieve reading, but only with explicit instruction.

Teaching by whole language adds efficiency to the reading process, but used alone is definitely deficient, because phonics is so fundamental to linking sounds and symbols. Furthermore, children who may be prone to reading difficulties benefit primarily from more extensive teaching of phonics. If children recognize distinct parts of spoken words they are more likely to be able to associate them with the letters that represent the sounds, and have greater success at learning to read.

DYSLEXIA

CHILDREN WHO HAVE TROUBLE discerning sounds belong to a much larger category of people known as dyslexics. The term "dyslexic" has been overused and oversimplified, and has incorrectly come to mean people who simply transpose adjacent letters or numbers. Clinically, dyslexia encompasses a far wider range of language disorders. Some dyslexics have difficulty processing sound, others have trouble processing the visual word, and still others find it difficult to extract meaning from printed words. Some who are impaired on the sound pathway can still recognize whole words, but can't decode or sound out unfamiliar words or pseudo-words. Others who are impaired on the sight pathway can decode words using phonics but can't recognize whole words as quickly. As much as 20 percent of the American population can be diagnosed as dyslexic, if all the variations in reading difficulties are counted.

As a result, a huge number of theories for a cause, and even more theories for educational "cures," have been suggested. One of the main questions that researchers are still exploring is whether the abnormal functioning results from a faulty reading system that is trying to do the best it can or whether a different part of the brain has substituted for the damaged system and is operating in a fundamentally different way.

Clinically diagnosed dyslexics are born with several structural differences in their brains that make reading, sounding out words, or spelling extremely difficult despite their normal or above-average intelligence. The common cortex has six basic layers, with layer 1 having essentially no cells, but dyslexics typically have bunches of cells in layer 1.

The trouble seems to begin in the middle of pregnancy, when small spots of miswired nerve cells, called ectopias, develop in the language areas in the Sylvian region. Dyslexics have also been found to lack a group of nuclei in the thalamus that relate to sound and sight perception. Most researchers now believe that the propensity to develop these abnormalities is genetically transmitted, although the exact genes involved have not been identified. There is an intermediate step between knowing that a disorder is genetic (based on familial aggregation) and knowing what the gene or genes are. This is linkage to a particular chromosome. Dyslexia has been linked both to the 6th and the 15th chromosomes.

The miswired organization causes difficulties in processing input from fast-moving objects that the eye sees and rapid changes in sound that the ear hears. A study in Italy showed that ten-year-old boys with reading difficulties were slower at tests of fast visual processing than peers without reading difficulties. The study concluded that these boys may have global difficulty with all sorts of fast processing. Therefore, it is possible that this condition affects hand-eye and hand-ear coordination as well.

Many language impairments may stem from an early childhood problem in hearing quick changes in phonemes; normal children can discern the fastest combinations of adjacent letter sounds, such as the "p" and "a" of "pa," which are separated by only 10 milliseconds, but dyslexics with auditory processing problems can't. This leads to reading problems because phonetic writing such as English (in contrast to pictographic writing such as kanji) is learned by matching sounds to letters. For these children the higher-level processing of reading seems to remain intact, but it is learned slowly because it is harder for the child to sound out new words, as well as to spell words while trying to write. Studies have shown that part of the cause may be a relative impairment of magnocellular visual cells in the geniculate body, a way station in the thalamus as information passes through there on its way to the cortex.

There is new hope for dyslexics who can't process fast sounds. Paula Tallal at Rutgers University and Michael Merzenich at the University of California at San Francisco have designed computer-based exercises that train children's brains to register the quick changes in phonemes in normal speech. The exercises were first tried on dyslexic children under the age of eight, who needed up to 80 milliseconds between sounds to distinguish them. Tallal and Merzenich designed a series of games on a CD-ROM that the children played on a personal computer for more than three hours a day, five days a week, for four weeks. The initial games slowed down speech and stretched out sounds, making the phonemes easier to understand. As the children improved through extensive repetition, the program gradually sped up the phonemes. After four weeks the children were able to process speech at the normal rate. The brain's amazing ability to reconnect neural circuits based on environmental input allowed the children to gain an average of two years' worth of language ability during one month of exhaustive training. Furthermore, the results, which seemed to carry over into home and school settings, were so encouraging that in 1997 Tallal and Merzenich undertook a nationwide field trial. The technique proved very effective for dyslexic children who have fast-processing problems; note that fast processing is just one subset in a range of causes of dyslexia. Their results have led to a commercial program called FastForward, a new computer tool for speech therapists.

Dyslexia occurs much more frequently in boys than in girls, perhaps because girls' more efficient corpus callosum makes them better able to compensate for dyslexia, so that the disorder often goes undetected. Recent research by Glen Rosen at Harvard confirms this theory. When ectopias, the small points of abnormally located neurons, are mimicked in the brains of rats, males develop a deficit in the detection of rapidly changing sounds but females do not. One explanation is that in males the ectopias are associated with more abnormal neural connections than in females, and that this appears to be a consequence of male hormones.

The visual system may be to blame in other cases of dyslexia. Some dyslexics complain that they have trouble reading because the words "move on the page." When they stare at a lone word they can focus on it, but when their eyes scan words from left to right the words appear to shimmer or fluctuate, making them hard to focus on.

When neurologist Margaret Livingstone heard these reports, she wondered if the cause could also be a relative absence of magnocellular cells in the lateral geniculate body, the partner to the hearing system's medial geniculate—in this case cells responsible for vision, not hearing. She had a tough time in the early 1990s getting anyone to pay attention, since the field had become convinced that the source of the problems was all auditory. She finally persuaded Al Galaburda of Harvard to examine the lateral geniculate of postmortem specimens that he had used in finding the medial geniculate abnormality. Confirming Livingstone's brilliant intuition, they found a relative lack of magnocellular neurons in the lateral geniculate as well.

Implicating magnocellular cells for fast-processing problems in hearing or vision leads to all kinds of speculation about humans' evolutionary advantage. The other cells in the geniculate bodies, parvocellular cells, are slow processors, and evolved later in the brain than the fast-processing magnocellular cells, which are the main type in the lower primates and are good at picking up quick movements and thus necessary for survival. The parvocellular cells help us discern subtleties in tone, texture, shape, and color. These are the very characteristics that artists, musicians, and architects are best at perceiving and using, and studies show that there is a high rate of dyslexia among this population. Perhaps, in some people at least, the later-evolving parvocellular cells have gained ground in the geniculate body at the expense of the magnocellular cells. We don't need to see the quick movements nearly as much.

LANGUAGE AND COGNITION

THE CLOSE TIE that language has with cognition makes it fascinating beyond its role as a means of communication. Language seems to be interwoven with our innermost thoughts, feelings, and sense of identity. Some theorists go so far as to say that language is the same as cognition. Linguist Derek Bickerton argues that our words are our thoughts; there is no abstract thought without the ability to combine and create, which is what language allows. Others envision a neurological language of thought that is distinct from the communicative language system and must be translated into spoken and written communication.

Is there a separate language of thought, what Steven Pinker of MIT calls "mentalese"? Or are our thoughts in English, Thai, or Swahili? If these two functions were truly one there would be no abstract thought in animals, infants, and individuals who never acquire language or who lose it through brain damage. This overstates the case; words can be separated from thoughts. We seem to have the capability of performing complex acts to learn, remember, analyze, and retain a self-identity without language.

A significant amount of knowledge must be nonverbal, stored outside the neural regions involved in language. Several types of impairments provide proof. Perhaps the first was Broca's aphasia, documented 150 years ago. It describes stroke victims who cannot produce speech even though their speech apparatus, speech comprehension, and intelligence remain intact. Specific language impairment (SLI), where individuals show severe language disabilities with no other intelligence or developmental problems, also shows the separate identities of language and thought. Children with SLI lag behind the normal language acquisition schedule and have difficulties with grammar that often last into adulthood; recent research suggests they may have particular problems with verbs, and that verbs may play a central role in the language acquisition process. The disorder runs in families and may be controlled by a single dominant gene that causes a greater-than-normal symmetry between the hemispheres. SLI youngsters can compensate somewhat by using their intact reasoning abilities to figure out the language rules that other children gain effortlessly.

Recent MRI tests on SLI children and others with normal language skills show that while the planum temporale (part of Wernicke's area) and pars triangularis (part of Broca's area) are more often larger in the left hemisphere than in the right in individuals with normal language skills, the regions are more often similar in size, or reversed in size, in children with SLI.

There are other cognitive disorders in which higher intellectual thought is wiped out, yet language persists. Two severe forms of mental retardation, hydrocephalus and William's syndrome, result in extremely altered brains and low IQs, which make daily functioning difficult. Remarkably, these individuals can converse fluently. They have intact or even above-average grammar and language comprehension and production, but are often referred to as "cocktail party conversationalists" because of the lack of deep meaning behind their smooth speech.

In a different case, a study showed an unusual disorder in a man named Brother John, an epileptic who lost all language abilities during his seizures yet could perform many higher intellectual functions while the seizures were in progress. His case adds to the evidence that language is not controlled by exactly the same neural connections as cognition. Without any language, Brother John was still able to assess events, make plans, evaluate responses, assign meanings, learn and use complex skills, and retain his sense of self. It's possible that Brother John's cognition was developed from the language abilities he utilized most of the time when he was free of seizures. If that is true, then perhaps language acquisition is necessary to develop the ability to combine and create and to move from the concrete to the abstract, but once there is cognition it can persist in the absence of language function.

More insight into the link between language and other cognitive abilities comes from a 1997 study that showed that spoken language has an astonishing impact on an infant's brain development. Some researchers even say the number of words an infant hears each day is the single most important predictor of later intelligence, school success, and social competence. There is one catch—the words have to come from an attentive, engaged human being. Radio and television talk do not work. Further, it's not necessary to teach different languages or numerical concepts; all that is required is ongoing interaction and adult speech.

This constant patter may be the most important underlying factor in early brain development for good thinking and reasoning skills, according to Betty Hart at the University of Kansas. Hart and colleague Todd Ridley at the University of Alaska, co-authors of the book *Meaningful Differences in the Everyday Experience of Young American Children,* studied forty-two children born to professional, working-class, and welfare parents. During the first two and half years of the children's lives, the scientists spent an hour a month recording every spoken word and every parent-child interaction in every home. At age three, the children were given standard tests.

The children of professional parents scored highest, and spoken language was the key variable. A child with professional parents heard, on average, 2,100 words an hour. Children of working-class parents heard 1,200 words an hour, and those with parents on welfare heard only 600 words an hour. Furthermore, children with professional parents got positive feedback thirty times an hour, twice as often as from working-class parents and five times as often as from welfare parents. Affirmative feedback is very important. A child who hears, "What did we do yesterday? What did we see?" will listen more to a parent than a child who hears "Stop that!" or "Come here!"

By the time the children were age two, all the parents had started talking to them more, but by then the differences among the children were great, forcing those left behind to catch up, if they could. The differences in academic achievement among the children remained in evidence through primary school. Hart postulated that the key to brain development is the rate of early learning—not so much what is prewired but how much of the brain gets interconnected in those first months and years.

To be balanced, though, we must note that there is skepticism about the results of this and other studies that find such direct relationships between specific actions of parents and infants. Professional parents might have passed on a set of more "talkative" genes to their children. Also, they may have risen in the society because they are more talkative, and may pass on not only talkative genes but a talkative environment. Infants in low-income homes may have poorer nutrition, may have to compete with more siblings, and may be compromised by other factors. John Breuer, president of the McDonnell Foundation, wrote in his article "The Bridge Too Far" that too much is made of neuroscience and education, that we have no conclusive evidence yet of any of this stuff that takes into account all the possible influences on children. Michael Gazzaniga, a revered researcher, goes further in his recent book, *The Mind's Past,* and calls such conclusions that you can read to babies and make them learn language quicker "politically correct pseudoscience babble."

Although language and thought may not be one and the same, clearly they are tightly intertwined. Language seems to shape the way we understand reality and ourselves in the world. The syntax that allows for infinite combinations, meanings, and categories gives us a way to organize our thoughts and effectively communicate them to others. The ability to represent objects in the external world through symbols gives us a separate world in our brains where we can store, manipulate, and generate objects and ideas in new ways. Symbols allow for the internalization and reflection necessary for stepping away from the immediate context and referring backward and forward in time, allowing us to plan and actively shape our futures.

Language, as noted earlier, also serves as a delay function. It lets us rearrange our ideas and take the time to vocalize them instead of acting directly on raw emotion. People who don't take the time to express their thoughts in words often act on impulse: an angry husband who stomps off from an argument with his wife, bumping her chair as he goes, instead of putting his feelings into words, and Jack, the boy we met earlier, who talked so steadily that he had no time for the off-line thinking that allows for reflection and planning for the future before acting on impulse again.

The process of writing may provide an even greater delaying function, allowing for even more organization of ideas before action. So many of us claim that we can organize our thoughts better and learn new tasks more easily if we sit down and write them out on a piece of paper. Perhaps journal writing would be a useful rehabilitation tool to train our brains to slow down, think, and vocalize before acting on emotion or acting out in a social situation—the subject of our next chapter.

8

THE SOCIAL BRAIN

RECALL THE REMARKABLE mid-nineteenth-century case of Phineas Gage, the railroad man who survived an accident in which an iron rod was driven through his skull. Readers may remember that although he survived and returned to work, over time he suffered a complete personality change, and from being an even-tempered hard worker became an irresponsible, profane rogue who was often compared to a dumb beast with no social or moral sense. Using computer technology to reconstruct a three-dimensional image of Gage's brain from photographs to assess probable brain damage, Hanna Damasio, Al Galaburda, and Thomas Grabowski found some years ago that the regions most likely to have been affected by the penetration were the ventromedial portions of the frontal lobes, known to be crucial to decision-making. This area is also the funnel through which emotional information from the limbic system enters the frontal cortex. It is here that we become aware of what we are feeling, that is, gain the understanding that is crucial to allowing us to feel for ourselves and empathize with others. It seems that Gage's injury ruined his ability to make choices that would have benefited his social survival because his center for matching emotions with reason was destroyed.

While Hanna Damasio was struggling with whether this conclusion made sense, her husband, Antonio, began to treat a patient named

Elliot, a man who had at one time been a capable businessman and a good husband and father. Unfortunately, he had developed a meningioma—a noncancerous tumor—underneath the frontal lobes, right above the nasal cavities. It had grown fast and by the time it was diagnosed and removed it had reached the size of a small orange, compressing both frontal lobes upward from below. During the operation, some surrounding tissue from the ventromedial frontal cortex also had to be removed because it had been damaged by the swelling tumor.

After the operation, Elliot began to recover normal functioning. He moved and spoke as before, but began to make bad personal decisions, both short-term and long. He eventually lost his job because he could not manage his time or prioritize his work in any way. He would often get hooked on an insignificant detail, such as obsessing for an entire day about how to sort a stack of documents—by date, size, or name. He still had the same business skills and knowledge, but didn't seem to know when to use them or care whether or not he was failing. Unemployed, he tried numerous improbable and unsuccessful business ventures, forming partnerships with obvious con men. In time he lost all his money. He divorced his wife and married a prostitute, whom he eventually divorced as well.

When Elliot came to Damasio for treatment, he seemed very reserved and unemotional. Lab tests showed that the damage to his brain was greater on the right side. He performed well on measures of intelligence, showing a good grasp of current events, including politics and economics. Damasio tested Elliot's perception, short-term memory, long-term memory, learning, language, and math skills. All were normal. Elliot also did well on the Wisconsin Card Sorting Task, which measures the brain's ability to reason quickly, and on the Minnesota Multiphasic Personality Inventory, which provides a basic view of an individual's personality and general disturbances of personality. Even when he was tested on different aspects of social decision-making, Elliot chose many correct options to hypothetical problems, but acknowledged that, based on his recent experiences, he was incapable of making such decisions in real life.

Damasio decided to focus on Elliot's emotions. Elliot didn't seem to show any, recounting even his own sad story in a detached way. Damasio wondered if this deficit could be affecting Elliot's decision-making ability. To test this idea, Damasio showed him disturbing pictures of violent and sexual acts and played him similarly disturbing audio material, while he was hooked up to a machine that measures the skin's electrical conductance. Normal subjects who see and hear this material have very strong reactions, giving meaning to the expression "It made my skin crawl." Remarkably, Elliot had no response at all.

Damasio reasoned that, like Phineas Gage, Elliot had no emotional guide to direct his activities, no gut response that he was aware of to tell himself to get on with a task or drop it—he had no emotional compass to follow. He also could not plan ahead, because he had no sense of what he wanted or of what was important. Damasio reasoned that Elliot's lack of gut reaction was due to brain damage and was causing him to have poor judgment in the social realm. Being reprimanded and then fired from work, being cheated by con men, and associating with prostitutes didn't bother him, so he didn't choose to avoid these situations. Damasio became convinced that Elliot suffered from the same type of brain damage as Phineas Gage.

AIMLESS MEN

THE CASES OF Elliot and Phineas Gage indicate that lack of emotion leads to poor reasoning and ultimately to poor social judgment, even when factual intelligence is still intact. In fact, the Damasios have identified twelve other patients with similar prefrontal damage, and all show a similar lack of emotion and have a history of terrible social decision-making. This hurts their ability to function in daily life—with a few ironic exceptions. The reader may recall the patient who arrived at Damasio's office one morning after driving over treacherous, ice-covered roads. He described how one driver after another ended up in a wreck in front of him because they overresponded in trying to control their skidding vehicles, putting their cars into tailspins. He noted flatly how he, by contrast, did not brake when he hit a patch of ice but gently steered out of the tailspin. This was the logical response. Damasio saw that the man had no fear about crashing and so didn't overreact as so many others had.

Most of us have some trouble knowing what we want, but these people have the problem to a much greater degree. For example, they often fall madly in love and are guided in every way by the other person; for the first time in their lives, life is worth living, and they want only to live for the other person.

One patient of mine, John, learned to capitalize on his lack of social and emotional skills, but was nonetheless very unhappy. He is the patient I introduced in Chapter 6, who belonged to an underground blackjack club whose members were able to win in casinos by applying a system of card counting and never letting their emotions interfere with the cold calculations needed to gain a slight advantage over the house. John had had difficulty in social settings as a boy, and was aware that his mother was overconcerned and too much involved with him. When he made her cry, he felt awful that he did not care that she was crying. He felt he should care, but he did not. Why this was so was a mystery. He was brilliant in math and science and breezed through classes at MIT, but got stopped when it came time to choose a major: math or computer science. He could not make the decision, and could not finish his senior thesis on the math model of predicting the shuffle in blackjack.

John was hired by a New York City brokerage house to trade stocks on the stock exchange. The pressure was high and he did very well, only he got bored with it. He returned to Boston and became head of the aforementioned blackjack club, and went on to clear $100,000 a year playing blackjack. Beating the casinos, which he felt were taking advantage of the common person, was the only thing that stimulated any kind of emotional interest for him.

People like John fail at sustaining social relationships because their emotional and social brains are deficient. Indeed, these two brain functions often work closely together, and in John's case there is a problem where they intersect. Without emotions as a guide, the person cannot form social relationships.

Today, with therapy and medication, John can pay attention to the rising of feelings within him and grab on to them. Following our discussions he felt relieved. He could begin to look at the way he felt through new glasses. He recently reported to me that he felt sad about a fight his sister and father had had. This surprised him. When his mother sent him a box of his favorite cookies, he actually felt a warm glow and thought how nice it was for her to take time to do this for him. Then he felt like a monster because he had not cared enough about her in the past to feel this before. He had not been able to attend to her needs before; there wasn't enough stimulation in it for him.

John and other people who have trouble forming social relationships and are little moved by emotions belong to a wider group I call the aimless men. They have milder cases—shadow syndromes—of what befell Elliot and Phineas Gage. I currently have two patients with similar brain complications and similar stories. Both of them are loners and both have a terrible time making decisions or plans because they are never able to determine what they actually want to do. Both are bright on factual matters, yet both did poorly in school because they never knew what was important and could not see why they should try hard to do well. If it came easily to them, then so be it; if not, they did not get upset despite the fact that their parents were very much invested in their doing well.

One of them, Jerry, has an excellent memory and recalls going to kindergarten and then first grade, which he repeated once. At this early age he was already unsociable. He did not see the point in all the things that other kids did, and wondered why they were so excitable. He was not. He saw himself on the playground alone, wondering what these kids were doing, running about and yelling. He had no interest. He also said, "I could see other people saying things and wanting things, but I did not find anything coming out of myself." He did not talk to other children in the classroom setting or even at recess. In the classroom, he did not talk to anyone because he was afraid of reprisals from the teacher for talking out of turn, and even though he did not care much about the threat he felt that this mild possibility outweighed any pleasure of talking. He was soon ostracized and grew up alone and friendless. Yet he never wanted to do anything. He could not get into schoolwork since he really did not see the point in trying. His one passion, which was short-lived, was skiing. He was not very good at it, but he at least felt alive bombing down the slopes. This lasted for a season; then he abandoned it.

Another man, Frank, came to see me because at the age of thirty-seven he had no life—no friends, no possessions, no career. He felt bad about not having a life but could not seem to do anything to change it. He had a difficult time feeling anything at all. He had a graduate degree in business administration and was working as a data entry person, but he had no desire to work in either field, only doing what he did because he fell into it. He was intellectually capable, as his degrees proved, but they came about mainly by default. His sister had suggested he get his MBA, so he did. He needed everything spelled out in literal detail. He maintained minimal contact with his family and was always envious of how his siblings and parents seemed to enjoy the holiday gatherings. To him they were merely an obligation.

Frank did not know how other people felt about things and was disturbed that he did not feel strongly about anything. His siblings and parents were very successful, and the contrast to his own life was so glaring that he eventually became depressed. He thought he should be better than he was. He had always had a hard time knowing what to do about friendships, but he did not pick up on cues from others and this bothered him. He knew it was expected that he should have friends, but he didn't bother to try to make any. He was just as happy watching TV. He had always preferred being alone to being with a possible friend, even though he knew he was expected to try to make friends.

Neither of these aimless men were addicted to substances; there seemed to be no point to it. They both responded somewhat to stimulants since they both had flagging attention systems, a result of their wavering interest and low motivation, which kept them searching for the next stimulus. They were shadow Elliots; they could not make decisions or plan anything for the future, because they had no intensity of feeling one way or the other. At this point, treatment possibilities for the Elliots and those with shadow syndromes of their condition are sketchy, but would include psychoeducational material on building social skills and social groups, motor coordination training such as tai kwon do and tai chi, and a variety of medicines for panic and attention.

A NEW VIEW

IN MY YEARS OF EXPERIENCE seeing patients of all sorts, I've become fascinated with the role that the physical brain plays in our being social animals. The newest neurological findings even suggest that the brain itself is a social organ; in the womb, neurons in the developing brain become functional only if they connect with other neurons. The brain's most primitive regions—the cerebellum and amygdala—are the very ones involved in the brain's social processing. Indeed, the amygdala has neurons within it that only fire in response to other people's reactions. Furthermore, evolution shows that the brain has changed itself to survive, adapt, and improve the success of its host person in a group of people.

Nonetheless, traditional psychologists and neurologists have been slow to acknowledge that social behavior is, at least in part, a brain function just like memory or language. The more I see the pieces put together, the more I am convinced that there is indeed a social brain. The pieces have long been identified, but we do not think of them as constituting a holistic function. Neurologists and neuroscientists have shown that damage to the cortex can affect one's ability to be empathetic, that problems in the cerebellum can cause autism and its social ineptness, and that deficits in the right hemisphere can make it difficult to understand life's overall picture. Together, these parts and others make up the social brain.

This simple declaration may seem heretical to some. But over and over again as we consider the different brain regions, we'll see links between their health and people's social abilities, from the simple motor skills that allow people to stand at a proper distance from an acquaintance to the highest functions that sustain moral decisions. There must be direct links between certain brain functions and socialness. How is it, for example, that so many geeks—highly intelligent people who've created our exciting and intricate cyberworld—are unable to function in personal, one-on-one relationships?

If we can understand how the social brain works, we can begin to find ways to treat people whose behavior crosses the limits tolerated by our social society. More important, we might find a way to give otherwise isolated and anguished people the ability to make friends, get along with co-workers, and form intimate relationships. Even though we typically think of these emotional, psychological, or moral capacities as learned, the existence of a social brain indicates that our social skills also have a partly biological basis.

HOPE FOR THE SOCIALLY AWKWARD

BEFORE GOING ANY FURTHER, it's important to point out just how fundamental effective social behavior is to living a successful life. Why are we involved with others? Why do we care about others? How do we recognize friendship and intimacy? These types of philosophical questions touch upon the very core of our lives. The answers are not simply a matter of whether we had happy, stress-free childhoods or proper sex education. The answers can be better derived from an understanding of how the social brain functions.

Humans evolved as social animals, which has given us an enormous evolutionary advantage, allowing one organism to benefit from the accumulated knowledge of another. For example, after a toddler becomes aware of how her mother takes care of her, she immediately turns around and does the same for her baby doll. As is true for every other brain function, some people have particular talents for perceiving, understanding, and engaging in social behavior. For other people, social deficits can be the foundation of psychiatric disorders. There is significant evidence that training in social skills can help individuals overcome antisocial personality disorder, autism, and even everyday shyness.

As John Donne said, "No man is an island." Yet many people become isolated because they aren't able to behave appropriately in social situations. Social cognition is the ability to cherish a friend, understand and empathize with another person, and communicate one's own needs in an effective way. Our highest human virtue is our connection with other humans, and social activity is basic to our health and happiness. Our brains are preprogrammed to look for other humans from the moment of birth, and continuing social interaction with parents and peers is essential for normal development throughout life.

All the different parts of the brain that we've discussed are involved when we interact with other people. We need attention, perception, and memory to recognize another person and recall what we know about them and our past dealings with them. We need emotion to interpret the feelings and intentions of others. We need motor skills and language to respond in socially appropriate ways.

One of the most intriguing aspects of the social brain is that lower and higher functions are equally important to successful behavior. Kids brutally chide their clumsy peers, saying "You can't even walk and chew gum at the same time." Well, social relationships require a lot of simultaneous walking and chewing of gum, that is, they require us to be able to pay attention to many stimuli and respond with many actions all at the same time. A simple conversation with one other person at the water cooler requires us to be able to maintain the right distance, a neutral posture, appropriate body language, good eye contact, and a proper balance of speaking and listening—all physical, "lower" skills that have nothing to do with intellect. The innocent child who can't walk and chew gum at the same time may indeed have difficulty handling these multiple motor demands, and so be unable to conduct himself well in social situations. I had two patients from a leading engineering school who, independently, reported that they had had a terrible time when they first tried to join in a pickup baseball game as children. Each one had been uncertain about everything—being in a group, taking turns at bat, hitting and fielding, and even understanding what the point of it all was. They were subsequently ostracized, and felt overwhelmed with the shame. As time went on, each of them tried new games less often.

This kind of early experience, of course, is the beginning of a slide down the slippery slope to social isolation. A child in this frame of mind may soon be mocked by other children, be called names, be given grief, and be left behind. Parents and teachers may conclude that the child is alone because somehow he doesn't want to make friends. Nonsense! The unfortunate outcome, however, is that by the time the child becomes an adult, after years of insult and rejection, he may well not want social relationships because he has been bruised so much by others along the way.

The hope is that research can find ways to improve a faulty social brain. There is already evidence that practice can help people overcome at least some of the motor deficits I've just described. Remember Temple Grandin, the autistic woman who learned how to approach people properly, without bowling them over, by walking through a supermarket's automatic doors over and over until she got the steps down? She overcame a social problem that was really a motor problem. As noted earlier, in the last few years therapists working with autistic children have greatly expanded the use of a similar technique called motoring through. A daughter holds on to her mother's leg, and then the mother walks the child through a situation, again and again, until the right physical schema is imprinted on the neural firing patterns in the child's motor cortex.

As we saw in Chapters 1 and 5, such repeated patterning strengthens neural connections in the brain by recruiting neighboring neurons to help in the task. Motoring through excites more pathways, which causes the mental task to be linked to the physical task. That's the great promise of understanding the social brain. If people who can't walk and chew gum at the same time, who are supposedly antisocial, who are geeks, who can't make friends, who can't form close relationships, who can't sustain the intimate bonds involved in being a husband or wife or father or mother, can strengthen the neural connections in the social brain, they can overcome the debilitating problems that don't seem to respond to counseling, drugs . . . or ridicule.

The beauty of training the social brain is that it can be approached from so many different angles, and the more that are tried, the stronger the neural connections will become. Individuals with faulty social brains can improve their behavior by seeing something done properly, hearing it, walking through it, and acting it out in various situations. They can deconstruct the process—break it down into small parts—then practice each one and begin to put them back together again. Just as children use different approaches to learn how to add 2 plus 2—counting on their fingers, drawing pictures, moving blocks on a desktop, and walking in pairs from the walls of a classroom into its middle—the more approaches taken to improve social skills the better.

The lesson is that practice can make perfect. Some schools are realizing this, and are beginning to put class time aside, even if it's as little as 15 minutes a week, to help children learn how to be friends, how to recognize and talk about different feelings, how to handle anger or pain, and how to express what they like and dislike. Teachers will act out situations—such as one child picking on another in the playground—and ask the children how each part of the episode made them feel and how they think the teachers could have acted differently. In another exercise, they ask the children to act out an incident, such as meeting a new child on the bus. By making believe that they are saying hello and trying to become friends, the children are actually practicing the techniques they will use on the real bus later in the day. This learning is invaluable to good development of the social brain.

Modern society has canonized successful social relationships as the ultimate in psychological adaptation, and much of psychology and psychoanalysis is based on this premise. But there is a definite neurological component to this exalted function, and the possibility of correcting the brain's social neurology has been largely ignored.

DEVELOPMENT OF THE SOCIAL BRAIN

THE SOCIAL BRAIN is not a single entity found in any one place. Rather it comprises a combination of different structures and systems working together in harmony. There are critical periods of development for the social brain just as there are for other brain functions. The ability to recognize a human face is an innate aptitude that is present in a human infant's brain from the first days of life and allows the infant to imprint faces quickly, just as a gosling has the ability to imprint its mother's shape immediately after emerging from the egg. Like all brain functions, this can go awry; thus the funny farm stories of a newly hatched gosling that follows the farmer's daughter or a sow, thinking that she's its mother. Does this mean we come into the world expecting to see human faces and ready to respond with our own prewired facial expressions? Yes!

Imagine a nine-month-old baby, ready with a smile when he sees his mother come into the room. The mother's social awareness prompts her to respond with an answering smile. The baby's smile broadens and then so does the mother's, building back and forth until they both erupt into joyful laughter. Many researchers suggest that the mother is teaching the infant emotional regulation through this process, which is imprinted on the baby's developing anterior temporal cortex. The mother and child are in a positive feedback loop in which the child is learning about emotional expression. The interaction goes beyond simple imitation; mother and child actually synchronize their emotional states. At this stage of development the mother and infant are so closely tied that the infant cannot distinguish between his own internal state and his mother's influence.

While it is important to learn about emotions and emotional regulation through this period of intense closeness, it is also important for both mother and child to develop boundaries, to learn how to individuate. During an enjoyable social interaction with his mother, an infant may experience too much physiological arousal and become uncomfortable. He will look away. The mother can effectively maintain an interaction with the infant by knowing when to pay attention and when to withdraw. This is how the baby learns the rudiments of social communication.

Babies pass through another critical period when they learn that other people are distinct selves. In normal development, this "theory of mind" begins to appear in one-year-olds with the practice of "shared attention," the interaction that occurs when a mother and child begin to share observation of outside objects and events. The mother says, "Look at the red truck," and the infant will look at the mother, follow her gaze, look at the truck, and look back at the mother. The child comes to realize that the mother has a separate mind that can deliberately communicate information.

Uta Frith and her colleagues at the Medical Research Council's Cognitive Development Unit in London identified this construct of theory of mind as part of what is not functioning properly in autistic children. They checked the theory with the famous Sally-Anne cartoon series. In this experiment a child watches the interaction of two dolls, Sally and Anne. Sally has a marble, puts it in her basket, then leaves the room. Anne takes the marble from Sally's basket and puts it in her own basket. When Sally comes back into the room, the researcher asks the child, "Where will Sally look for the marble?" Children with autism fail this test of imagination; they cannot navigate the social environment by trying to put themselves in someone else's shoes. The autistic child will choose Anne's basket, which indicates that such children do not have a theory of mind and think that Sally, who was out of the room, has the same information that they do. They do not grasp that the other minds out there are different, with different experiences and different thoughts. That is why these people are often so literal, honest to a fault, and completely devoid of cunning. They think that everyone knows everything that they do, so why bother lying. They can get into trouble with the various bureaucracies in life because they assume that everyone always tells the truth and that honesty prevails.

Temple Grandin says that she didn't realize until her forties that people use their eyes to communicate social cues. Many autistic people can learn social routines that will help them compensate for these kinds of deficits. They can learn to look at a face and to watch for body language—but it is not that simple. For example, one autistic person who had learned to look a person in the eye when talking overdid it with me, gluing his eyes to mine persistently, which would ordinarily register as a social faux pas. Thus, while learning can be a great help—eye contact did improve this individual's communication with others—it can never completely take the place of the social integration that occurs in normal development.

As social development continues, children learn not only how to behave with their parents, but how to get along with their peers. Facial expression, tone of voice, and body language are all important social cues that the child must learn in order to navigate the environments of day care, school, and playground. As the child becomes an adult, new social skills become necessary for finding a mate. The human body and brain are prewired for this eventuality and will respond with the appropriate hormones and neurotransmitters. Eventually, the cycle will be repeated when the former infant becomes the parent of another new baby who will look up with another social smile.

WHY ARE WE SOCIAL?

HOW DID HUMAN BEINGS become the social, interconnected creatures that we are? We can find important clues by looking back to the hunter-gatherer groups of our ancestors. Social behavior seems an odd subject for evolutionary analysis. At first glance, Darwin's basic tenet of "survival of the fittest" doesn't seem to include a mother's love or the selfless acts of heroism featured on the evening news. As we will see, however, effectively propelling our genes into the next generation involves important interactions with both our nearest relatives and the unknown stranger.

All human features evolved in the environment of our hunter-gatherer ancestors. There was a set of basic problems that the brain evolved to solve: how to relate to parents, secure food, find a mate, take care of helpless children, cooperate with kin, and defend against enemies. These basic survival skills are just as important to twentieth-century humans as they were to Cro-Magnon man.

The evolution of social behavior starts with the interactions between a mother and her offspring. A mother who could lead her children to the best food sources or alert them to the danger of predators in the area created a better chance for their survival, and through them the perpetuation of her genes. Communication, through both language and emotion, is important to adaptation. Children who were effectively taught in this way would live longer and reproduce more. Social behavior directed toward the opposite sex was also important in order to maximize reproductive viability.

Kin selection goes a step further. It involves the idea of showing deference, and if need be, of sacrificing one's own survival to help blood relatives survive and pass on their genes to the next generation. Survival of the genetic line is more likely when this behavior helps an individual's kin to reproduce because relatives share many of the same genes, in siblings up to 50 percent. Helping your kin at a calculable cost to yourself is a benefit to those 50 percent of your genes. Robert Wright, in *The Moral Animal,* describes this well with an example of two brothers, Bill and Bob. Bill is drowning in a river. If Bill and Bob are from the "Nice" family and have a genetic predisposition to help each other, Bob will jump into the river to save his brother. There is some risk that they will both die, but there is a greater likelihood that they will both live. In this way, Bob is preserving the genes that he shares with his brother. If Bill and Bob are from the "Mean" family, however, Bob won't jump in, because he's not motivated enough to overcome his fear of self-destruction. He lacks the right combination of genes that code for altruism. In a less dramatic way, differences in the evolution of a genetic line may contribute to sibling rivalry in a family that runs a business, eventually bringing the business down with it, versus a family in which cooperative siblings develop the business for the family's common good.

The next degree of social behavior is that which is directed toward nonfamily and strangers. Reciprocal altruism—the "you scratch my back, I'll scratch yours" philosophy—is found in all human cultures. Even chimpanzees participate in this type of exchange. The problem with altruism is that it lowers the chances for an individual to reproduce while enhancing that of another; you give of yourself—your food, energy, stores—for the good of another. So why does this trait persist? First, the emotions of love and trust that evolved through the mechanism of kin selection sometimes extend to close nonfamily individuals such as members of the tribe or neighbors. Second, reciprocal altruism has a unique adaptive economy. Often it is easier for an individual to participate in an exchange relationship than to do everything for himself. Once you open a door for yourself, it requires only a bit of extra effort to continue to hold it for the person behind you. This little effort is rewarded with a lessening of your own burden when someone else holds a door open for you. This economic principle, which surfaces as little politenesses, helps make civilized society possible.

Critics will point out that cheating on reciprocal altruism can provide a greater benefit than actually following through with the altruistic act. Yet even if cheating exists in a population, selective cooperation will still work. This is the basis for the golden rule of all major world religions: "Do unto others as you would have them do unto you." In 1981 Robert Axelrod and William Hamilton created a computer program called *Tit for Tat* that operated on this principle. They entered it in a competition with several dozen other programs that used cooperative, cheating, or exploitative strategies. The programs each interacted with the others about 200 times in the attempt to determine which one did the best in approaching the "prisoner's dilemma," a classic test of survival based on cooperation. *Tit for Tat,* which followed the rule "Cooperate on the first move and on subsequent moves do whatever your partner did on the previous move" won hands down, and had the simplest software coding of any of the programs.

Perhaps human brains have algorithms that are specialized for detecting cheaters. Researchers have shown that we have an "error catcher" mechanism that requires an intact anterior cingulate and orbitofrontal cortex and that allows us to pick up on errors or shifts in expectations in our interpersonal interactions. Without the awareness that sleight of hand is being used or that a cunning move (masking the shift in logic) has been made, we would have a hard time detecting cheaters. This "error catcher" can be co-opted into the social realm and could be the brain mechanism that malfunctions in paranoia—where everyone is cheating us. An ability to detect cheaters would be one indication that the brain is innately prepared to participate in social life and the give-and-take of everyday existence. In an interesting study that shows this predisposition, John Tooby and Leda Cosmides of the University of California at Santa Barbara found that people think more logically about social content than about other kinds of information. We can sort out social situations more effectively than problems dealing with, say, an abstract math proposition. What's more, Cosmides found that people are not as good at identifying altruists as they are cheaters, and not as good at seeing violations of social contracts when the violations do not involve cheating. The brain's error-recognition ability is biased with some foreknowledge of how people usually act: it is important for you to be able to identify the cheating.

THE CEREBELLUM AS COORDINATOR

REGIONS THROUGHOUT THE BRAIN contribute to its social capabilities, from the "lower" areas in the back of the brain to the "higher" ones in the front. At the base of the brain is the cerebellum. Taking in visual, auditory, and somatosensory information, the cerebellum provides a coordinating function for body movements and possibly some mental processes, such as cognition and attention. It also has connections to many parts of the brain involved in attention and is intimately involved with the higher functions, setting the timing and rhythm and other aspects of language, memory, and emotion.

The cerebellum has only recently been implicated in the normal functioning of social behavior. Traditionally, this "little brain" was recognized only for its role in motor control and balance. But new research has shown that the cerebellum is important as a mediator in cognition. To perceive an object or event, we must pull together the various sensory qualities and any relevant memories or thoughts in a carefully timed way. When we see an object, we determine it to be a "chair" based on its outline, color, and position. We may also simultaneously associate it with the place that the cat seems to favor for sleeping. The cerebellum assists in delaying or accelerating these associations, and regulates attentional states.

Coordinating associations and attention is essential to entering into a relationship with another human being. Communication, conversation, and graceful social interaction all depend on being able to pay attention to another person and to one's own internal states, and to alternate easily back and forth between them. Stroke victims with cerebellar damage struggle for the rest of their lives with simple physical maneuvers like walking up and down stairs. Instead of being able automatically to put their feet down in the right place on the stair step, they have to consciously think about where to put their feet. And like autistic patients, they also find it harder to shift their attention quickly from one thing to another.

Eric Courchesne devised a study in which subjects look at a computer screen containing two empty square boxes with an X between them. The subjects were asked to focus on the X, and to press a button as soon as they saw that a light had been turned on in one of the boxes. Both autistic patients and cerebellar patients took much longer to register the light than normal subjects. The lesson is that those with autism and cerebellar damage are slower to pick up on and react to new stimuli in the environment, obviously making it harder for them to manage social interactions, which are characterized by constantly changing stimuli. Like putting our feet where we want them without having to think about it, our ability to put our attention where we want it without having to think about it is coordinated by the cerebellum.

Courchesne has also found that while a normal baby can shift his attention from the parent's nose to an eye or to the mouth in a fraction of a second, the autistic baby may need as many as five to six seconds to make these shifts. When we imagine ourselves in the position of the autistic baby, autism makes more sense: if it takes five to six seconds to shift your gaze from your father's nose to his eyes, you are not going to see your father's face as a coherent image. You are going to see disparate parts of a face that do not combine, in memory, into a meaningful whole. They will be stored, simply, as face-pieces.

Personal testimony supporting Courchesne's conjecture comes from high-functioning autistic people, many of whom have spoken of "visual dropouts," of not being able to see all of a tree, or see the tree as a whole. An autistic adult might simply see a branch, or a leaf, or a bird perched on a branch. But the tree as a tree, as a coherent entity, does not jump out at him.

When it comes to social competence, this inability to shift attention can have devastating consequences. Social information, the look on a mother's face, her tone of voice, is fleeting; it happens in a moment and is gone. The autistic baby, locked into whatever stimulus has captured his gaze, cannot move his eyes up to his mother's face quickly enough. If the baby is staring at a puppy and the mother smiles, he will miss her smile. By the time he can attend to her face, her expression has changed. If he pulls the puppy's tail, he will miss his mother's frown. Her "No!" and his tail-pulling will not form a coherent whole in his memory. They are separate pieces of reality, disparate fragments in a life that does not add up.

Thus the autistic baby misses a vital first developmental step: he fails to develop "joint social attention"—two people attending to the same thing at the same time. By the age of fifteen months a normal baby will look at something his mother is looking at, or ask her to look at something he is looking at. But the autistic baby does not do this, and one result is that he is difficult to entertain. If his mother says, "Look at the kitty," he does not look. Nor does he ever ask his mother to look at a kitty with him. When the mother says, "Look at the red truck," the baby with autism will be looking at his mother's ear, as she has already turned her head and he has not. The skill of shared attention is a major component of social communication and one of the building blocks of language and conversation. It may be very difficult for infants with autism to decipher which of the rapid and unpredictable incoming stimuli to attend to. Not being able to coordinate attention might account for their social impairments.

Autopsies on autistic persons show that almost all had cerebellar malformations and that there was significant loss of Purkinje neurons, which provide the only pathway for information leaving the cerebellum. Not being able to coordinate basic cognitive functions could underlie the behavioral and social impairments of autism: uneven memory, insistence on sameness, repetitive behaviors, and more. This deficit might also create a chaotic, incoherent world. Children who are unable to shift attention between behavior, events, and others' verbal or facial responses may never learn to interact with people. They will never be able to participate in the shared attention that is so essential for bonding with caretakers and for developing language.

While this deficit is a prominent feature of autism, it may well be that milder forms of social awkwardness, such as clumsiness, shyness, and nerdiness, may derive from less obvious cerebellar abnormalities. Indeed, the "social klutz" is just that, awkward, uncoordinated, out of step, lacking social graces, all of it driven by an inability to properly pay attention, share attention, and coordinate the many simultaneously incoming and outgoing signals.

This was the case with Jeff, a thirty-four-year-old multimedia consultant who worked with start-up computer companies. He was extremely good at his job, and financially well-off as a result. He was by no means autistic. Yet he could not put together a social life. He was a virgin and had been in and out of therapy for years to try to understand his sexuality and his lack of connection with other people. His life confounded him.

Jeff had been a lonely child, unable to make friends. Other children teased him. His parents played it down, saying that if he only had one good friend, that was a lot. But even the typically shy or chided child ends up acquiring the social skills needed to make a few good friends. Jeff made none.

Jeff came to me for help. In talking about his awkwardness in social situations, he mentioned that he could not dance. It wasn't that he danced poorly or was afraid to try. He physically could not do it. He had joined a group that put on small plays, in an attempt to connect. There was one group dance number, and Jeff could not learn it. Finally, he asked the director to humor him. Jeff placed his body behind the director's, his front to her back, his knee to her knee, and moved in step with her through the whole thing. Finally, he got it.

That's what tipped me off: motoring through. Cerebellar damage. Lack of coordination and social grace. Jeff's social deficits were subtle enough that most therapists would not suspect there was anything physically wrong with him, especially with his brain.

In talking further, Jeff told me about some group therapy sessions he had gone to. When Jeff started to talk he would continue nonstop until someone in the group objected. Again he felt rejected, until one day the therapist explained to him that what he was doing was rude and wrong. He hadn't even realized that he was doing it, and rather than feeling rebuffed, he was glad the therapist had pointed it out. He viewed it as useful information and asked the others to interrupt him if he did it again. He could use constructive criticism and was eager to receive it, as long as he was told that's what it was.

Jeff also told me he could not make small talk. He could not deal with cocktail parties. He could not schmooze. He hated it. He gave the facts and expected them back, and that was that. He could not attend to all the rest. More than once he blew a first date by instantly providing a detailed account of his psychiatric history. He also could not engage in romance or foreplay, which led to sexual rejection and confusion. All of it left him bewildered.

The evidence added up to a clear conclusion: Jeff had an impaired social brain. The happy ending for him is that even though he has a shadow syndrome of autism, he is not seriously impaired like a true autistic person. He can correct his behavior. He needs things spelled out, he needs people who can direct him how to act in social situations, but once he is told what to do he can understand it, practice it, and master it.

So it is for many others whose social brains are impaired.

INSIGHT AND SOCIAL BEHAVIOR

AS HUMAN BEINGS have the largest and most fully developed frontal lobes of all animals, these are considered "the organ of civilization" or "the seat of abstract intelligence." The frontal lobes are also important to insight, one of the primary capacities that separate us from the apes. Insight is how we know that we are ourselves, and what particular abilities and weaknesses we have. As noted in Chapter 3, on attention and consciousness, insight depends on working memory, which allows us to know what we feel and have felt and thought and done, all at once. Keeping all this in mind is also what allows us to rehearse and plan.

Patients with right-frontal-lobe lesions typically lack knowledge about their own deficits. They are not aware of the significant loss of social graces that often accompanies this handicap, and are unaware that they are different from others in this regard. Problems in the frontal lobe are also implicated in ADHD, where individuals often lack the ability to see that their behavior may be unacceptable and should be changed. Insight helps us learn the personal responsibility that is essential for social interaction. Insight into ourselves engenders empathy with others, as we imagine what another mind may be thinking about itself and the world around it. Empathy and a "sense of the other" are fundamental to the humanness that makes civilization work.

Imagine what a person would be like if she didn't have a system to guide the development of social integration through reinforcement, empathy, and attachment. She would be impulsive, self-centered, and aloof from the needs and desires of others. These characteristics are seen in people with antisocial personality disorder (APD) and in some with ADHD, and differences from the normal in frontal-lobe structure and activation have been found in people with both conditions.

Researchers also now maintain that the frontal lobes are responsible for working memory, and not having an operational working memory makes it impossible to have a meaningful conversation. People with ADHD may have a problem with this type of memory, which can result in a forced conversational style that jumps from topic to topic. The other participant in the conversation ends up feeling that he is not being listened to or paid attention to. The frontal lobe is also responsible for the temporal organization of behavior, which allows us to navigate the complex and ever-changing social world around us. Obviously, these types of communication and organization problems will interfere with social relationships.

Intact frontal lobes are also important for learning new behaviors. Patients with frontal-lobe damage can use previously learned skills and carry out temporal sequences of behavior, but learning new skills and behaviors is very difficult for them. The exciting news is that with learning and practice through social-skills training, people to whom socializing doesn't come naturally can route neuronal connections around the deficient frontal-lobe area and acquire new abilities. For people who have no social skills, acquiring even one helpful script can be an "in" to a bewildering social world. A patient who can present the same cheerful smile and friendly hello for everyone, even though he doesn't have any conversational interest or skill beyond that, at least has a beginning. Other social scripts can follow.

FEELINGS AND THE FRONTAL LOBES

THE VENTROMEDIAL CORTEX, a section of the frontal lobes, is responsible for the emotion that colors our decision-making processes, especially in the personal-social realm.

Contrary to the popular notion that decision-making requires a "cool head," it is feelings that point us in the right direction and help us make moral, personal, predictive, and planning decisions. Feelings are generated when the brain perceives the varying physiological states of the body. The body as represented in the brain is the basis for what we call "mind." Imagine if we didn't have emotion to guide us in our reasoning. Unfeeling robots, we wouldn't have that instinctual gut reaction to tell us quickly and accurately that we prefer Tuesday over Wednesday for an appointment, that Mary is more trustworthy than Marty, that we should skip that doughnut, or that a potential new job is worth taking. Many decisions like these don't have a reasoned, rational basis, and even those that do are still largely made by what our gut has learned from experience and how it guides us.

Damage to the ventromedial cortex is what caused the social problems for Phineas Gage and Elliot. Other historical cases also show the link between such damage and loss of social skills.

The neurotransmitter serotonin may be implicated in the results of damage to the ventromedial cortex. Serotonin has been shown to inhibit aggression in primates and encourage social behavior. Monkeys with good social behavior have more serotonin receptors in their ventromedial cortex than monkeys with poor social behavior. The production of serotonin is also mediated by social circumstances. Vervet monkeys who are at the top of their social hierarchy—as well as college fraternity officers who are at the top of theirs—have been found to have more serotonin in their brains than their rank-and-file counterparts.

Antonio Damasio draws the conclusion from historical, clinical, and animal cases that damage to the ventromedial cortex consistently results in deficits in reasoning and feeling abilities, especially in the realm of social relationships. Connecting the decision-making, emotional, and physiological monitoring processes requires comprehensive knowledge of the social system and strategies for using that knowledge to make decisions. Further, information from different brain areas regarding everything from the social system to emotions needs to be held in "mind" for a certain amount of time in order for decisions to be made, which requires good working memory. All of these frontal-lobe processes, as well as proper regulation by serotonin, are essential to survival.

EMOTION AND THE AMYGDALA

AS WE MOVE FROM THE LOWER and upper brain into the middle we find the amygdala, a central component of the limbic system—the system of emotion and motivation that enables us to participate in the social world.

The amygdala's influence on the whole brain is very large notwithstanding its very small size. This almond-shaped structure consists of about a dozen different clusters of neurons that have different functions. Each tiny nodule is connected to a broad range of brain regions, and a complex mix of neurotransmitters and hormones acts upon it. The amygdala has been found to regulate autonomic, endocrine, somatosensory, and motor functions, as well as reproduction, memory, sleep, and orientation. As such, its influence on emotions, particularly fear and aggression, is important for the social brain.

Among other functions, the amygdala is the investigator into the ambiguous. If something is different or seems amiss, the amygdala fires to find out what has happened. It is especially attuned to social ambiguity. It responds quickly to a fearful face, even more quickly than it does to an angry one. A fearful face indicates that there is danger, but not what or where the danger is. An angry face indicates that there is danger coming directly from the person who is angry. Tests at the University of Sheffield in the United Kingdom show that people who have normal intelligence but amygdala damage have good recognition of basic emotions such as happiness, surprise, fear, sadness, disgust, and anger, and various facial expressions, yet cannot recognize fearful faces and have trouble seeing the potential danger when meeting a stranger.

Research with monkeys from the 1930s on has shown how important the brain's emotion circuits are to social behavior. For example, the social behavior of monkeys released into the wild with only their amygdala removed changed dramatically. Rhesus monkeys in particular did not survive past several weeks, falling prey to aggression from other monkeys. Research on the connections of the amygdala shows that it quickly triggers autonomic and endocrine responses to social stimuli, from sexual advances to group cooperation. The frontal cortex, responsible for the brain's most complex processing, has the heaviest projections to the amygdala, and the two work together as part of the network that is the social brain.

Output from the amygdala has two components. The first feeds back to the sensory cortex and thalamus areas sending messages to it. The second goes to the hypothalamus, which initiates the autonomic and endocrine responses. Meanwhile, the central nucleus of the amygdala sends outputs to the brainstem areas that regulate heart rate and breathing. These connections are important to social behavior because they influence the perception of another animal and modulate the body's reaction.

As an example, when a dog growls and bares its teeth, it is displaying an emotion. But that emotion is only meaningful when it is perceived by another animal and correctly interpreted as "Stay away!" This "affective display" has both an internal and an external component for the growling dog. Internally, his heart starts to race and adrenaline starts to flow. Externally, his facial expressions and posture communicate his meaning. The amygdala is closely involved in this behavior and in the other dog's—or person's—interpretation and response to it.

In similar but much more complex ways, the amygdala helps regulate our production of and response to higher social stimuli. Its connection to the anterior cingulate gyrus, for example, appears to have a role in speech. Connections between the amygdala and hippocampus support memories of previous social situations. The amygdala, temporal lobes, and posterior medial orbital cortex, part of the ventromedial cortex of the frontal lobes, constitute a connected system that is important for attaching emotional significance to stimuli.

The amygdala and structures around it, called the extended amygdala, send information to various parts of the temporal cortex to help the neurons there encode and remember an engram—a representation of something to be remembered—and to change, regroup, and recalibrate. It sends cholinergic neurons up to the temporal cortex, making those neurons more likely to fire and fire more easily, so that it is easier to encode a memory.

A lot of panic, particularly social panic, is driven by the amygdala. This may beset a woman who walks into a cocktail party feeling that she doesn't know anyone, and is nearly overwhelmed by the lights, smells, and sounds that all eventually converge in the amygdala. The amygdala goes on alert, looking for danger or the ambiguous, and responds by flashing, "Threat to survival. Stop. Watch out. Be on alert. Keep wary of strangers." If a link to a memory of previous panic is formed, or the new threat or discomfort is high enough, the signals turn on the nucleus basilis, a structure in the extended amygdala, which sends acetylcholine to the sensory cortex. This raises the potential to fire away in the sensory cortex, thus making the likelihood of encoding the event and storing the memory of it much greater. This distress is fed back to the amygdala, and the sensory cortex is activated further. The whole system jazzes up. There is a reverberating circuit effect, a runaway reaction. Fear begets further fear, and soon the woman runs out of the room, sweating and paranoid. This motor activity then helps her calm down.

Alternatively, as the woman's anxiety heightens, someone may approach her who has a friendly expression and a quieting tone of voice and give her a reassuring touch on the arm. This raises the woman's serotonin level, which dampens the network, tells the cortex to quiet down and the amygdala to reduce its vigilance for there is no longer a threat.

When we see a fearful face we are instantly thrown into overdrive to search for more information. But this hyperalerting can also take place in nonthreatening circumstances; a man may get a whiff of perfume that reminds him of the eighth-grade teacher he adored (and had the hots for), and become energized and readied for action.

This kind of disconnect can also be seen in the hothead who loses his temper and then quickly says he is sorry when he realizes he has gone too far. Too much of his cortex has been hijacked by the amygdala and thus he can't put on the cortical brakes of reason. His emotional system is activated before the facts are really in, and he reacts before the situation is clear.

If the frontal lobe can't intervene fast enough in these kinds of situations, we latch on to our emotional response and shut off any further investigation of possibilities. In this way, emotion affects social cognition; we get swept away by our feelings, which overrule our better judgment. We "go emotional" and become less cognitive, less logical, since the frontal cortex's reasoning and decision-making apparatus is overwhelmed by the emotional response.

FACE RECOGNITION

THE AMYGDALA PLAYS a crucial role in face recognition. Being able to recognize faces is an important part of the human repertoire of social behaviors. For one thing, it is essential for survival, a key to determining whether a friend or a foe is approaching. It is also essential to maintaining social relationships, and even social status (another kind of survival), by identifying others of a higher, lower, or similar status and regulating behavior accordingly, and impairment here can be devastating.

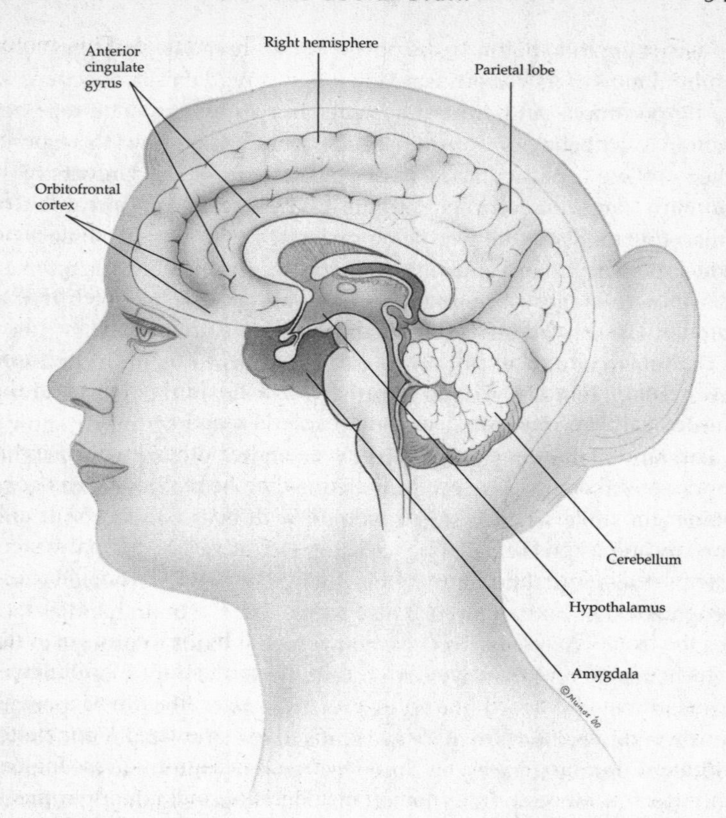

THE SOCIAL BRAIN The cerebellum is the "rhythm and blues" center of the brain and helps keep our actions, emotions, language, and memories running smooth and seamless. It is crucial for enabling us to do the social dance gracefully. The limbic system, especially the amygdala and hypothalamus, gives us the pressure to seek others and helps add intensity to social bonds through hormones and direct neuronal input. The anterior cingulate gyrus directs our inner response to others and keeps us willing and interested in being with them. The orbitofrontal cortex is the error catcher and with its partners, the anterior cingulate and the ventromedial cortex of the frontal lobe, is crucial for empathy and evaluation of the genuineness of the words and intentions and comments of others.The right hemisphere helps us pick up on nonverbal cues in speech and gesture as well as in facial expressions, while the right parietal lobe is crucial in giving us an overall picture of ourselves in space and how we relate physically to another person.

Face recognition points to the power of the amygdala and represents a culmination of its various functions. The amygdala's connections to the hippocampus and arousal systems and its ability to tie together memory and behavioral responses are what tell us how to respond when we see a particular face. Leslie Brothers of the University of Southern California recorded outputs of single cells in the amygdala of a macaque monkey that was being presented with various images on videotape. They found one cell that fired specifically when the monkey was shown images of monkey and human faces with which it was familiar. The cell did not fire when the faces were unfamiliar.

The image of a face, as I noted earlier, seems to be imprinted in a baby's brain. Just as a gosling imprints the shape of the mother goose immediately after hatching, newborns, minutes after birth, will show a preference for looking at a correctly arranged picture of a face as opposed to a picture with random features. At two to three months an infant will smile when it sees a balloon with eyes painted on it and stop smiling when the balloon face is rotated away.

Thus it seems that humans are hard-wired or driven to perceive and recognize other human faces. It also seems that the brain handles face imagery as special visual information, which it tends to process in the right hemisphere. When we scan a face, we make a lot of eye movements directed toward the eyes and mouth of the other person. Research shows there are more eye scans in the left visual field. This is odd, even unique to faces, because we scan most stimuli in a symmetrical way. Information from the left visual field goes to the right hemisphere, which suggests that the right hemisphere has a mechanism for recognizing faces, similar to the left hemisphere's for recognizing words. Tests using chimeric figures—faces with the left and right halves made from two different people—support this conclusion. Split-brain patients, who don't have any communication between the hemispheres, consistently identify the chimeric people as the one whose half-face is on the left.

A deficit in the ability to recognize faces is called facial agnosia or prosopagnosia, derived from the Greek words for face (*prosopon*) and not knowing (*agnosia*). Prosopagnosia seems to be a result of an impairment in the medial occipitotemporal cortex of the brain, due to stroke or brain damage. Although bilateral damage usually causes the full-fledged syndrome, damage to the right hemisphere alone is far more debilitating than damage to the left. Prosopagnosiacs can sometimes use cues such as a mustache or birthmark to identify a face, and most can recognize expressions. It's the specific link between the face and its identity that's the problem. More often these patients have to wait for the person to speak, as they have no trouble recognizing the identity of familiar voices.

The opposite of the inability to recognize faces is a disorder called Capgras's syndrome, also caused by damage to the temporal lobe. In this heartbreaking malady, patients recognize the faces of family and friends, but there is no emotional input connected with the faces. They often then assume that the members of their family have been replaced by impostors. Although they usually retain many other cognitive abilities, they can become delusional and paranoid when they try to explain what has happened to their family.

One man, Arthur, described in V. S. Ramachandran and Sandra Blakeslee's *Phantoms in the Brain*, is a good example of Capgras's syndrome. He had suffered extensive head injuries in an accident, but had fully recovered following a three-week coma and seemed completely his old self. However, he thought that his parents were impostors. They looked like his parents, but he knew they were not—he didn't have his emotional compass to guide him. He did not feel the warm glow he had associated with their faces; he had no emotional input that confirmed who they were. His temporal lobes were fine; he could recognize his parents. But his amygdala had been damaged and he could not feel they were who he thought they were. He thought he was really speaking to his parents when he talked to them on the phone—that is, without the usual emotional cues of vision operating. (By the way, the reason our pets do not respond to our voices over the phone or on video is that they operate solely on emotional cues and context, and thus do not recognize an owner's voice without a three-dimensional reference.)

Arthur's father even tried to trick his son, saying that the couple who were masquerading as his parents had returned to China and that his real parents had come back. Arthur believed this for a while, but still did not feel any warmth toward them and eventually took up the idea again that they were impostors. He would say, "The man looks identical to my father, but he really isn't my father." Or, "That woman claims to be my mother; she looks like her but is not." Arthur's deficit was in the ability to link the emotional response with the face recognition. So his deficit was not in the amygdala per se, but somewhere in the temporal lobe; not in the frontal lobes, but in the connection between the amygdala and the temporal.

Fuller understanding of the process by which we recognize both the face and the identity of the people we know will increase our understanding of the social brain.

Facial expressions constitute one of the primary methods of communicating information in the social realm. As with face recognition, both the production and recognition of facial expressions are hard-wired in specific areas of the brain, notably the occipitotemporal lobes and the amygdala. Again, the right hemisphere seems to play a more important role; patients with left-hemisphere lesions have some difficulty recognizing certain expressions, but patients with right-hemisphere lesions have an even harder time.

Creating facial expressions is an innate skill. Infants respond to certain stimuli with prewired facial expressions. However, unlike adults, who create fairly specific faces for each emotion, infants use the same face for many different situations. As the brainstem and the cortex develop, muscles and coordination strengthen, and recognition of different emotions improves, a baby's facial responses become more specialized.

FREE WILL AND THE ANTERIOR CINGULATE GYRUS

ANOTHER REGION IN THE CENTER of the brain that is critical to social function is the anterior cingulate gyrus. Part of the limbic system, it has many connections to other brain areas and plays a role in social behavior, emotion, and motor functions. It also receives more inputs from the thalamus—the sensory filter—than any other cortical region.

Electrical stimulation of the anterior cingulate gyrus can cause changes in the autonomic system, including heart rate and breathing. Involuntary vocalizations, visceral symptoms such as nausea and vomiting, and automatic movements of the hands and mouth can also result. In some people, damage to or loss of this area due to stroke, tumor, or surgery has no apparent effect on personality or behavior. In others, it decreases social awareness. Monkeys with a damaged cingulate gyrus seem to treat other monkeys as if they were inanimate objects, and walk right over them. Other animal research indicates that the cingulate gyrus plays a role in regulating stress. Disturbances in this area due to lesions, epilepsy, chemical imbalance, or surgery can cause mild modifications or major upheavals in social and emotional behavior. These can range from apathy, impulsiveness, disinhibition, aggression, psychosis, sexual deviancy, obsessive-compulsive behavior, and impaired social judgment to the simple desire to change hobbies or reading habits.

The anterior cingulate gyrus also seems to be involved in regulating the emotional content of physical pain in three ways: by determining the emotional meaning of the pain; by initiating a motor response to the aversive stimulus; or by learning how to predict and avoid the pain. Some people have found relief from chronic nerve pain through surgical lesion of the cingulate. They report that they still feel pain, but that they do not experience any emotional reaction, such as despair.

The anterior cingulate gyrus is also important to social behavior in its control over free will. Sensory neglect—where patients ignore the sensory information coming from whole areas of their bodies—has been seen in individuals with lesions in this area. The most stunning case is that of akinetic mutism, where a patient becomes completely mute and motionless, as if in a coma, but is still conscious. Antonio Damasio had a thirty-five-year-old female patient who had had a stroke in the anterior cingulate gyrus. For a month she had no spontaneous speech or movement, although she would answer questions, or pull at the bedsheets when questioned. When she recovered she reported that she did not talk spontaneously because she "had nothing to say." She "felt no will" to speak.

Conversely, an impediment to the functioning of the anterior cingulate gyrus can produce a lack of control over free will, as in Tourette syndrome. These patients have involuntary obsessions and compulsions in movement or vocalizing, sometimes characterized by physical tics and unprovoked streams of foul language. People with Tourette have been shown to have less activation in the anterior cingulate gyrus and related areas, such as the caudate nucleus in the basal ganglia. Tourette syndrome reveals the integrating function that the anterior cingulate gyrus performs for thought, motivation, emotion, and movement.

NONVERBAL CUES AND THE RIGHT HEMISPHERE

DOING THE "SOCIAL DANCE"—engaging in the right physical moves in concert with others—is a very important aspect of social skills. For example, as noted earlier, judging where our bodies are in space compared to those of other people so that we stand at the appropriate distance from someone while having a conversation is a basic social skill that is not easily taught. The right hemisphere, particularly the parietal lobe, is responsible for analyzing external space and the body's position within it. The parietal is the "where" area of sensory perception, as mentioned in Chapter 2. It is also responsible for pulling together a complete perception of the spatial and social components of the world. Studies of lesions in the right parietal indicate that it is involved in attention, music, body image, body scheme, face recognition, and the physical act of dressing. Further, the entire right hemisphere plays a role in the attentional system and in feeling and displaying emotion.

Research has shown that adults with right-hemisphere lesions respond with indifference to emotionally disturbing events, and similar damage in children is associated with chronic difficulties in social relationships. Often, the nature of the problem is linked to one or more nonverbal learning disabilities, such as what I call social dyslexia, which causes people to mistakenly interpret or misread others. Social dyslexics can decode and recognize words easily, but have poor comprehension of what they read. It is hard for them to pick up meanings and innuendos. One patient told me that she had a good sense of humor and began our session by telling me a joke. However, when I later referred to a friend she was mad at, and jokingly said, "You might want to beat him about the head and shoulders," she immediately responded by saying that that would be an overreaction.

People with this kind of brain deficit may use words well in conversation and love to talk, yet they do not get along comfortably with others because they cannot pick up on nonverbal cues like body language, facial expressions, or tones of voice. Though they are intelligent, they can seem dense or obtuse owing to the lack of information they cannot extract from the environment. In short, people with nonverbal learning disabilities cannot put things in context or manipulate them well.

Nonverbal learning disorder, now known as "right-hemisphere deficit syndrome" (RHDS), results from impairment to the right hemisphere during early development. Some 50 percent of such individuals had prenatal problems or problems at birth. They are often "difficult" babies—sleepy, inactive, uninterested in social interactions, yet at times overaroused and hypersensitive. In the first six months there are problems with eye contact. At one year problems with joint attention become apparent, and they show a lack of understanding of facial expressions. At two years they fail to develop the theory of mind like other children. When they begin to explore their environment as toddlers, they show one of two patterns: fearlessness or timidity. The fearless toddlers don't seem to have a clear conception of their bodies in space and are therefore accident-prone. The timid ones prefer to label objects rather than handle them. Both of these traits, we may speculate, are due in part to a deficit in the ability of the right hemisphere to put the child in space and in the proper dimensional context.

Adults with RHDS may show a lack of awareness of other people's interest in conversational topics. They may also have flat voices, little emotional expression, difficulty reading others' emotional signals, a limited vocabulary of emotion words, and an inappropriate sense of interpersonal space. There also seems to be a genetic predisposition for this disorder; 50 percent of patients have family histories of social problems.

The social-skills problems that right-hemisphere damage creates are often mistaken for the social problems of other disorders. Many people with RHDS have been diagnosed with ADD. There is a difference, however, between the impulsive and sometimes intrusive social behavior of ADD and the unaware social behavior of RHDS. Individuals with a right-hemisphere deficit may also be misdiagnosed as autistic, but they are more likely to have fluent spoken language and don't exhibit the stereotypical movements that characterize autism.

One patient of mine, J.C., is an extremely smart librarian. Talking, reading, and writing at an early age, he was a child genius, and yet throughout his entire life he has had a hard time with meaning. He is interested only in the literal sense of things. I once suggested increasing the dosage of a medication he was taking, and he balked, saying that that was not indicated on the bottle. He stops all conversations short with "I got it" when he understands the literal meaning of what is being said. He wants to communicate only through words; he refuses to make eye contact; he takes no joy in human encounters. Although he is an honors graduate from one of the world's leading universities and has received many academic accolades, he has an impossible time deciding on all sorts of details for himself, and is extraordinarily dependent upon authority figures. Neuropsychological testing confirms that he has an underperforming right hemisphere. Despite his brilliance, J.C. is unable to relate to others in any intimate way.

LANGUAGE

THE HUMAN CONDITION is a complicit interaction between culture and individual minds, one shaping the other. I co-evolve in reading books and writing this one; I am stimulated by what I read and this causes my own book to evolve into something other than exactly what was there when I started. In the same way, people are continually co-evolving, and, fortunately and unfortunately, language is the primary medium of this exchange.

One of the right hemisphere's most critical contributions to the social brain is its role in the social aspects of language. Language, consciousness, and social behavior probably evolved concurrently, each driving the others and causing them to expand. Social interactions provided the opportunities to learn new behaviors from others, increasing intelligence. Expanded social behavior patterns, in turn, created a more complex environment that had to be navigated. Language developed as one navigational tool for this journey.

Language, especially syntax, is an essential part of human intelligence and social behavior. Syntax gives us part of our ability to plan ahead. Emotional prosody, or the tone of voice in which words are spoken—the emotional content of speech—is also important as it often provides important cues as to other people's moods and intentions. It also contributes to our understanding of humor and metaphor, higher forms of social communication.

As noted earlier, although the brain's left hemisphere plays the dominant role in deciphering the content of language, the right hemisphere seems to hold the most influence over the production and comprehension of emotional prosody. Patients with right-hemisphere lesions don't understand metaphors; when people in one group were asked to choose a picture that matched the phrase "give someone a hand," they picked a picture of a platter with a hand on it. These same patients also had impairments in regard to understanding humor; when given a choice of endings to fun stories, they often chose bizarre conclusions instead of appropriate punchlines. They tended to laugh more at humor that relies on the totally bizarre and unexpected rather than on a subtle switch of meaning, which in fact underlies most of what we encounter as humor. They miss nuances and have a hard time understanding metaphors used in conversation; they will think that the metaphor or hyperbole is real and literal and be surprised by it.

Language and nonverbal learning disorder point to the importance of the right hemisphere in the social brain. Simply understanding the deficits of the right hemisphere can help people who are so impaired find ways to change their social behavior. One patient of mine who has right-hemisphere problems used to get highly insulted at humor in her workplace. She took everything far too literally. Only by realizing that this was happening and analyzing her past bouts of anger could she begin to see her way out of getting furious in response.

WHEN IT DOESN'T WORK: ANTISOCIAL PERSONALITY DISORDER

BEING SOCIALLY CAPABLE depends on smooth coordination between the different brain regions involved in various aspects of social behavior. But when one or more regions fail, handling social situations can become a nightmare. This is what happens to people with antisocial personality disorder (APD), who are branded as simply socially inept but may actually have brains that are structurally different from the norm. Some of the compromised behaviors in APD can be seen in the activities of many of us "normal" people, too.

APD would be the clinical description for the behavior of Phineas Gage and Elliot. A person with this disorder may seem to lack a conscience or moral sense regarding other people. Those who have milder cases may even be charming and likable, but may use these characteristics to manipulate others. More often, people with APD may be impulsive, irresponsible, and sometimes violent.

Research has shown that underactive frontal lobes may be at the root of this disorder. Some studies have indicated that antisocial people are also physiologically underaroused; their heart rate, skin conductance, and EEG readings are lower than those of socially normal people. This underarousal could cause such individuals to seek inappropriate stimulation or make it difficult for them to learn the rudiments of social behavior. They don't have access to normal emotional cues to help regulate their relationships. Further research has shown that many antisocial people may have prefrontal dysfunction and dysfunction of the region in the left hemisphere responsible for language—or of the corpus callosum between the hemispheres.

Adriane Raine at the University of Southern California conducted a study of underarousal in 101 fifteen-year-old boys. He tested them for skin conductance, heart rate, and EEG measures and correctly predicted 75 percent of the criminal and noncriminal outcomes when they reached age twenty-four. Another study found abnormal EEGs in 50 percent of violent offenders. Raine's explanation was that much of the underarousal was caused by a combination of genetic and environmental factors, among them a compromised structure of the frontal lobes, a lack of neurotransmitters or neurons or both in the region, a paucity of connections to the area, and a lack of practice in using the area.

Overly aggressive people fall into the antisocial category of misbehavior. Raine used PET imaging to measure glucose metabolism in the prefrontal cortex of forty-one homicide defendants who were pleading not guilty by reason of insanity. The study had a control group of the same size of age- and sex-matched people. Raine found that the killers had a deficit in the frontal cortex and abnormal concentrations of norepinephrine, dopamine, and serotonin in the area, but that the effects could not be picked up on typical psychiatric or neurological exams. The preliminary findings provide initial indications of a network of abnormal cortical and subcortical brain processes that may set up a predisposition to violence in such individuals.

Dysfunction in the left hemisphere, which results in a loss of control over impulsivity, is another possible cause of antisocial behavior. Lesion studies indicate the possibility that violence can also result from damage to the left temporal cortex, amygdala, and hippocampus.

AUTISM

APD IS A COMPLEX FAILURE of the social brain. Autism, on the other hand, is a developmental disorder that strikes at the heart of that which makes us human: our ability to love, laugh, and encounter others. People with autism have a range of symptoms that can include deficits in IQ, delayed or nonexistent language, self-injurious behavior, and repetitive, stereotyped movements. Some autistic people can have amazing talents as well, such as an almost photographic drawing ability.

Leo Kanner of the Johns Hopkins Children's Psychiatric Clinic described and named autism in 1943. He listed four traits: a preference for aloneness, an insistence on sameness, a liking for elaborate routines, and some abilities that are remarkable compared with those of the average person. Autistic people can be very upset at the slightest change in routine. You might recall the autistic character played by Dustin Hoffman in the movie *Rain Man;* he needed to watch certain TV shows accompanied by certain rituals, such as Judge Wopner of *People's Court* with cheese balls and toothpicks.

Today autism is the diagnosis when a person has impaired social skills, impaired language, and significantly restricted interests. There are many ways that these symptoms and their subsets can be expressed. Whatever the individual mix, however, all autistic people share a profound deficit in social abilities. This lack of awareness of the social realm is a terrible handicap, but for scientists it is a window into the workings of the social brain.

Autism occurs in one or two of every 1,000 births. Boys are affected two to four times as often as girls. For years autism was considered a mental illness without a biological cause, even though there was no supporting empirical evidence. However, more recent research has shown that autism does have a genetic basis. Identical twins are more likely to both be autistic than fraternal twins. Autism is 50 to 100 times more likely to occur twice in the same family than would be expected by chance. Structural differences in the brains of autistic individuals—such as a smaller cerebellar vermis or uncommon cell structures in other parts of the cerebellum, hippocampus, and amygdala—have been widely reported in postmortems, and the deficits that autistic people show in tests of planning, initiative, and imagination are similar to those suffered by people with frontal-lobe damage.

As mentioned earlier in this chapter, one of the most troubling aspects of autism is the lack of a theory of mind. For example, autistic children cannot understand pretense and do not pretend when they are playing. It is unclear why this is so. Perhaps they do not have an inborn mechanism that allows them to delay long enough in their thinking process to shift to a different level of meaning, or have difficulty shifting from one line of thinking to another. They are bound by where they are. They are unable to put themselves in another person's place. It is as if the ability to manipulate ideas and see them from more than one perspective as well as a powerful integrating force to seek meaning are missing. However, some autistic persons can learn social routines that hide their deficits, and they can excel in areas such as music or math that don't require social interaction.

Other theories link autism to cerebellar damage, which would interfere with smooth coordination of shifts of attention. Most recently, another group of theorists, Lynn Waterhouse and colleagues at Trenton State College, identified four neurofunctional deficits in autism. The first is canalesthesia, a problem with the hippocampus, the memory center of the brain. In this dysfunction the neurons in the hippocampus have too great a cell-packing density, which results in abnormal fragmentation of both current events and long-term memories. The second is a problem with the amygdala that disrupts the ability to assign emotional significance to events as they happen. The third is asociality, a problem with the oxytocin system that leads to less bonding and affiliative behavior. The fourth is extended selective attention, a problem with the temporal and parietal lobes that causes overprocessing of primary representations, where the brain gets stuck on some detail. Causal relationships between symptoms have not been established, though, and the variety of symptoms in autism makes it difficult to find a single cause of this disorder.

Regardless of which theory prevails, all of them take into account that autism results from deficits in several areas of the social brain. Some scientists think this could be the result of damage to the developing embryo's brain. Patricia Rodier of the University of Rochester reported that in 33 percent of mothers who, before its use was banned in pregnant women, had ingested thalidomide between days 24 and 27 of pregnancy, their children developed autism, and that fetuses exposed at other times did not (thalidomide is a tranquilizer that was used to help prolong pregnancy and was banned because it was proved to cause birth defects). The period between days 24 and 27 of pregnancy is exactly the time that brain neurons are just starting to form.

Other evidence supports injury in the first trimester, particularly to the brainstem, as the cause of autism. Many autistic people have problems with eye movements, facial-nerve palsy, and hearing problems that are indicative of brainstem injury. All of this research can give us insight into how to help people with autism and how the social brain works in everyday situations.

THE BIOCHEMISTRY OF LOVE AND INTIMACY

LIKE OTHER BRAIN FUNCTIONS, the workings of the social brain are greatly affected by hormones, drugs, and even the food we eat. We may not realize it, but we are adept at changing our social brains. We drink tea and coffee, with the stimulants they contain, at our social gatherings, and automatically accept the obligatory glass of wine or martini upon walking into a cocktail party. All these are useful drugs to help the brain function more socially—to a point. Smoking clubs, chocolate parties on college campuses, and even fitness centers induce effects on the social brain. We use alcohol and marijuana to help calm our anxiety and improve our mood and social assertiveness. We use chocolate, tea, coffee, and nicotine to improve our attention so we can attend more fully and easily to others and be good social partners. In psychiatry we use Prozac and other SSRI drugs to treat depression, panic, and anxiety, and they have the added benefit of increasing social assertiveness. When we treat adult ADHD patients with stimulants, their spouses often remark on a tremendous improvement in the couple's relationship; the ADHDer becomes more present, a better listener, and thus more social.

Perhaps we will soon learn how to adjust hormones and alter genes to modify the function of the social brain. This kind of research has been led by animal studies on hormones that are involved in bonding, monogamy, and parenting.

A lot of research into the social brain concerns the two primary phases of romantic love: attraction and attachment. Attraction begins with intrusive thinking about another person. Eventually, there is a fixation on the loved one and the characteristics that make him or her special. Many people report experiencing a range of emotions at this stage, including elation, hope, shyness, uncertainty, and fear. There also may be a sense of helplessness and a feeling that the relationship, though unplanned, was meant to be. Tests at this stage have shown an increase in the brain of a chemical known as phenylethylamine (PEA), which is related to the amphetamines. This might account for the "rush" at the sight of the loved one, or other physiological responses such as decreased appetite or fitful sleep. This attraction phase usually lasts from eighteen months to three years. This may be because neurons in the limbic system eventually become habituated to phenylethylamine. A crush may subside or a love affair may settle into routine, all because the brain becomes desensitized to its own internal "upper."

Attachment, the second phase, is associated with feelings of calm and peace. A person becomes sure of the love relationship as a support, and the loved one's presence provides comfort. The brain has been shown, at this stage, to increase its production of endorphins, which are chemically related to morphine. Oxytocin and vasopressin also play a role in the attachment phase between a male and a female. The increase of these hormones or an increase in the sensitivity of receptors to these hormones at the time of the birth of a child may be the trigger that initiates the attachment phase. As the parents begin to nurture their child, they also begin a phase in their relationship in which they nurture each other. Perhaps oxytocin and vasopressin also play a role in this phenomenon.

Animal models of the two phases of social bonding—attraction and attachment—encourage the theory that the emotions of love were selected for in the process of evolution. Some animal species mate for life and show physiological distress when separated from their mates. Others exhibit the attraction stage through elaborate courting rituals, but may stay with the mate only through conception or birth of the young. Studies on animals have shown a rise in vasopressin and oxytocin levels in the blood and limbic system during sex.

Some researchers feel that there is a third stage of romantic love: detachment. The animal model is that of a bird leaving the nest. Human beings break up or divorce. This phase may be caused by excessive feelings of safety. Brain receptors to the endorphins of attachment may become desensitized. Obviously, not all couples experience this stage.

Research has shown that 90 percent of all Americans marry at one point in their lives. Anthropologists have learned that most people in the world are married to only one spouse at a time. If a culture approves of polygamy, it is usually limited to a man being allowed to have more than one wife. These statistics on monogamy do not take account of adultery. In the evolutionary environment there was an advantage to staying with a mate to take care of the helpless young, but this did not preclude the opportunistic strategy of extramarital copulation.

Research on divorce patterns throughout the world has shown that: (1) most divorces occur around the fourth year of marriage; (2) the most common age of divorce is the twenties; (3) most divorces occur when there are no children or only one child; and (4) most divorcees remarry during their reproductive years. This evidence reinforces the theory that the pair bond is formed to conceive and take care of offspring. After a period of not conceiving, couples may find themselves less attracted to each other, probably because of the infertility. If the couple did conceive and bear a child, they might still find themselves restless four years later, because at this age the infant needs much less care. Furthermore, a change of mate helps to produce more genetically varied offspring that may have an advantage in the survival of the fittest.

Thomas Insel and colleagues at the National Institute of Mental Health have been looking for years at how the hormones oxytocin and vasopressin influence the social bonds between animals. These hormones, which are very similar, differing in only one link in an otherwise identical chain of nine proteins, are both made in the hypothalamus but they have drastically different effects on the body. They also play a central role in attraction and attachment.

Hormones are essential to the feeling and expression of romantic love (as anyone who has gone through puberty knows). The female hormones, estradiol and progesterone, have been shown to contribute to women's sexual interest. A woman's peak level of testosterone, at ovulation, also correlates with increased sexual interest.

The hormones oxytocin and vasopressin, which are produced by the pituitary gland in the brain, have been previously associated with pregnancy and nursing in women and are important to sexual activity in both sexes. Oxytocin in women causes the milk ducts in the breast to contract so that milk is ejected, allowing the infant to nurse. It has been found to cause smooth muscle in both males and females to contract and is released in both sexes during orgasm. It may play a role in sexual satiety, as well as being an important factor in social bonding. It also facilitates sexual behavior in both sexes, and in males it has been shown to increase erection.

While oxytocin is an important hormone and chemical messenger of the body, it also serves as a neurotransmitter in the brain. It is involved with nearly every type of animal bonding: parental, fraternal, sexual, and even the capacity to soothe one's self. It has played an essential role in the evolution of social behavior, particularly of mammals. Its first evolutionary role may have been in helping to forge the bond between mother and infant. But its ability to influence the brain has since been co-opted to help us form alliances and partnerships, hastening the evolution of advanced cognitive skills. This, in turn, has improved both the quality and the length of human life.

Vasopressin, while essential to pregnancy, also acts as a neurotransmitter, and increased levels have been correlated with male sexual behavior.

Pheromones constitute another group of chemicals involved in sexual behavior, acting as messengers from one organism to another. Animals use pheromones to determine when a mate may be ovulating and ready for sex, but there is as yet no conclusive evidence as to the role these chemicals play in human beings. Some researchers say that the way women's menstrual cycles synchronize in an all-female living situation may be attributable to pheromones. One study showed that androstenol from male sweat increased women's social interactions with men. Some perfume companies are marketing pheromone-enhanced perfume, betting that these chemicals are important to human sexual behavior.

There are also specific areas of the brain that are involved in sexual behavior. The medial preoptic area has been shown to be critical to male sexual behavior. In animal studies, stimulation here engenders copulation behavior. Conversely, copulation results in increased activity in this area. Destruction of the medial preoptic area eliminates male sexual behavior.

The ventromedial nucleus of the hypothalamus is essential to female sexual behavior. The hormones estradiol and progesterone act here. This area has connections to the periaqueductal gray matter of the midbrain, which also is important to female sexual behavior. The periaqueductal gray matter is connected to the reticular formation of the medulla and the spinal cord, which is activated in the lordosis response, or female sex posture. The medial amygdala has connections to the olfactory bulbs and plays a role in pheromone perception. The smell rules as the olfactory system gets directly connected to the amygdala. The pheromones, which are the powerful ingrained guide to action among a species, go right to the entry point of the limbic system.

HOME, HOME ON THE RANGE

NONE OF THE NEUROCHEMICAL ASPECTS of sex is uniquely human, nor are the three primary forms of love, so it is possible to investigate the neural basis of social attachment in animals. Interestingly, the greatest insight into our understanding of social attachment has been derived from studying a mouse-sized rodent known as the prairie vole.

The validity of relating research on the sexual behavior of animals to that of humans is a platform for fierce anthropologic debate, but the parallels between the sexual and social habits of prairie voles and humans are intriguing. Most men and women are monogamous, wedding only one individual at a time, but only 3 percent of all mammals are monogamous. Prairie voles manifest many of the classic features of monogamy. Breeding pairs share the same nest and have frequent contact with each other there; males and females participate proportionately in parental care; offspring remain sexually suppressed while within the new family group; and intruders of either sex are rejected. Following the death of one of the pair, a new mate is accepted only 20 percent of the time. When prairie voles mate, they mate for life.

Regardless of divorce, humans intend to pair up for life, establish a home base, build networks of family and friends, and bear and nurture children. Current relationships are perpetually taxed with various social and economic constraints, suggesting that divorce or extramarital affairs may not be a reflection of biological deviance. Clearly, the topic of human monogamy warrants endless debate, but for the purpose of this discussion, we will assume that human beings, like prairie voles, strive, at the very least, to be a monogamous species.

In addition to their highly sociable, monogamous nature, prairie voles offer a second virtue for research: the possibility of comparative studies. Enter the montane vole. It looks remarkably like the prairie vole, but is generally found in isolated burrows, shows little interest in social contact, and is clearly not monogamous. Male montane voles demonstrate little if any parental behavior, and female montane voles often abandon their young between 8 and 14 days postpartum. Like the good and the bad twin, the prairie and the montane vole share similar genetic backgrounds but differ substantially in their social personalities.

For these reasons, prairie and montane voles provide intriguing natural subjects for studying the neural substrates of social attachment. The two species differ remarkably in the receptor distribution of oxytocin and vasopressin. For example, after delivery, when the female montane vole demonstrates brief parental behavior, the pattern of oxytocin binding changes to resemble that of the prairie vole. The differences in oxytocin and vasopressin receptor distribution between the prairie and montane voles may contribute vastly to their distinctive social behaviors.

The formation of a pair bond requires the generation of a preference for the mate over a stranger. Prairie voles that demonstrate pair-bonding reliably choose to sit next to their mates rather than other voles, and they continue to show preference for their mates even after weeks of separation. Not surprisingly, female montane voles are ambivalent about mate preference. Additionally, after mating, male prairie voles become highly aggressive toward other males, to serve the sole purpose of mate guarding. Male montane voles do not demonstrate an initiation of aggression after mating.

Various tests indicate that oxytocin and vasopressin have been adapted for different roles in male and female prairie voles. It is not difficult to generalize from the social pair-bonding seen in prairie voles to the more highly evolved social attachment seen in humans. For example, once in a monogamous relationship, most human couples prefer to spend intimate time with their mate of choice rather than with a stranger (although promiscuity varies from human to human). It is also not uncommon to witness a husband or boyfriend retaliating against an unwanted intruder if that intruder poses a threat to his relationship. Clearly, humans form enduring, selective bonds, but the role of oxytocin and vasopressin in this process remains largely unexplored.

Emil Coccaro and colleagues in Philadelphia reported recently that there was a correlation between the levels of vasopressin and life-long aggression in personality-disordered patients. C. F. Ferris and colleagues at the University of Massachusetts Medical Center have reported that there seems to be a reciprocal relationship between vasopressin and serotonin, and that serotonin may inhibit vasopressin-induced anger by stopping its release in the hypothalamus. Thus in experiments on prairie voles and hamsters we use serotonin-enhancing drugs to treat aggression, and the level of vasopressin goes sky-high in prairie vole and hamster fathers. This may have a direct effect on paternal protective and aggressive responses. Given the lack of any real data, it is fun yet folly to speculate that the increase in vasopressin in the father at labor and delivery—and the rise of vasopressin in a bonded relationship—make for a more aggressive father. And since this acts a bit in the amygdala, one could go on to speculate that domestic violence and paternal protectionism seen in aggressive workaholics and the like are fed by the increased amount of vasopressin.

Oxytocin and vasopressin influence several other forms of social attachment, including parental behavior. When administered to rats, oxytocin facilitates the onset of maternal behavior, and blocking its transmission results in a significant inhibition of maternal behavior. As in the montane voles, the oxytocin receptors increase in two key limbic brain regions at the onset of maternal behavior in rats. The changes in receptor binding can be rapid and profound; some studies have demonstrated a 300 percent increase in hypothalamic binding within 72 hours of birth.

In humans, oxytocin is secreted into the bloodstream during parturition and nursing. Researchers suggest that females without this dramatic increase in oxytocin neurotransmission may exhibit normal labor but fail to experience the motivational changes essential for maternal care. Clearly, however, human maternal behavior does not commence at parturition and does not require either labor or nursing.

It is important to note that oxytocin is only one link in a complex chain of neurochemical reactions necessary for maternal behavior. Steven Thomas and Richard Palmiter of the University of Washington have suggested that mothers who feel no attachment to their newborns may lack a neural pathway that is normally triggered by the act of giving birth. From female mice they removed a gene labeled Dbh, which produces the enzyme dopamine beta-hydroxylase, the precursor for norepinephrine. Interestingly, the survival rate for the offspring of the knockout moms was only 35 percent, but it increased to 85 percent when the pups were given to foster mothers with normal Dbh genes. Also, when the knockout mice were given injections of norepinephrine during birth, their nurturing skills increased and about 75 percent of their pups survived.

What about paternal instincts? When injected into the lateral septum of male prairie voles, vasopressin has increased the time males spend with their pups. Additionally, when vasopressin antagonists were injected into the same region, decreased paternal care was observed. These results are supported in related studies by Geert DeVries and colleagues at the University of Massachusetts at Amherst. DeVries demonstrated that testosterone appears to regulate vasopressin synthesis and paternal behavior in male voles. He found that testosterone had the dual capacity to make male prairie voles aggressive toward outsiders, but friendly and nurturing to their pups.

SOCIAL SUCCESS

CLEARLY, BIOCHEMICALS IN THE BRAIN influence the ability to engage in sex, love, bonding, and child-rearing—all fundamental social behaviors. Anthropologists have found that romantic love is universal among all cultures, which provides strong evidence that it is biologically based. From an evolutionary perspective, feelings of attraction and love may have evolved as a way for two adults to bond, conceive, and provide the long-term care that a helpless human infant needs. In the ancestral environment, two adult humans were necessary to protect an infant from predators and provide enough food. Today, we know that two adults together provide a better and more stable learning and growing environment for the developing infant brain. Since evolution has found it most fit for humans to be in long-term relationships, the human brain needs and wants to interact with other people. It longs to fall in love and works at maintaining a primary relationship.

The clear influence of the brain over our ability to sustain love relationships is one strong indication of the power the brain has to influence our social being. The capacity to get along with another person requires a strong social brain as much as any other intentional or psychological action. Constructive social behavior requires a healthy social brain—an activation of neurons every bit as intricate as the mechanisms controlling language, movement, or emotion.

For as much as individuals must be able to fight or flee, they need sociability. It is necessary for human survival. Children who are not held or given love when young may grow up disturbed, scared, or dangerous. Adults who isolate themselves from the world are more likely to die at comparatively young ages. We have a central dependence on others. We are designed for group living. If we can begin to understand how the brain affects social functioning, we will have even more success as social creatures in the future.

9

THE FOUR THEATERS

THE DAY AN INFANT IS CONCEIVED it begins to perceive the external world, and also becomes aware of its own internal states, such as hunger, tiredness, and discomfort. As the baby is born and develops, it begins to pay attention to light, then voices, and more and more stimuli, external and internal. As children grow and learn they begin to acquire a conscious understanding of what they are perceiving. Along the way, the two hemispheres of their brains, their male or female characteristics, and their major brain functions—movement, memory, language, emotion, and social ability—mature. Ultimately, every child becomes an individual with a unique identity, a special set of behaviors, and his or her own problems—mostly mild, but in some cases serious enough to qualify as mental disorders.

Despite all this activity going on in the brain, the treatment of mental disorders—extreme and mild—has centered upon the lone issue of "affect": a person's emotional state. Since the earliest days of psychiatry, every diagnostician has inevitably asked some version of the question "How do you feel?" Feelings are what hurt the patient, and the therapist is drawn to them, wanting to fix the hurt. But feelings are not

This chapter represents thoughts developed over many years and contributed by Robert Kirkpatrick and Jeffrey Sutton.

the cause of the problems but the result, the outcome of an enormous amount of brain activity, including perception, attention, consciousness, and the brain functions.

A person's emotional state is an important therapeutic consideration, of course, because that is the means by which the brain communicates to itself qualitative aspects of experience. But as we have seen in earlier chapters, the many advances in neurology indicate that there is almost always a biological root, or at least a biological contribution, to every "mental" disorder. Depression, rage, or anxiety may be the symptom that drives a patient to seek treatment, but the cause of the suffering is often a neurological deficit, which cannot be reached, much less changed, by conventional insight-oriented psychology or psychoactive drugs.

If a clinician asks a patient, "How do you feel?" he is only working at the surface. Instead, the primary diagnostic question is "How do you perceive and comprehend the world?" or, more generally, "How do you know?" In this approach, the clinician investigates, together with the patient, how the latter perceives the world, relates to the perceptions, becomes conscious of them and understands them, and how he or she brings the many brain functions to bear on this understanding. The clinician acts as a co-explorer of the physiological factors that might be shaping psychological aspects of the patient's experience.

This view leads to a radically different but simple model for analyzing human experience, which delves far below the emotional surface of feelings while recognizing that emotion conditions the entire process. The model consists of four "theaters" of exploration, which flow and feed back into one another: perception captures incoming stimuli; attention, consciousness, and cognition filter and process these perceptions; the brain functions then work with this information and affect how subsequent information is perceived and processed; the final result is behavior and, ultimately, identity. By logically investigating each theater, clinicians—and people themselves—can find the fundamental cause of difficulties and design lasting cures.

WHY A NEW PERSPECTIVE?

THE TRADITIONAL EFFORT to uncover the hidden trauma supposedly responsible for suffering is largely being replaced today by searches for neurotransmitter imbalances, aberrant genes, and altered brain functions. Where we once spoke of superego, ego, and id, we now speak of serotonin, gene sequences, and neural networks in various brain regions. Nevertheless, mental health practitioners continue trying to treat affect directly, as if it were the illness itself, rather than attempting to investigate the ways in which it might be a consequence of a patient's underlying disorders. Furthermore, they continue to insist on a Pasteurian notion of illness: one pathogen, one antigen, one cure.

The pressing desire to provide immediate relief also leaves many doctors and patients addicted to the search for a convenient, comforting diagnostic label and a hot new drug to cure the affliction. The hunt for a single villainous gene for each behavioral problem is just as intoxicating, and the media have fostered unwarranted hopes among the general public, which now plainly expects miracle cures for suffering. This trend is unfortunate and distressing.

The true measure of a drug's effect is not the degree to which the target symptoms are measurably relieved, but the degree to which the patient taking the drug feels that it has improved his or her life. In an era when Prozac is taken nearly as often as aspirin, clinicians find it tempting to attack any emotional or behavioral problem with a prescription. Too often symptoms are relieved but not cured, and the relief comes at the expense of other dimensions of the patient's life.

Sedating the overactive brain with medication may calm a patient's fears and worries, but it often impairs cognitive ability. Stimulants may lengthen a person's attention span but compromise creativity. Lithium can stabilize mood swings, but it cripples short-term memory. Neuroleptics can dampen the vivid and terrifying hallucinations of schizophrenic psychoses, yet they often erode motivation and narrow cognitive capacities. Although psychiatry has finally achieved recognition as a medical science, drug therapy is still a crude and primitive tool for treating an organ we are only beginning to understand. Which faculties and sources of satisfaction should a psychiatric patient expect to surrender for the sake of therapeutic convenience?

Drug therapy is not a process that uncovers the true, basic cause of the problem and allows a patient to participate in his or her own treatment. Virtually all aspects of personality, emotion, and cognition vary continually along a spectrum between extremes. An individual can exhibit any combination of mild ADHD, OCD, mania, paranoia, depression, autism, or other "pathological" traits to some degree, yet still lead a normal, productive life. Each trait may actually be an adaptive behavioral mechanism to compensate for a neurological deficit.

Almost 50 percent of the U.S. population, at least once in their lives, experience some form of what is traditionally classified as mental illness. But only when one particular neurological and behavioral combination comes to dominate a person's ability to live a productive life do clinicians assign the diagnostic label of a "disorder." That's because the clinician can't find evidence of the milder forms of difficulty—the shadow syndromes—without the patient helping to analyze subtle problems in daily life that would open the door and cause them, together, to say, "Aha!" Although it is vital for any practitioner to recognize the patterns into which human behavior tends to fall, the science of diagnosis and healing is still very much a personal art.

Every brain is a unique expression of the interaction of a range of neurological dynamics, and we need a new, multifaceted paradigm to define mental disorders more effectively. The task in assessing any individual is to determine which factors, both neurological and psychological, contribute directly to his or her pathology. The clinician must attempt to find out how patients experience the world in a very real, physical sense.

HOW DO YOU KNOW?

BY TRACKING EXPERIENCE from perception to identity, we achieve two valuable objectives. First, patients become active participants in the search for a biological source of their misperceptions. Second, the approach frees them from a great deal of shame and self-blame, as they come to see the neurological origins of their challenges. Much of the shame, guilt, and self-loathing that patients—and all of us—experience stems from a belief, inadvertently echoed by psychoanalysis, that the failures caused by a person's deficits stem from problems of morality. Nonsense!

Be clear that I am not attempting to add to the growing list of forgiveness syndromes; right action is still the ultimate goal. If, however, from the first day of therapy, the clinician explores the quirks and intricacies of how patients process their experiences, much of the criticism that patients unknowingly direct inward might instead be transformed into constructive, inspiring motives for self-awareness, personal responsibility, and change. The simple act of pinpointing a physical disorder is often sufficient to provoke a remarkable transformation in a patient's life. Feelings may be the cause of distress, but they also cover up the problem. Furthermore, in light of the paucity of our understanding of the brain, it seems more productive for clinicians to replace confidence in diagnostic categories with curiosity and a knowing humanity.

Be clear, too, that I am not dismissing psychology. Psychological counseling is still needed in many cases, but only after the biological cause and effect has been considered, if not altered. Even if emotions are largely the cause of a problem, those emotions are created by the physical firing of neurons in the brain. Furthermore, even if an emotional trauma is the root cause of a problem, that trauma reorganizes the brain's circuits—a biological result that can be turned around.

Similarly, we must remember that the brain's biology is influenced by a person's environment. The external influences a person is exposed to—cuddling or abuse, talking or silence, mentoring or scolding, support or ridicule—alter the brain's network of neural connections, which are otherwise guided by genes. Indeed, a recent analysis of more than 200 studies of intelligence concluded that nurture edges out nature in determining even a person's IQ. Environmental factors, especially during early development, have a greater impact on IQ than genes do.

By attempting to elucidate how a patient physically apprehends the world, the clinician may achieve a more accurate, empathetic, and effective understanding of the patient's experience. A systematic way of looking at the brain's adaptation to experience permits the therapist to forge a more immediate, guilt-free bond with patients, most of whom have never considered the possibility that the means by which experience comes to them might be the cause of their problems. When patients become investigators, they are able to participate in a way that greatly enriches and empowers the therapeutic process. The main advantage of the theaters approach is that it offers the therapist a heretofore hidden source of explanatory power.

THE FOUR THEATERS OF THE BRAIN

THE THEATER METAPHOR IS regaining popularity, as it suggests the unification of many small parts and pieces that acting alone are virtually nothing but acting in concert give form and shape and creation to something. Consider memory, for example. The focus of memory research these days is on the storage process—how a memory is constructed and how synapses change, as if that were a separate process from how a memory is retrieved. It isn't. Memory storage and retrieval are integrated. Memory consists of both processes. Memory is the retrieval action as well as the physical neurons that hold the memory's puzzle pieces. As Endel Tulving suggests, the wind is both the energy and the air molecules that are moved along.

The theater metaphor is useful because evidence indicates that consciousness creates access to many knowledge areas in the brain. It is the stage for it all to happen. It is about the structure and function of perception or memory or language together, and the call for it to function and to act.

The four theaters lie along a neurophysiological river of the mind, with each theater further downstream from immediate experience than the one before it. Sensory information enters the first theater, perception, and flows through attention, consciousness, and cognition; the information then flows through the brain functions, such as language or social ability, and into the fourth theater: who the perceiver has become. This river of the mind can be envisioned as having a downstream flow from our constantly changing immediate experiences to the slow-to-change domain of personality and behavior.

This is not a simple linear process. There is also an upstream flow. Any factor affecting the third theater of brain functionality, for example, can flow back up to affect consciousness and perception, as well as down to affect behavior. As the brain is an interconnected network, a complex system, an influence on any part of it affects all the other parts, so any clinical treatment of any theater will have an effect on the entire stream.

Consider an actual river. The mighty Hudson River is a mile wide and a hundred feet deep as it flows along the western shore of Manhattan Island. But it begins as a small stream in the Adirondack Mountains, two hundred miles north. This initial stream, perception, flows south toward Albany, where it combines with the Mohawk River and becomes a larger stream, consciousness. The river continues down into the broad Hudson Valley, where local tributaries feed it, creating a still larger watercourse, functionality. As this now major river reaches the northern tip of Manhattan, a salt-water estuary flows into it, giving it its fourth and final character, and its ultimate identity, as it passes midtown and Wall Street and empties into New York Bay.

As in the brain, there are feedback loops. If the weather is hot in the New York City area, it will warm the river water there, and the heat will flow up north. The salt water from the estuary is moving back that way, too. Analogously, if changing conditions affect one theater, that in turn will affect the others; if weeds and algae grow in the slow-moving portions of the midriver in the Hudson Valley, they will subsequently affect the oxygen content of the water upstream and down.

Understanding this model makes clear the folly of finding a problem in one theater and trying to fix it by addressing just that theater. If fish are dying in the Hudson Valley section of the river, scientists looking there might well find pollution in the riverbed. Public works departments might then spend millions of dollars scraping mud from the river bottom, only to find later that the fish continue to die. Why? Because the source of the pollution is far upstream, released in small quantities by small industries, accumulating as the waters flow.

It bears repeating that we experience life as a seamless entity, not in stages. The river analogy of the four theaters is a conceptual framework that helps us understand and treat the biological base of psychopathology. To that end, though, it is extremely useful in determining the nature of an individual's behavioral and emotional responses, to life in general, and from moment to moment.

My hope is that with the metaphorical framework of the four theaters we can begin to base modern mental health care on a more holistic method of healing. By starting upstream and working their way down, psychiatrists stand a better chance of identifying the true etiologies of the disorders they seek to treat and of engineering the environment in such a way that patients' internal shortcomings are met with self-forgiveness and effective steps to correct their problems. Let's first look at each of the four theaters and then consider how they work in concert as a system.

The First Theater—Perception

THE BRAIN DOES NOT mechanically store the information that it acquires. It is changed forever each and every time it interacts with the world. Each time, it becomes the information. Perception is the gateway through which we receive information from our five senses and from our internal awareness. Perception is the beginning of all experience. Problems in this theater often begin early in physical development, but the trouble they cause commonly manifests itself much later and much further downstream. A deficit in perception has the potential to radically change the course of cognitive, emotional, and social development, just as a marked change in the stream up north in the Adirondacks would significantly affect the wider river further south. Abnormalities in the perceptual apparatus are particularly likely to impair a substantial range of neurological functions.

A perception problem lies so far upstream from the major foci of our lives, such as emotional distress, social difficulties, or destructive behaviors, that it is often difficult for the individual or the clinician to identify, or even imagine, a perception problem as being the root cause of a major life disturbance. It is even more difficult because there is no objective standard of "normal" perception. A lifetime of misperceptions leaves many patients ashamed of their limitations and fearful of repeating failures. Perception is the starting point for diagnosis, because mental life develops primarily in response to the information that the brain apprehends.

The Second Theater—Attention, Consciousness, and Cognition

WHEN WE ATTEND to a perception, we become conscious of it, and then we think about it or react to it. The second theater encompasses a person's conscious experience of the world. It is how a person represents the world to himself or herself, from moment to moment, and how he or she interprets events within it. Upstream problems in perception gradually result in cognitive deficits, as the brain tries to learn from and adapt to a lifetime of faulty information. Consciousness is subject to its own impairment, too, particularly difficulties with attention and working memory.

The quality of one's inner awareness can deteriorate as poorly formed perceptions fail to provide the structural basis for well-coordinated attention shifts, and as ill-formed cognitive networks lead to confused internal representations of the world. This leaves the brain trapped in a state of constant "noise" and starved for accurate information. Exploring a person's conscious awareness is the most direct means to gain insight into his or her experience of the world. Understanding a person's reasoning abilities, social intelligence, and cognitive survival tactics is essential to treating any brain and mind problem.

The Third Theater—Brain Function

THE THIRD THEATER comprises the primary functions of the brain, movement, memory, emotion, language, and the social brain, all of them affected by differences between the brain's hemispheres. These processes directly influence the moment-by-moment experience of life, yet adapt relatively slowly. The brain functions both contribute to conscious experience and are gradually molded by it. The brain's networks modify their connectivity as external and internal environments feed back to them, altering long-term memory, arousal, feelings, reactivity, rhythm, timing, motivation, emotional liability, and a host of other traits as we age through life.

Brain functions lie downstream from the first two theaters in our model because they emerge and develop in response to whatever kinds of experiences enter consciousness from perception. A second theater filled with mental noise can distort functions in the third in numerous ways, including ADHD, OCD, autism, anxiety disorders, and others.

Clinicians are just now learning that this downward flow from the second theater to the third could be particularly relevant to the treatment of substance abuse. They are finding, for example, that many people who overuse alcohol and marijuana do so because the drug quiets the noise in their second theaters. For others, cocaine or nicotine improves attention and short-term memory. In both cases, the drug suppresses the noise that otherwise flows into the third theater from the second, enabling such people to focus better. This temporarily improves their brain functions, particularly language and memory. The obvious problem, however, is that the drug also interferes with the normal communication among neurons, leading to physical and behavioral problems each time it is used, and to long-term destruction of the brain overall, both of which wipe out any temporary gain from the isolated improvement in noise levels.

Like the other theaters, the third can create its own challenges, among them problems with language, sense of timing, startle response, levels of arousal, and physical and social graces. Many of these are far too complex for today's pharmacological and psychological treatments. For example, a developmental deficit in motor coordination in the cerebellum can impair attention, mood, reasoning, language, and even the ability to approach people in social situations; we currently have no drugs or talk therapies that can correct the problem directly. Similarly, various abnormalities in the language centers of the brain may alter not only how people experience emotional states but how well they communicate those states to spouses, supervisors, or therapists.

Problems in the third theater can be even more confounding when they are compounded by faulty information flowing in from the first two theaters. Patients suffering from anosognosia are unaware of, and even vehemently deny, symptoms of paralysis, blindness, or other gross impairments. It is difficult to imagine that anyone could fail to perceive and so plainly deny such obvious facts, yet the condition is very real.

There are also milder forms of this problem. While dyslexics are often quite aware that their reading abilities are impaired, "social dyslexics" may blunder through life with no clue that they are lacking interpersonal skills. Psychiatrists miss the clues just as often, because the source is frequently faulty visual perception or a lack of social rhythm owing to poor motor coordination. Either these deficits are not addressed in treatment, or the person ends up being "diagnosed" as having some unfixable defect of character.

The Fourth Theater—Identity and Behavior

THE FOURTH THEATER constitutes the "output" of the brain: one's decisions, behavior, and historical sense of self. It is the sum total of neurological and psychological traits that, at any given moment, constitute who a person has become. This theater is, in a sense, the space occupied by the life narrative that individuals tell themselves—and their care-givers. It is also where modern psychiatry has spent most of its time, for it is where we confront early traumas, lowered self-esteem, fantasies, phobias, behavioral troubles, broken marriages, character disorders, and personality. A patient's quest for accurate self-knowledge begins in the fourth theater. The obvious problem, of course, is that a life of long-compromising influences flowing in from upstream alters one's own self-observation, self-esteem, sense of self, and memory.

Problems in the fourth theater are the ones most readily apparent to ourselves and others, so not only are they the ones most likely to motivate people to seek treatment, they are often the sole focus of investigation and treatment. Many clinicians never even look at or through the other theaters of the brain. Personality is not a cause of problems; it is rather the expression of good and bad influences from the earlier theaters. Although inquiring why a patient has sought help is certainly the first step in diagnosis, characteristics in the fourth theater are most often symptoms of underlying pathology. When a patient asks, "What is wrong with me?" it is essential that the clinician suspend the desire to seize upon a diagnosis based on behavior. The patient wants to be rid of the behavior or the feeling state of the disease. It is painful. The clinician wants to help relieve the pain. So both lock onto fixing the behavior and feelings, and become trapped in a region of vague speculation, while the river of upstream influences keeps flooding in, unexamined and unaltered.

The last theater is where the various biological processes of the brain fuse with life's experiences. The resulting persona is complex. But improving patients' abilities to understand what is going on around them, and how their lives may be compromised by more straightforward difficulties in the first three theaters, can go a long way, and this practice must be brought to the forefront of psychiatry. We have devoted so much effort to deciphering the relationships between trauma and misery and genes and behavior that we have overlooked the mediating processes of comprehension. A patient's understanding of the world, or lack thereof, can make all the difference between a life of success and a life of failure, between meaningful existence and despair.

WORKING FROM THE FIRST THEATER—THERESA

THE POWER OF WORKING from the first theater down through the last can be seen clearly in the case of a woman patient of mine, Theresa. Examining her situation also helps clarify how the four theaters flow and feed back, and how a small perturbation in one theater can build to a major disturbance in another.

Theresa grew up with parents who placed great emphasis on athletics and education. As girls, she and her sister Pauline, who is a year younger, would head to the backyard with their father to play catch. Both girls threw well, but Theresa just could not seem to catch the ball. She also had a hard time keeping up with her classmates in reading. The parents made sure the girls had eye exams every year, starting early in life, and both girls were always told that their vision was normal.

Normal—for stationary objects. Theresa had 20-20 vision when looking at letters on a wall chart, but she and I would determine much later, when she was an adult, that she could not see fast-moving objects. She lost sight of them. She had a problem with the fast-processor of the vision system in her brain. It was not a focus nor a retina problem; it was a problem with the region of the brain where visual information is processed. This also made it difficult for her to focus on letters and words as her eyes scanned across a line of text on a page. Although the letters were stationary, moving over them from left to right—the act of reading—put them in motion.

Theresa's inability to track objects properly—a perception or first-theater difficulty—caused her attention and cognition system, in the second theater, to be off. She could not attend to the moving ball, and so could not catch it. The same was true when she walked into a classroom full of children; people and objects were moving all about and she couldn't track them. As a result, she couldn't pay attention to any of them, so she simply stood there and didn't join in. Her difficulty in reading made even the simplest classwork a struggle. Emotion began to color all these experiences. Her father put pressure on her to "try harder" to catch the ball. Her classmates chided her because she stood on the sidelines, too slow to pick up on the fast action of the playground. Her teachers admonished her because she fell behind in reading.

By the time Theresa was a teenager she spent most of her time alone. Walking into a new situation caused her to panic. She skipped school parties because they overwhelmed her. She felt awkward almost all of the time. Years of physical and social inactivity, during which her motor function as well as her emotional and social brain—all in the third theater—were underused, left them underdeveloped.

A decade later Theresa came to me for therapy. As a personality she was tame. She was shy, retiring, and didn't say a thing. Initially I had to drag every word out of her. But after a period of investigation and working backward, we pieced together how she had become that way. We started in the fourth theater: identity and behavior. She was a panicky person. She missed so much of what other people saw—and knew that she did—that she was always trying to reanalyze and reconfigure the situation, checking on things over and over, afraid she had missed something because she hadn't seen it all.

On the surface, Theresa behaved like a "nervous Nelly." She had all the characteristics: she seemed fearful and she avoided new situations. Other people even called her that, acting on her emotional system and convincing her that she indeed was a chronic worrier. But then she told me something that lit a light bulb over my head. When she was eighteen years old she had inherited some money and decided to take a trip around the world—by herself! If she really were such a nervous Nelly, she would never have done that. It simply didn't match with a phobic personality. Although Theresa seemed to be a fearful person, in her own mind she was not afraid. Further, she didn't avoid new situations for lack of interest. She was, in fact, extremely interested in new things, but shunned them because she felt out of control in many environments.

Another behavior also provided a clue. Driving, for Theresa, was a nightmare. Even as an adult her eyesight was fine, so she had no reason to suspect that there was something wrong with her visual perception. But there was: tracking moving cars requires fast perception, and Theresa had difficulty perceiving moving objects.

Starting from the first theater, perception, we began to explore and explain Theresa's troubles from childhood on in a new light. A perception problem, never diagnosed because it was subtle, is what caused her to be uncoordinated and have trouble reading. These second-theater problems, in turn, led to her inability to engage in sports, learn, or navigate social situations. She wasn't lazy or a slow learner or unmotivated, but she was labeled that way, and suffered from constant feelings of failure and rejection, which altered her emotional and social brain functions in the third theater. It was these altered functions that, ultimately, caused Theresa to become shy, reserved, and withdrawn—the behavioral problem of the fourth theater.

By the time Theresa came to me she felt really bad about herself. She felt she couldn't accomplish anything or manage any situation. Her self-image, her identity—also the fourth theater—was terrible. As an adult, Theresa's sister Pauline was outgoing and popular, a real doer. She was close to Theresa in age, had had the same upbringing, and had gone to the same schools. The only difference was that she didn't have Theresa's perception problem. These two women had developed into completely opposite personalities, partly because of a slight perception problem way back in the first theater.

Slowly, Theresa began to realize she wasn't all the things she had been labeled for so many years. Understanding that she had a perception problem made all the difference. She wasn't uncoordinated; she literally couldn't see the ball. She wasn't stupid; she couldn't focus on the letters. She wasn't uninterested in social gatherings; she had trouble processing all the stimuli in the room. She wasn't panicky; she was just constantly rechecking the environment. She wasn't a dull, ponderous person; she was just slow to react.

Theresa came to realize that, in truth, she had a billion thoughts going on, new ones, different ones, interesting ones. She realized that she was smart after all. She also realized that she was motivated to learn new things and that she was an adventurer. She wanted to be outgoing, and now she knew she could be. What she would have to do is find new ways to handle situations that require fast-processing.

A PROTOTYPE OF INTERDEPENDENCE—CLAUDIA

AS INFORMATION FLOWS TO and fro in the brain, the imagination strives to explain it. This lays the categorical and emotional groundwork for how future information will be perceived, and produces permanent alterations in the structure of the brain. No perception enters the brain without being interpreted to some degree, and every experience occurs in the context of all four theaters. As information enters the second theater, it trains and restructures the cognitive networks, for better or worse. Cognition, in turn, reaches back upstream to influence perception, for the predictive mechanisms of the brain play a huge role in determining what we attend to and, thus, what we perceive.

The third theater, brain function, is the product of a whole range of factors, yet it also exerts an upstream influence. For example, arousal directs the tone of the second theater, fueling its attentional and cognitive states. Arousal and the startle response assign an emotional value to every perceived stimulus, even before it enters conscious awareness.

The fourth theater is the enduring aspect of the whole system, the continual reevaluation of the historical self, which affects the three other theaters. As treatment moves into this theater, concrete neurological descriptions of control, activation, and cognitive function are replaced by more abstract notions of will, morality, and personal philosophy. Behavior, personality, and way of life all affect the kinds of perceptions, mental challenges, and stresses a person experiences in daily life. No one theater in a living brain has an existence independent of the other three.

We can see how this interdependent system works by looking at Claudia, who was finally diagnosed with ADD when she was a fifty-two-year-old grandmother, following the diagnosis of four of her seven grandchildren. Claudia is a salt-of-the-earth type, working as an interpreter for the deaf and ready to help anyone at a moment's notice. She has raised four now-successful children. Yet she claimed she saw life "as a blur." She felt like she was "living in pea soup," moving through her day against a constant pressure that pushed back at her. Life, she said, was like a B-movie, with odd camera shots, bad lighting, and a discordant sound track, particularly when she wasn't working. Claudia's secret dream was to join the Peace Corps and get an assignment on Cook's Island. Known for its poverty and profound human misery, it is reputed to be the most difficult mission the Corps has to offer. She knew that, once there, she would be needed twenty-four hours a day.

Claudia loved her work with the deaf, and with her kids grown, felt she could work sixty hours or more a week. Apparently, the more stress she was under, the clearer her world became; the blur, the noise disappeared, and she became focused and highly productive. Yet, after twenty-two years of signing for the deaf, she had never managed to pass her certification exam. In the course of the testing she discovered that she, herself, had an auditory processing problem, although it had remained undiagnosed. This had made her job all the more difficult all those years. She also came to find that she had poor fine-motor coordination, a common symptom of ADD and dyslexia, and thus was even further challenged in the act of translating what she heard into Sign. Yet it was the very difficulty and complexity of the task that made her feel most alive and cleared her mind of noise. She once told me, "You know, you talk about people with ADD having to live life 'on the edge,' but I have to be standing on the blade!"

For Claudia, an impaired attention span made life seem somehow remote and indistinct, a blurred state of consciousness in the second theater. Her faulty auditory apparatus—a misperception of reality in the first theater—compounded the problem, for her consciousness in the second theater was trying to predict events based on misinformation. Downstream, Claudia's brain tried to compensate by relying progressively more on intuition, which, though not a negative development per se, compounded her impulsiveness.

Still, Claudia was able to hyperfocus to a degree most non-ADDers rarely experience. When given a task of sufficient complexity and challenge, such as signing, despite her auditory and fine-motor deficiencies, all distractions vanished from her ADD consciousness. Her mind laser locked on to the task at hand, blocking out the noise, giving her a calm intensity that was so gratifying that she was driven to seek greater and greater stimulation, such as sixty hours of work a week and the assignment on Cook's Island.

There is, however, an extremely positive aspect to Claudia's feedback–feed-forward phenomenon, which suggests that some component of her psyche can be a creative adaptation to the problem. A constantly shifting attentional system produces a hyperawareness, which, while it causes individuals to miss details and fail to follow through on tasks, often renders them brilliant at imaginatively reconstructing entire gestalts from partial perceptions and ambiguous information. People like Claudia are often highly creative and empathic, and have a knack for making unobvious but keen associations. The flip-side of their tendency to go down the wrong streets is a constant striving to find new and different ways to do things. It is this positive aspect of the brain that contributes further to the success of the theaters approach, for while these patients are typically poor self-observers on the social level, they are drawn to complexity, and the exploration of their own neurological peculiarities presents an irresistible challenge—a true inner force for a cure. There is a plus-minus phenomenon going on with Claudia, as there is for us all. The downside is a disorder, or simply a weakness, and the upside is a special talent, or strength.

BETTER DIAGNOSES AND TREATMENT

THE EXPERIENCES OF Theresa and Claudia demonstrate that the theaters model offers a powerful new way for individuals and their clinicians to assess and alter behavior. But it is important to point out that the metaphor of the river provides only a partial handle on reality and fails to convey fully the nonlinear plasticity of the brain in response to positive or negative feedback. In the real world, each stimulus changes the way the brain interprets subsequent stimuli. Prediction, desire, and imaginative interpretation all play a tremendous role in determining how the brain will respond to a given situation.

Similarly, the notion of four theaters is necessarily a simplification. It provides for a diagnosis that integrates neurological, psychological, and behavioral domains. This integration is absolutely necessary if we are ever to devise a truly holistic approach to treating the brain: perceptual dysfunctions may be diagnosed through neurological testing and cognitive disorders through educational testing. Deficits in basic brain function must be diagnosed neurologically, while affective disorders must be addressed through behavioral and psychological consideration of the patient's history. Perhaps one day there will be practical tests for deficits in social skills. Indeed, late in 1998 research was completed that suggests autism might be predicted by careful observation of early movements babies make in their cribs.

These tasks become less daunting for the clinician if he or she pulls the patient into the hunt. Although the workings of one's own brain are generally quite transparent to self-observation, the appropriate questions open huge insights and multiply the powers of observation. The four theaters provide the clinician and the patient with a convenient tool that may be extremely useful in determining what, exactly, has contributed to and is shaping the patient's subjective experience.

Modern medical practice tends to regard patients' self-evaluations as too tainted by subjectivity, but this is a grave error. The attitude has arisen partly in response to the rise of "pop psychology," the culturally learned, popular explanations of mind and brain problems that are far too shallow and generic to be applied to any individual. However, if patients are told what to look for and how to look, they may become aware of parts of their subjective experience that are inaccessible to a therapist. Such revelations may well provide the key to devising a unique combination of interventions.

As happened with Theresa, once patients learn to see themselves in terms of a more neurologically based paradigm, they are able to free themselves from the self-imposed stigma of moral failure. The freedom is liberating! It clears up many kinds of self-deception and brings unrecognized symptoms, as well as positive traits, out into the light of day. When the patient has nothing to defend against, the defensive behavior often simply disappears. Shame and doubt are abandoned in pursuit of a gentler, more forgiving kind of insight . . . and a more positive self-image.

The brain is a multitude of processes that may be in need of redirection. The theaters model offers a broad-based yet practical approach, as even minor changes in environment, basic brain functions, and self-knowledge can have profound developmental consequences. It is an unfortunate fact of modern mental health care that a great many people afflicted with first- and second-theater disorders end up in psychotherapies that limit themselves to the fourth theater of behavior and identity. It is quite beyond the average patient's ability, within the framework of these insight-oriented therapies, to pinpoint the true source of unhappiness and frustration, which leaves him nowhere, except in all-encompassing depression at finding no visible source for his pain.

The pharmacological approach to helping these patients is just as ineffective. Prozac is hardly a remedy for the self-blame, lost opportunities, and intellectual insecurity of a lifetime compromised by unrecognized perceptual and cognitive deficits.

The limitations of current therapies are particularly evident in the treatment modes for children and adolescents, who are in stages of life in which there are enormous biological, psychological, and social changes. Both drug and talk therapy are usually unrewarding, even in cases of juvenile depression. Young bodies are too sensitive to the chemicals, and patients of that age are cognitively unable to fully and accurately articulate their emotional experiences. If taught to examine the way they experience the world, however, children are quite capable of observing their deficits, learning to cope with them, and, most important, forgiving themselves for their shortcomings before there is damage to their self-esteem. Without knowledge of the biological bases of such disorders, people often feel a sense of shame at their behavior, which can, in time, ripen into self-hate.

Out of a well-meaning desire to relieve patients' suffering, we in psychiatry are still encouraged by the germ theory of illness: one problem, one culprit, one cure. While modern medicines may occasionally have an almost magical effect on a person's life, the time has come to move beyond our rather arrogant clockwork model of the brain, within which a clinician has only to isolate some faulty cognitive gear and repair it. We have to begin to think of the brain as a self-organizing ecosystem, one of such staggering complexity and delicate balance that almost any aspect of a patient's life may be relevant to a diagnosis or essential to treatment. Genetic predispositions and early traumas are significant considerations, yet who a person becomes depends on many other factors as well. Finally, as is true in any highly complex system, a single element has limited meaning if considered in isolation. When assessing a human being, almost everything merits co-examination.

It will be the clinician's duty in the new century to help and to teach patients to explore whatever modifications of neurochemical, behavioral, psychological, and environmental factors might compensate for an illness. This includes devoting more time to looking for what is good in patients' lives, for the strengths and talents that are not yet being fully realized, and for the secret pleasures and sources of happiness that they have never allowed themselves. Many disorders are so emotionally damaging that patients never discover activities at which they might be particularly gifted and from which they might derive significant joy, challenge, and self-satisfaction.

Discovering creativity in one's self can be a highly effective component of treatment. By thinking solely in terms of pathology, as if our Hippocratic duty required only that we restore the patient to some former, imagined state of perfect health, we fail to notice traits that provide not only a path to recovery but a means to progress beyond it. The brain's processes can be utterly transformed by self-discovery and the right pursuits in life.

We in psychiatry continually risk mistaking our labels for the disorders themselves. No disorder ever cleanly fits its diagnostic criteria. We still know very little of the relationship between microscopic and macroscopic events in the brain, let alone how they contribute to psychological phenomena, and our levels of description at this point are determined by the limitations of our measurement techniques. While the four theaters are simply a convenient construct through which to examine the human condition, such a metaphor may provide a valuable pathway by which to proceed.

10

CARE AND FEEDING

IT HAS BECOME OBVIOUS that we can actually change our brains. By altering the external environment of our surroundings or the internal environment of our bodies, we can take better advantage of our strengths and amend our weaknesses. The possibilities for change are bounded only by our imagination, our willingness to assess our brains accurately through self-reflection, and our commitment to do some hard work.

If you've read every chapter in this book, you've learned about dozens of specific actions that might be taken at specific times in people's lives to train their brains—from exposing infants to music and lots of talking to challenging the elderly with crossword puzzles and debates on current events. But many of the fundamental tools for the care and feeding of the brain are everyday matters. Physical and mental exercise, proper nutrition, and adequate sleep will help anyone gain cognitive clarity and emotional stability. Other important practices come from the realms of spirituality, meditation, even the pursuit of one's passions. In all these we take advantage of the brain's great plasticity—our power to reconfigure our brains.

One necessary precursor to change, though, is often a change in attitude. The spectrum of mental disorders, from everyday anxiety to chronic schizophrenia, are not moral failings or the result of bad upbringing or a failed personality. Mental problems, from hot temper to laziness, from chronic worry to excessive drinking, all have roots in the biology of the brain. That doesn't excuse the behavior. But it shifts the focus of responsibility: we and our care-givers have to examine our biological lives as well as our social and emotional lives.

Understanding that biology lies at the root of mental disorders can be extremely liberating. We don't blame ourselves or feel ashamed if we have a bad back or a physical illness, so why should we impose blame and shame on ourselves when the problem is mental? If our back is bad we may alter our external environment by getting a new chair, and strengthen our internal environment by doing back exercises each day. The same view should be extended to mental problems—those we have, and those other people have. Simply lifting the burden of blame and shame can be a major first step toward a cure—for the patients, their families, and their care-givers.

TIPPING THE BALANCE

CRITICS SOMETIMES CLAIM that a focus on "ordinary" measures like exercise and diet is too simplistic to affect unordinary behavior. Not so. The brilliantly simple evidence from exciting new areas of physical and social science—complexity theory and tipping points—shows how powerful such universal factors can be in affecting the brain-body system.

Complexity theory finds meaningful patterns in what looks like chaos. The idea has been applied primarily to large physical systems, such as weather patterns. Each water droplet, each puff of air has an influence, and together the billions of influences compound into a grand, additive result. Edward Lorenz, a scientist at MIT, calls this the butterfly effect: even the tiny flap of a butterfly's wing on one side of the world could potentially build to a hurricane on the other. And yet, even though there are virtually an infinite number of inputs into this complex system, there are fairly predictable outputs: a clear day, clouds, rain, hot or cool temperatures, and an occasional hurricane.

Scientists are applying one aspect of complexity theory to explain human phenomena. They call the small change that can precipitate a large effect a "tipping point." At some critical juncture, an isolated change tips the whole system in a different direction. Take a jar of water, cool it and cool it, and it remains water down to 35 degrees, then 34, then 33. But reduce the temperature by one degree more and suddenly the water turns to ice. The temperature was consistently reduced, but at one point the whole system suddenly changed. This way of analyzing systems is now being used in many fields, from economics to ecology. What causes a safe middle-class neighborhood to slip over the edge and become a dangerous slum? What turns a city district with a high rate of crime and teenage pregnancy into a stable neighborhood? In the latter case, perhaps new citizen activities such as neighborhood watches, the DARE program, and a zero tolerance for crime build upon each other until one day the neighborhood reemerges as safe.

Arguably, the most complex system known to science is the human brain. Given a hundred billion neurons, each with possibly thousands of synaptic connections bathed in multiple chemical transmitters, it is difficult to imagine the sheer volume of interactions that go on. It may be hard to determine what small factor could have been in the wrong place at the wrong time to contribute to a mental health problem—like the tiny pebble kicked off a path that creates an avalanche farther down the mountain.

The beauty of tipping points is that a complex system can just as easily tip up as down. A small and seemingly inconsequential action can lead to successful treatment of a disorder. Perhaps a lonely, depressed young man decides one day to go for a walk. Then he starts walking a few times a week. On one occasion he tries running for a while. Then he begins to run more often. Each time he can go a little farther. He soon notices that he is sleeping and eating better, that he has more energy and looks better in the mirror. Like the movement of the butterfly's wing, that isolated first walk may have started a sustained chain of events that are now building toward improving the man's self-esteem and increasing his energy, affecting his entire brain-body system. One day, feeling physically and mentally stronger, the man finds a new job. He begins to make friends, starts to laugh—and turns his life around.

This man's extraordinary metamorphosis began with an ordinary walk. There are many tools right at our fingertips for changing our mental health, both in correcting our problems and simply in becoming the kind of person we want to be.

PHYSICAL EXERCISE

EVERYONE KNOWS THAT EXERCISE is "good" for you. However, not only is it hard to start, but everyone has different recommendations to follow. Is three times a week for 30 minutes enough? At what intensity? Aerobic or anaerobic? Weight-bearing or not?

The first step in starting an exercise program that will help your body and brain is to do something active—anything. Research shows that just a brisk walk will bring health benefits, from reducing the risk of heart attack, hip fractures, diabetes, and colon cancer to lowering weight and blood pressure. Exercise, especially swimming, has also been found to help make arthritis less painful.

Obviously, exercise improves the body, but some of the most exciting new findings point to the positive effect exercise can have on mental health. First of all, it increases the amount of blood that gets to the brain. And it has been shown that it can augment the number and density of blood vessels in the areas that need them most: the motor cortex and cerebellum. Thus, in addition to its well-known effect of increasing blood vessels and improving circulation in the heart, it seems that exercise or activity has a similar effect on the brain. The more we use it, the more we stress it, the better our circulation is, and the more fit that part of the brain becomes.

Better self-esteem is also a known effect of exercise. Women who regularly participate in sports and exercise have a more positive self-perception than those who do not, according to a 1996 survey done for the Coors Brewing Company. Telephone interviews of 505 women age eighteen and older found that 80 percent of them believed an active fitness regimen heightened their sense of well-being, and 52 percent said it improved their social life.

Different types of exercise seem to have different effects on mental health. There have been over 1,000 studies on the effect of exercise on depression. Short sessions of vigorous aerobic exercise, usually in a program that lasts for several weeks, seem to be the most helpful for mild to moderate depression. A 1994 Harvard study found that men who burned more than 2,500 calories a day in aerobic activity were 28 percent less likely to develop clinical depression than less active men. It is hypothesized that prolonged, strenuous exercise raises the brain levels of endorphins, the naturally occurring opiate-like substances that decrease pain and enhance feelings of comfort and well-being. However, with our advancing understanding, we now see that the three major neurotransmitters—norepinephrine, dopamine, and serotonin—that have preoccupied researchers concerned with mood, cognition, behavior, and personality are all increased by exercise and are strongly implicated in its mood-elevating effects. Dopamine, the key neurotransmitter involved in the feelings of reward, motivation, and attention, is increased both in the intensive phase of exercise and in prolonged exercise (at least in rats). Similarly, serotonin, the neurotransmitter of the '90s made so famous by the Prozac craze, is also elevated by both intensive and regular exercise programs. Thus one could say, using all the poetic license one can muster, that exercise increases the neurotransmitters, which help with mood regulation, anxiety control, and the abilities to handle stress and aggression and to become more attentive and social.

Exercise that involves learning complex movements can impact our brains in other ways. Modern dance, basketball, and the martial arts involve a host of coordinated movements, and practicing them causes more connections to grow between neurons. Exercise that forces us to improve balance and coordination may not only help overcome clumsiness but also reduce shyness and even enhance the ability to make friends; these kinds of exercises strengthen neural networks in the cerebellum, which is the area responsible not only for balance and physical coordination but also for coordinating our social interactions. There is also the psychic benefit; a dance instructor involved in one of the studies observed that as adult students gained control over their bodies, they also gained control over their lives.

New research indicates that these kinds of exercise also affect the basal ganglia and corpus callosum, sharpening memory and increasing the capacity to master new information. While some deterioration of these areas can occur with age, it is not inevitable. Maintaining memory and repairing weakened brain connections depend on stimulation. As most of us approach our fifties, our minds slow down. This generalized slowing affects almost everything we do, from our reaction time to how quickly we learn or recall. Part of the reason is that the body has become less efficient at delivering nutrients to the brain. The antidote: exercise, to get more nutrients up into the brain. Several studies show that older men who have stayed in shape do better on mental tests than those who have not; indeed, they do just as well as men thirty to forty years younger. Aging brains may also decline in function owing to lower levels of dopamine crossing the synapses. Physical exercise elevates these dopamine levels.

MOVEMENT

PHYSICAL EXERCISE IS a special case of a larger activity: movement. As we have seen, motor function is crucial to all the other brain functions. The many connections we are finding between motor and cognitive functions suggest that any sort of physical activity can improve our motor function and therefore our cognition. For example, we often experience anxiety because we have no possible motor schemata to solve a problem, no movement we can make to cope with the perceived situation we face. Outwardly we "freeze" with anxiety, while inwardly we "churn." Without a motor response, there can be a buildup of these physiological changes that can cause additional anxiety. Movement provides the physiological release that we need to bring our bodies back into balance. As human beings, we are meant to be moving.

Stagnation is a problem for many people in today's society. Computers and television have further contributed to our already sedentary lifestyle. Inactivity can hasten the decline of our mental abilities. We don't have to run a marathon to benefit; simply moving is good for our bodies and our brains. When we combine movement with a learning activity such as dance, we get extra benefits because new connections form in the brain as a result of learning a new task. As new routines and dance steps are introduced, the brain produces a greater number of connections between its neurons, which creates a brain that is better able to process more information.

Patients with Parkinson's disease who practice tai chi on a regular basis have seen improvement in their physical and mental symptoms. Coordinated movements as in tai chi and dance increase concentration, will power, and balance, and also enhance the ability to cope with physical and emotional stress. Hatha-yoga, an ancient Indian technique used to develop physiological, psychological, and spiritual potentials, has lately become popular in Western countries as a method for counteracting stress. It consists of active exercises, postures, and regulation of breathing, combined with relaxation and meditation. Hatha-yoga has been found to decrease excitability, aggressiveness, and somatic complaints and to enhance emotional and life satisfaction.

Routine physical activity may also help us solve problems we're stuck on. When we start moving, we activate the brain, which can help us in many undefined ways. For many people, the act of pacing helps them think; for them the old adage is true: "I think better on my feet." Since the health-conscious '90s, more and more people have been walking several miles a day, finding that this motor activity not only makes them feel more fit and invigorated, but allows them to sort out things that are on their minds or solve vexing problems. Many people claim they do their best thinking or come up with their most creative ideas while performing some fundamental physical activity like walking or jogging.

The reason is that the primary motor cortex, basal ganglia, and cerebellum, which coordinate physical movement, also coordinate the movement of thought. Just as they order the physical movements needed for moving, they order the sequence of thoughts needed for thinking. Fundamental motions like walking and running trigger the most deeply ingrained neural firing patterns in these brain regions. It may be that as this happens, it causes the brain to establish fundamental firing patterns among complex thoughts, helping us to find a solution or generate a creative idea.

One offshoot of movement and exercise is massage, or more fundamentally touch. Modern medicine has lost touch with the benefits of touching. However, clinical studies have shown that full-body massage has a positive effect on medical conditions such as hyperactivity and diabetes. Massaging the temporal artery can ease migraines. Massage can help asthmatics breathe easier, boost the immune function in HIV-positive patients, improve autistic children's ability to concentrate, reduce anxiety in depressed adolescents, limit agitation in Alzheimer's patients, and even lessen apprehension in burn victims about to undergo the extremely painful procedure of debridement, in which all of the dead or infected tissue is removed.

It doesn't matter what style of massage is used. Whether it's Swedish massage, reflexology, or shiatsu, it's all rubbing, and that's what matters. It works in part by stimulating nerve pathways that tell the brain to lower the levels of the stress hormones cortisol and epinephrine. It may also affect the brain directly; Tiffany Field at the Touch Research Institute, University of Miami School of Medicine, has reported that tactile stimulation appears to increase the activity of the vagus, one of twelve cranial nerves that influence heart rate, speech, and the absorption of food. Studies show that touch helps babies born prematurely to grow faster, helps ADD kids pay more attention, increases attentiveness and responsivity in autistic children, raises feelings of calmness and well-being in adolescents with ADHD, and reduces anxiety and pain in children with juvenile rheumatoid arthritis.

Touch may even help couples rekindle their sex life. When you were a teenager, the mere touch of a girlfriend's or boyfriend's hand on yours in a movie theater could send you reeling. When couples consult with Mark Schwartz, a psychologist at the Masters and Johnson Institute in St. Louis, on how to improve their sex life, he tells them to go home and caress each other all over, but not on the genitals or breasts. And intercourse is not allowed. As they tease each other in this way, they drive each other wild—just like the teens in the movie theater—and reignite their desire for actual sexual contact.

Touch, movement, exercise—these physical aspects of our lives—are extremely important to our mental health. Our brains originally evolved to direct complex motions, such as are needed in the constant competition for food and mates. Much of what the brain does is still quite similar; we should learn to view all brain functions as descendants of motion. Motion is involved in almost every aspect of human experience: thoughts move from one topic to another, emotions stir us deeply. Language is essentially a complex semantic dance by the mind and the tongue, a sophisticated form of motion that allows us to manipulate the contents of the world without laying a hand on them. To improve our brains we have to move our bodies, take action, get going.

MENTAL EXERCISE

WE SHOULD ALSO purposely exercise our minds. The value of pursuing mental fitness has come up repeatedly in the preceding chapters. Young children who study piano in addition to routine academics end up doing better on standardized tests. The nuns of Mankato, who keep on doing crossword puzzles and debating politics into their eighties and nineties, live far longer and have far fewer and milder cases of dementia such as Alzheimer's than the general public. And even within their own ranks, the nuns who have attained higher levels of formal education and who constantly challenge their minds in old age live longer and resist brain disease better than those who are less well educated and spend less time pursuing cognitive activities.

Mental exercise strengthens and even renews neural connections, keeping the brain flexible and resilient. For years scientists believed that the irreversible loss of neurons was inevitable in old age. However, actual evidence indicates that there is not a great deal of loss. PET scans show that the frontal lobes of a twenty-five-year-old and a seventy-five-year-old glow equally bright after the same memory test. Decline in old age is caused primarily by the lack of mental exercise.

New mental tasks increase neural connections and help the brain become more adaptive to future events, whether these include coping with the death of a family member or mastering a new computer language. You have the best chance of growing connections between your axons and dendrites by tackling activities that are unfamiliar to you. Learn to play the piano, to tap-dance, to repair small engines. Study philosophy, converse with provocative people. Like the nuns of Mankato, if you constantly challenge your brain to learn new things, you may develop more neural connections that help you delay the onset of Alzheimer's disease, recover from a stroke, and live a longer life. And your life will be more interesting. It's never too late to start: studies show that the adult cortex retains its basic plasticity. You can indeed teach an old dog new tricks.

The lesson here is that we have the power to change our brains. The human brain's amazing plasticity enables it to continually rewire and learn, not just through academic study but through experience, thought, action, and emotion. As with our muscles, we can strengthen our neural pathways with exercise. Or we can let them wither. It is wise to remember once again one of the brain's most basic principles: Use it or lose it.

Every time we choose to solve a problem creatively, or think about something in a new way, we reshape the physical connections in our brains. The brain has to be challenged in order to stay fit, just as the muscles, heart, and lungs must be deliberately exercised to become more resilient. Some stress makes us tougher in the face of future adversity. There is even research that shows that exposure to reasonable challenges during childhood alters the balance of brain chemicals so that children are able to respond better to stress later in life. Houston psychologist Emmy Werner found evidence for this when she went looking for trouble in Hawaii. For years she studied the offspring of chronically poor, alcoholic, and abusive parents to understand how failure was passed from one generation to the next. To her surprise, one-third of the children ended up leading more productive lives than their parents. Many social scientists now suggest that while we must continue to study children who fail, there may be much more to learn from children who succeed despite adversity. Such children, researchers find, are not simply born that way. The presence of a variety of positive influences in their lives often makes the difference between a child who fails and one who thrives. The implications are profound; parents, teachers, volunteers, peers, and all those who are in contact with children can create a pathway to resiliency. Werner later studied women who overcame adversity in their adult lives. She found that several factors made the difference: at least one person who gave them unconditional love and acceptance; a sense of faith in themselves; the willingness to seek support; and finally, hope.

For children, cognitive cross-training may be a creative way to harness the brain's plasticity. While music lessons have value of their own, they may also help children develop stronger math and spatial skills. Talking to babies often, and exposing them to foreign languages in their early years, aids their cognitive development. If you are socially or politically inclined, you can fight for arts programs in schools and communities. Budget-cutters often see the arts as "frills," but research shows that these programs are more than just an adjunct course of study; they are an important part of intellectual development.

As noted before in this book, we must nonetheless be careful not to place too much emphasis on a single fix. Several years ago, when research showed that babies might be able to discern spoken language while still in the womb, many pregnant women began talking incessantly to their bellies. They played foreign-language instructional tapes on their stereos and strapped the headphones to their abdomens. Obstetrician René Van de Carr opened Prenatal University in California, for women from their fifth month of pregnancy on. He led mothers in exercises in which they counted and spoke loudly to their fetuses, accompanied by coordinated pats on the abdomen to make the fetus pay attention. The intent was to foster the child's future intellectual growth. When asked on an ABC news show whether he believed that parents who didn't follow the program were somehow letting their fetal babies' brains lie fallow, Van de Carr said, "I wouldn't want anybody to feel guilty. I think they may be missing an advantage."

Perhaps, or perhaps not. While studies do show that some fetuses respond to the sound of their mother's voice, they may not hear much more than a distant, muffled sound, blurred by the aqueous womb, the whooshing of blood through the placenta, and the beating of the mother's and baby's hearts. Some scientists on the ABC program maintained that the expectant mothers at Prenatal U. were wasting their time and money, since no research has shown that a fetus's brain benefits from this kind of treatment. One scientist went further: "The fetus has so much work to do in maturing and getting its brain developed, that it doesn't need this other type of stimulation that may interfere." Another said, "The stimulation might be interfering with the fetus's need for rest. The fetus spends about 90 percent of its time sleeping, when it is growing, recharging its batteries, and conserving its metabolic energy."

The theory of mental exercise can be applied to several of the brain's functions, notably memory. By challenging yourself to remember, you keep synaptic pathways alive. Constantly challenging yourself to learn will also maintain an efficient process of long-term potential in your brain, so that your working memory remains spry in old age. As you age you can use these kinds of mental activities to counteract some of the degradation of the frontal lobes, which are responsible for working memory. Remembering your memories is a wonderfully beneficial cognitive exercise. Perhaps that is why so many older folks regularly recount stories of the past—to keep them alive not just for others, but for themselves.

Memory changes due to advancing age are largely the result of lesser efficiency in processing, not of some pathological breakdown in the brain. Aging is unavoidable; severe memory loss is not. Several factors may slow the onset of memory problems encountered in Alzheimer's disease and other forms of dementia. The intake of antioxidants such as vitamin E and selegeline may help the brain fend off cellular attacks on the neurons. The use of nonsteroidal anti-inflammatory drugs (NSAIDs) such as aspirin and ibuprofen may be helpful too, since inflammation is thought to be a cause of the plaques and tangles found in the brains of Alzheimer's patients. Corroboration comes from a review of numerous epidemiological studies on people with arthritis; early indications are that those who have taken anti-inflammatory agents for years to reduce the symptoms of arthritis may have a decreased risk of Alzheimer's. But researchers have issued a strong warning against using these studies as a mandate to start downing aspirin or ibuprofen daily. None of the anti-inflammatory drugs was ever designed for long-term use, and there could be serious side effects such as damage to the stomach lining, internal bleeding, and interference with kidney function—conditions that could be much more life-threatening than Alzheimer's.

Can drugs reverse loss of memory? Our decade of the brain has certainly renewed the pursuit of pharmaceuticals intended to do so. For example, Gregory Gray at the University of Southern California found that vitamin B_{12} supplements miraculously initiated the full recovery of 8 out of 13 senile subjects in his study. In Italy, another study concluded that Alzheimer's patients who took 2 grams of carnitine daily for a year showed a significant improvement in long-term memory and verbal ability.

John Morley, chief of geriatric medicine at the St. Louis University School of Medicine, discovered that beta amyloid, a component of the nerve-cell membrane, will actually damage other neurons. In a leap of intuition, Morley tested and found that beta amyloids were overabundant in the brains of Alzheimer's patients. Consequently, he suggests that beta amyloid blockers may be a good future treatment for that condition.

Whether drugs can actually enhance a healthy memory is another matter. The notion of a pill that can help a synaptically challenged fifty-something remember where he left his keys is quite a temptation for hustlers, con artists, and quacks. In early 1997, for example, ginkgo biloba became all the rage. An extract from the tree of the same name, it was being widely sold in health-food stores as a herbal supplement. Research showed that it improved blood flow to the brain. Salespeople therefore insisted that it would bring more oxygen to the brain, which "of course" would improve memory. Well, oxygen plays no special role in coding memories. Besides, our brains get all the oxygen they need naturally. If a person had enough oxygen debt in his brain to actually impair his memory, he'd have a much bigger problem, namely, being on the verge of a stroke. There are some studies that show that test subjects who took ginkgo biloba did better in certain brain tests, but other studies show no improvement. In short: we don't know enough about this substance to recommend its use, much less make sweeping claims about its benefits.

Just the same, recent research indicates that one class of chemical compounds called ampakines may enhance some processes involved in forming memories. Gary Lynch at the University of California at Irvine injected ampakines into the brains of laboratory rats, and observed that they remembered their way through mazes much better than untreated rats. In very preliminary tests, young men with normal memories who took ampakines performed up to 20 percent better on some standard short-term memory tests, and men in their sixties and seventies as much as doubled their scores.

Only extensive trials will tell if ampakines and other memory drugs are really effective for humans. Time and again "cognition enhancers" that worked wonders on lab rats and even on small human test groups have fizzled in controlled clinical trials. Many turn out to be no more effective in stimulating the brain than a nice cup of espresso.

The clear message you should derive from the benefits of mental and physical exercise is that the worst thing you can do to your brain is to be content living a passive life. The habit of passivity is pervasive in our culture, from longing for miracle cures to watching television for hours to being politically apathetic. Physical and mental action is fundamental to maintaining mental health.

ROOTS AND SHOOTS

WHAT WE INGEST is also fundamental to how we think and feel. The brain is an incredibly active furnace, consuming 25 percent of the glucose and oxygen we take in. It burns glucose as its sole fuel, and yet it has no storage site for it. What little glucose is present in the brain at any time can be depleted within 5 to 10 minutes if replenishment is not available. The brain also needs a range of other nutrients. Researchers are making great strides in discovering how different foods can help us stay younger, get smarter, and feel better.

The relationships between what we ingest and what goes on in our brains are at times straightforward and at other times complex. Poor diet is common in people who suffer from depression and other mental disorders. Even minor nutritional deficiencies can be associated with changes in mood. Memory loss, confusion, depression, and other mental disorders in the elderly, once attributed simply to aging, can also be consequences of a poor diet.

One significant relationship is that food intake affects the formation and activity of neurotransmitters. It is no coincidence that people turn to pasta, desserts, and other carbohydrate-rich foods when they feel down in the dumps. Carbohydrates increase brain concentrations of an amino acid called tryptophan, which is the building block for the neurotransmitter serotonin. Consequently, increased tryptophan means a higher level of serotonin in the brain, which in turn relieves depression, insomnia, and irritability by promoting feelings of calmness and serenity. This finding was illustrated in a study conducted by Katy Smith at Littlemore Hospital, Oxford, U.K. She gave half of a group of women with a history of clinical depression a drink containing tryptophan and half of them a drink without it. The women rested for several hours and were then assessed for mood changes. The group without tryptophan had significantly lower moods.

Even cholesterol is getting some play in mood regulation. Although we have put the hex on cholesterol, we still need it, at least in moderate amounts, because it is necessary to make brain stuff such as the myelin that surrounds and protects nerve fibers. In recent studies, rapidly lowered cholesterol levels have been associated with depression, anxiety, panic disorder, violence, and suicide. When we go wild in altering food intake, in this case absurdly lowering cholesterol levels, we risk upsetting the equilibrium we need.

Central nervous system stimulants such as caffeine can also affect mood. One to two cups of coffee may combat fatigue, increase alertness, and improve work performance, but higher doses often result in agitation, headaches, nervousness, or decreased ability to concentrate.

Deficiencies in vitamins B_1, B_2, B_6, and B_{12}, niacin, folic acid, and vitamin C are often found in psychiatric patients and are thought to contribute to a cycle of depression, lack of interest in food, progressive malnutrition, and increasing mental or emotional disorders. Vitamin deficiencies seldom occur singly and are almost always accompanied by inadequate intake of other vitamins, protein, iron, and minerals. Even small deficiencies in normal people can promote personality and mood changes, insomnia, aggressiveness, and impaired reasoning and judgment. Clearly, a better diet rich in vitamins and minerals, with supplements if necessary, is the road to reversing many ailments of the brain. This may sound as old as the ages, but a balanced diet is the best brain medicine.

Eating foods rich in certain nutrients at the right time of life may also increase intelligence. For example, a lack of iodine in the mother's diet during pregnancy may lead to hypothyroid conditions that can cause severe brain damage in the unborn baby. The teen years are vulnerable to iron deficiency because of increased growth, and the onset of menstruation in women. A Johns Hopkins University study of teenage girls showed how important iron is to proper brain function. Half of a group of girls who had a slight iron deficiency were given supplements, while the other half were not. The group with supplements improved their performance on tests of memory and learning.

An important subset of the roots and shoots we ingest is the medications and drugs we take. Doctors and drug companies typically concern themselves with how a psychoactive drug affects a patient's body and brain. But instead of just asking, "What does this drug do?" the physician should ask, "What does this drug mean to the patient?" The issue that is overlooked is the patient's attitude toward the medication he is taking. What is his impression of this particular drug's therapeutic value? What is his desire and expectation concerning side effects and treatment outcome? The complexity of any pharmacological intervention is compounded by the thoughts, moods, and attitudes of the patient. As drug therapy strives to become more accurate, it will become even more dependent on understanding psychology. Matching pharmacological treatments to symptoms, therefore, must be recognized as an essential component in the treatment of almost any brain disorder.

Direct application of neurotransmitters to the brain might one day be part of the chemical arsenal. Why do children who suffer from brain damage often recover fully, while adults with the same damage are permanently incapacitated? The University of Wisconsin's Ronald Kalil found that the brains of young cats are bathed in neurotrophic factors, which help the brain reorganize or rebuild damaged networks. Adult cats' brains have far less of these chemicals. When Kalil soaked tiny sponges in neurotrophic-factor solution and placed them in the brains of adult, brain-damaged cats, the adult brains began to repair themselves.

Using drugs or even natural substances to enhance the performance of our brains brings us smack up against our puritan roots, and we feel that we are cheating. Yet we daily use other environmental tools to help us function better. As more options become available, the debate will continue as to what we should or should not use and whether it ought to be the goal of our lives to become more efficient and more content through such means.

TECHNOLOGY

DRUGS ARE JUST ONE WEAPON in the larger technological arsenal society has to combat mental illnesses and otherwise improve the brain. When the cell phone is ringing, the microwave is beeping, the computer has crashed, and the fax modem is on the fritz, modern technology can seem more like a curse than a convenience. Scientists and engineers, however, are developing many ways to use technology in healing.

For example, the body's own electrical signals are being harnessed to help paralyzed people communicate. Scientists at Emory University have successfully implanted an electrode in the motor cortex of two such patients. Over several weeks, neurons grew onto the electrode, creating a connection between its circuits and the brain's circuits. Now when the neurons fire, the signals are captured by the electrode and transmitted through the skin to a receiver on the surface of the scalp. Once the patients are trained, they can control the pattern of the impulses, which can be used to drive a cursor on a computer screen. The two patients thus far have been able to conduct basic communications in this way. In the future, it might be possible to use this kind of signaling to command a wheelchair, a prosthetic limb, or even the actual muscles in a paralyzed limb.

Technology is also being used in brain-training—from children with learning disorders to soldiers in the military. Anyone who has seen a video game can imagine how a virtual-reality simulator could be an effective tool for training pilots or tank drivers. Some of the really exciting research, however, is in using computer games to help children with dyslexia. As described in Chapter 7, Paula Tallal at Rutgers University and Michael Merzenich at the University of California at San Francisco have developed software for children who have problems processing certain combinations of letter sounds, which makes it difficult for them to read.

The hit interactive toy from Japan, Tamagotchi, is showing the possible power that interactive technology may have on our psychology. Tamagotchi is a cyberspace pet, a little figure on a key-chain screen that needs to be loved, played with, and fed over days and weeks or it will die. Children watch the pet hatch from an egg on the screen and then care for it as it signals its needs, as often as every ten minutes during peak times. If the child is neglectful, the pet dies. The toy can be reset, but keeping the pet alive is still an incredible emotional experience for many children. Could this type of technology be used to teach social skills or behavior?

Computers are also being used to simulate talk therapy and assist in relaxation and meditation. A software program called the Therapeutic Learning Program, recently developed at UCLA, operates like an interactive self-help book to assist patients in grappling with their depression and anxiety. Sessions last 30 minutes once a week for 10 weeks. Patients tell the computer their problems, and the computer develops a strategy to help them cope with those problems. This type of program should be used only with the supervision of a mental health professional, but it allows patients a degree of self-control and independence to proceed at their own pace without interruption.

There are also many Internet sites and computer programs that are devoted to teaching relaxation, meditation, and even hypnosis. These programs vary in their degree of interactivity and also in their effectiveness. Other sites provide straight educational material on medical information of all types, not to mention lists of support groups, doctors, and organizations that address every kind of disorder. Accurate, timely access to information can be the patient's best means to a swift recovery.

Another application of new technology may help the 2 million Americans who have "essential tremor," a little-understood disease that causes persistent shaking but no other symptoms, and the 500,000 who have Parkinson's disease. A tiny electrode is implanted in the thalamus. From there a wire runs under the scalp down to the collarbone, where a pacemaker-sized "pulse generator" is implanted. The generator sends electrical waves to the electrode, which blocks tremors by emitting constant, tiny electrical pulses. In studies of 120 patients who received the implant, about half of those with essential tremor saw their shaking disappear. Others showed different ranges of improvement. The implant did not help the Parkinson's patients as much, but it may still hold promise.

One recent advance that seems straight out of science fiction is transcranial magnetic stimulation (TMS). This new technique for affecting the brain does not seem to have the sometimes harmful side effects of drugs or surgery. A small coil of magnets, shaped in a figure 8, is placed against the scalp beside the area of the brain that is to be affected. This device delivers repetitive bursts of magnetic fields for perhaps 20 minutes at a time, which can excite or inhibit synaptic activity in localized regions of the brain. Researchers at the National Institute of Mental Health have used this technique to treat the frontal-lobe hyperactivity found in obsessive-compulsive disorder. Twelve patients diagnosed with OCD were given the stimulation at several different points on their skulls. The patients found that stimulation to the frontal lobe on the right side of the head resulted in a reduction in compulsions and an improved mood. More research needs to be completed before we truly understand the possibilities of this new technology. Researchers are most excited about the possibility of using magnetic stimulation to vault people out of serious depression, a replacement for effective but damaging shock therapy. Tests have succeeded in arousing neurons in the left prefrontal cortex (behind the forehead), where electrical activity is often abnormal in depressed people.

As the magic of our machines increases and as we gain more and more understanding of how the brain works, we will be better able to use technological devices to help direct our lives, not just as the wonderful slaves that we now employ in the external world, but as intelligent helpers that will guide us to a better use of what we have in the "black box."

Years ago I proposed that *The Sharper Image* would soon be advertising the complete "Home Brain Gym." My personal gymnasium of the future would be a machine individually designed to maximize every day. It would probably be something like a Stairmaster, or a treadmill with all sorts of gadgets attached, and there would be a built-in computer monitoring and asking questions of whoever steps into the machine. Yesterday's activities and achievements would be reviewed, and a game would be played to assess current brain activity. All of the relevant history of the gym-user would be known by the computer, and, after performing a range of neurological evaluations of physical and mental alertness, mood, and so forth, as well as inquiring into the coming day's activities, it would design a comprehensive body-brain workout to tune up the user for whatever might lie ahead. If a particular system was functioning below par, the machine would center the workout around getting things back in balance. Accompanying the physical exercise, the right kind of light and the proper harmonies, rhythms, and scents would be projected to get the brain in tune. Puzzles, video games, and mental drills would be involved as well, in order to raise or lower the level of one system or another. The net result would be that a person could step out of the gymnasium not only physically but mentally primed for the day ahead.

SPIRITUALITY AND MEDITATION

AN OFTEN OVERLOOKED ASPECT of the treatment of mental disorders is the connection between the mind and the body. The mind, or free will, can impact on the physical manifestations of illness. The famous case of Broadway actor Yul Brynner, who long starred in *The King and I* and who seemed to put his cancer into remission for a decade using only his will to counteract it, made many people aware of the strong powers the mind can have over the body. Religious people, practitioners of yoga or meditation, and disciples of Eastern cultures are familiar with the powers the mind can exert on mental illnesses and their effects on the body—powers that are rarely examined by Western medical science.

Herbert Benson at Harvard Medical School is one medical researcher who has championed the mind's power in using the "relaxation response," the physical reaction of the body to relaxation by the mind. This is usually accomplished by the practice of meditation or progressive relaxation. Now Benson is looking into the effects of prayer. In his new book, *Timeless Healing: The Power and Biology of Belief,* he cites a study of 92,000 people in Washington County, Maryland, who went to church once or more per week. This group of people had 50 percent fewer deaths from coronary heart disease and 53 percent fewer suicides than the general public.

In 1998 there were several major media reports that looked into the possible links between spirituality and mental health. We are far from drawing any conclusions. As with medications, any efficacy of spirituality would be affected by a person's attitude toward it. Here the brain's structure may make clarity even harder to find. The two brain hemispheres provide two ways of looking at and understanding the world. The right attempts to see the whole, the left orders specific experiences. This leads to very different ways of approaching abstract concepts such as the divine.

The left hemisphere would favor a god of the Word. People with dominant left hemispheres would hold the Word as sacred—the Bible is the truth and the guide and its order rules. As Jerry Falwell tells it, "The writ is the writ" and the narrative carries the day. God is about victory, and heaven is a reward that is full of happiness. There is a point and an endpoint to it all.

Right-brainers hold experience to be most dear; they would want to insert themselves into the gestalt of God, and see and accept everyone and everything as part of the whole of God's creation. To them, God is simply oneness, and that is a continual mystery. The right-brainer's heaven is just a part of the oneness.

Unfortunately, proponents of these two views often clash with each other over which is the "truth." In trying to reign supreme, however, each side misses a fundamental point. Both would agree that God is something of which nothing greater can be conceived. However, for humans, all conception is limited by the language, context, and meaning capabilities of the brain—a limit that falls far short of being able to define something that is beyond all conception. Instead of arguing about and in some cases killing each other over our different conceptions of God, we might try bringing both deities—both hemispheres—together. Then we might at least find a better way to think about the Almighty, and about the universe and our place in it.

Powerful as the brain may be to conceive of such complex thought, it also deserves an occasional break. Our muscle analogy again: when overworked, the brain too becomes fatigued. Unfortunately for the brain, going to sleep doesn't provide absolute rest. The brain continues to receive stimuli from the environment and the body even during sleep, as well as generating a steady supply of dreams.

To promote health and well-being, we must help the brain rest. One way to accomplish that is through meditation. Meditation is a focused awareness, achieved by sitting quietly and turning one's attention inward. Thoughts, feelings, and perceptions are observed and then put aside. Judgment is suspended and attention is paid only to the object of the meditation. Some people choose a word or phrase to say aloud or to themselves. Others concentrate on looking at an object such as a candle or a picture. Logic and language are cast aside. Distractions are ignored. Gradually, the mind becomes quiet and deeper levels of awareness are reached.

The body has a physical reaction to this altered state of consciousness. Sympathetic nervous system activity decreases and metabolism slows down. Heart rate, blood pressure, and breathing rates fall, and electrical skin conductance and blood flow decrease. These reactions in the body have been found to help reduce high blood pressure, relieve chronic pain and migraine headaches, and soothe depression and anxiety. The brain's own electrical activity changes, too. Instead of supporting a decentralized storm of signals, large numbers of its neurons fire in a pleasing synchrony.

PASSION

FINDING OUR INDIVIDUAL synchrony in life is equally important to a healthy brain. It is crucial to letting our talents blossom, and important for getting us away from the addictive behaviors that abound in life.

Any activity that gives us a sense of purpose and accomplishment, that makes us feel glad to be alive, can help us care for and feed our brain. Many people put off doing what they love, or what they know they need to do for themselves, until later in life, trying to get the world's demands out of the way first. What a grave mistake! It is far better to make sure that part of our lives is consumed with activities that we can put all our hearts, minds, energies, and joys into at once.

Find a mission in your life. A commitment to a calling, a career, even a hobby focuses the mind and the soul. Psychotic patients report that they don't hear "the voices" while they are busy working. Surely we ordinary people can calm our own internal voices with some intent activities. Almost any form of work or concentrated effort can quiet the noisy brain. Work you love is more powerful still because it brings with it a sense of accomplishment, pleasure, and well-being. Passion heals.

Remember one important point: In pursuing your passion, the actual doing is what matters, not any measure of success. A diet of constant, stimulating activity is the best prescription for our troubles. It keeps the brain in a state of constant change, flow, confirmation, and anticipation, thereby reducing the noise, fragility, self-doubt, and stagnation with which we all have to contend.

ACKNOWLEDGMENTS

A book grows in many ways; this one evolved. This book developed in distinct evolutionary niches with quantum jumps into the next phase, as did the brain. Almost five years ago I proposed to Jill Mullen, my research director, that we write a much-needed primer on the brain for mental health professionals. This seemed a more relevant project than yet another drug study. Jill gulped and took on yet another impossible task, with appropriate fear and trembling. Over the next three or so years she led our ever-changing research group. She helped everyone to stay focused on producing a guide through the "black box." Many years and two thousand pages later, the primer is awaiting publication. I owe a huge debt of gratitude to these people: Gail Garber, Rosemary Kwa, Carolyn Brenner, Susan Bullivant, Robert J. Kirkpatrick, Julie Maynihan, Jessica Smock, Jessica Ratey, Susan Herzberg, Tika Shah, Robin Fox, Graham Spruiell, David Hoffman, and Jennifer Chadwick. Our work and learning together became the basis for the next evolutionary jump.

Also, in this early phase, I owe a special note of thanks to Robert Kirkpatrick and Jeff Sutton, who helped mold my "Four Kingdoms of the Brian" essay that I wrote in 1993. They both were instrumental in helping me to create a more up-to-date and scientifically correct version. Their work resulted in a series of papers titled "Four Theaters of the Brain," which became the basis of Chapter 9.

The next phase grew out of the first and was promoted by my editor at Pantheon, Linda Healey, who had edited my previous books that I coauthored with Edward Hallowell and Catherine Johnson. When I told Linda about my brain project, she suggested that I think of doing a smaller and even more readable version stuffed with vignettes that would try to instruct the public at large—to explain the machinery in the black box to everyone. Jill Neerham, my agent at Palmer and Dodge, provided encouragement and direction, and with the help of Susan Heath and Robert Kirkpatrick I was able to get started. We agreed that the work would be better filtered if we engaged a ghostwriter to tell the story in a nontechnical manner. I owe much to Mark Fischetti, who filled this role so well. With the chapters from the primer as a base, I proceeded to try and teach him all about the brain, using incidents from my own life and practice. Mark is a gifted science writer and editor, but had sparse knowledge of the brain, so it was both a learning and a writing task. I owe him much for his constant pressure to look for a clearer way to present an idea or come up with a more concise story that fit the concept better. He was truly an artist at work.

For the final phase of fine-tuning, I drafted two of my colleagues, James Hopper and Carl Albert, who generously agreed to read and criticize the finished work. I am forever grateful to Dr. Albert Galaburda, who added depth and precision to the subject matter of this book. He agreed to be my neuroscience editor and correct my mistakes and challenge my wild metaphors. He read it line by line and commented and cajoled me to change this fact and then that one. Al provided a solid backdrop, and I will be forever grateful for his interest, and his support in making this kind of book accessible to the public.

I owe a debt of gratitude to many who influenced my life and led me to my passion and thus this book. To Frank Elliot and Norman Geschwind, who blazed the trail for all of us who care about the brain and behavior. To Ned Hallowell, who began my career as a writer for the public. To Catherine Johnson, who added so much confidence and support. To Ben Lopez, Bart Hershkovitz, Larry St. Clair, Allan Hobson, George Valliant, Jules Bemporad, Bessel van der Kolk, Marcel Kinsbourne, Frederick Schiffer, Polly Heninger, Brenda Bemporad, Douglas Watt, Martha Stark, Robert Stickgold, Paul Wender, and my patients—the best teachers. To my wife and children, who provided support, love, and editorial suggestions.

Finally, I would like to thank my new editor, Dan Frank, and his reliable assistant, Jennifer Weh, whose skill and gentle prodding brought the book to its conclusion.

SUGGESTED READING

1. DEVELOPMENT

The Mozart Effect. Don Campbell. Avon Books, New York, 1997.

Search for the Tourette Syndrome and Human Behavior Genes. David E. Comings, M.D. Duarte, Calif., Hope Press, 1996.

The Growth of the Mind. Stanley I. Greenspan, M.D. Addison Wesley, Reading, Mass., 1996.

Living with Our Genes. Dean Hammer and Peter Copeland. Doubleday, New York, 1998.

Developmental Neuropsychiatry, vols. 1 and 2. James Harris, M.D. Oxford University Press, New York, 1995.

Brain Development and Cognition: A Reader. Mark Johnson, ed. Blackwell Science, Malden, Mass., 1993.

Galen's Prophecy. Jerome Kagan. Basic Books, New York, 1994.

Developmental Psychobiology: An Interdisciplinary Science. George F. Michel and Celia L. Moore. MIT Press, Cambridge, Mass., 1995.

Principles of Neural Development. Dale Purves and Jeff W. Lichtman. Sinauer Associates, Sunderland, Mass., 1985.

Development Throughout Life: A Handbook for Clinicians. Michael Rutter and Dale Hay. Blackwell Science, Malden, Mass., 1994.

Developing Minds. Michael Rutter and Marjorie Rutter. Basic Books, New York, 1993.

2. PERCEPTION

Conversation with Neil's Brain. William Calvin and George Ojemann. Addison Wesley, Reading, Mass., 1994.

The Neurological Side of Neuropsychology. Richard Cytowic, M.D. Bradford Books, MIT Press, Cambridge, Mass., 1996.

The Human Mind Explained. Phyllis Greenfield. Henry Holt, New York, 1996.

The Three-Pound Universe. Judith Hooper and Dick Teresi. Dell Publishing, New York, 1986.

Wet Mind. Stephen M. Kosslyn and Olivier Koenig. Free Press, New York, 1992.

Inside the Brain. Ronald Kotulak. Andrews McMeel Publishing, Kansas City, Mo., 1996.

Figments of Reality. Ian Stewart and Jack Cohen. Cambridge University Press, Cambridge, 1997.

3. ATTENTION AND CONSCIOUSNESS

In the Theater of Consciousness. Bernard Baars. Oxford University Press, New York, 1997.

The Cerebral Symphony. William Calvin. Bantam Books, New York, 1990.

The Conscious Mind. David Chalmers. Oxford University Press, New York, 1996.

The Engine of Reason, the Seat of the Soul. Paul Churchland. MIT Press, Cambridge, Mass., 1995.

The Astonishing Hypothesis. Francis Crick. Scribners, New York, 1994.

Consciousness Explained. Daniel Dennett. Little, Brown, Boston, 1991.

The Prefrontal Cortex. Joaquin Fuster. Lippincott-Raven, Philadelphia, 1997.

Toward a Science of Consciousness. Stuart R. Hamerof et al., eds. MIT Press, Cambridge, Mass., 1996.

Complexity: The Emerging Science at the Edge of Order and Chaos. M. Mitchell Waldrop. Touchstone, Simon & Schuster, New York, 1993.

5. MEMORY

Memory's Voice. Daniel Alkon. HarperCollins, New York, 1992.

The Longevity Strategy. David Mahoney and Richard Restak. John Wiley, New York, 1998.

Committed to Memory. Rebecca Ruff. Crown Publishers, New York, 1998.

Searching for Memory. Daniel Schachter. Basic Books, New York, 1996.

Traumatic Stress. Bessel van der Volk, Alexander MacFarlane, and Lars Weisaeth, eds. Guilford Press, New York, 1996.

6. EMOTION

The Amygdala. John Aggleton, ed. Wiley-Liss, New York, 1992.

The Neurology of Thinking. D. Frank Benson, M.D. Oxford University Press, New York, 1994.

The Nature of Emotion. Robert Ekman and Richard Davidson. Oxford University Press, New York, 1994.

Emotional Intelligence: Why It Can Matter More Than IQ. Daniel Goleman. Bantam Books, New York, 1995.

The Emotional Brain. Joseph LeDoux. Simon & Schuster, New York, 1996.

Molecules of Emotion. Candace Pert. Scribners, New York, 1997.

Affect Regulation and the Origin of the Self. Allan Schore. Lawrence Erlbaum, Hillside, N.J., 1994.

The New View of the Self. Larry Siever, M.D., and William Frucht. Macmillan, New York, 1997.

7. LANGUAGE

How Brains Think: Evolving Intelligence, Then and Now. William Calvin. New York: Basic Books, 1998.

The Cerebral Code. William Calvin. Bradford Books, MIT Press, Cambridge, Mass., 1996.

The Symbolic Species. Terrance Deacon. W. W. Norton, New York, 1997.

Origins of the Modern Mind. Merlin Donald. Harvard University Press, Cambridge, Mass., 1991.

Nature's Mind. Michael Gazzaniga. Basic Books, New York, 1992.

The Language Instinct. Stephen Pinker. HarperCollins, New York, 1994.

8. THE SOCIAL BRAIN

Friday's Footprints. Leslie Brothers, M.D. Oxford University Press, New York, 1997.

Descartes's Error. Antonio Damasio. Grosset/Putnam, New York, 1994.

Phantoms in the Brain. V. S. Ramachandran, M.D., and Sandra Blakeslee. William Morrow, New York, 1998.

The Cerebellum and Cognition. Jeremy Schamann. Academic Press, San Diego, Calif., 1997.

INDEX

Page numbers in *italics* refer to illustrations.

ABC, 19, 366
abortion, spontaneous, 27
absolute threshold, taste and, 73–74
acetylcholine, 218, 313
Acredolo, Linda, 180
acrophobia, 235
adaptation, taste receptors and, 72–73, 74
ADD (attention deficit disorder), 27, 80, 125–26, 243, 255, 321, 350–51, 363
addiction, 118, 123–25, *125*, 126–27, 128, 243, 244
adenosine triphosphate (ATP), 88
ADHD, *see* attention deficit hyperactivity disorder
adrenaline, 173, 223, 225, 232, 313
affect, 336
affect-centered diagnosis, 7
"affective display," 313
"afterimages," 73
age, 360–61
  memory loss and, 217–20
ageusia, 74
aggression, 236, 238, 333
aging, brain and, 39, 43
agnosia, 269
agoraphobia, 233
akinetic mutism, 319
alarm hormone, 223
alcohol, alcoholism, 25, 27–28, 30, 118, 123, 124, 126, 127, 238, 243, 344
Ali, Muhammad, 44
ALS, 34
altruism, 303–4
Alzheimer's disease, 32, 39, 43, 44, 167, 185, 207, 217–18, 364, 366, 367
amblyopia, 36
American Psychological Association, 216
American Sign Language (ASL), 96
Amgen, 47, 169
amnesia, 202–3
  traumatic, 211–17

amphetamines, 117, 244, 245, 328
amphibians, 22
amputees, phantom limb syndrome of, 58–59, 86–87
amygdala, 63, *67*, 71, 117, 161, 172, 192, 210–11, 221, 244, *315*, 324, 325, 326
  anxiety and, 173, 222
  description of, 311–12
  emotions and, 174–75, 225–26, 227, 228–29, 311–14
  face recognition and, 314–18
  fear and, 232, 233–34
  functions of, 66, 68, 121, 312
  pain and, 88
  sadness and, 239
  social processing and, 295–96, 311–14
analog processor, brain viewed as, 5
Anderson, Douglas, 170
androgens, 245
anesthesia gases, 29
anger, 229, 236–38
  as basic emotion, 226, 232, 236
  control of, 236–37
  pathways and expressions of, 237–38
anosmia, 62
anterior cingulate gyrus, 75, 120, 121, 150, 152, 171, 193, 227, 229, 248, 313, *315*
  free will and, 318–19
anterior insula, 75, 76
antibiotics, 29, 74
antidepressants, 16, 29–30, 128, 241, 250
antioxidants, 366
antipsychotics, 118
antisocial behavior, 28, 196–97, 237
antisocial personality disorder (APD), 309, 323–24
anxiety, 61, 66, 126, 173–74, 222–23, 226, 236, 239
"anxiety gene," 236
apathy, 249–50
aphasia, 270, 272, 274, 277, 286
Aplysia, 191, 195
apnea, 77
apraxia, 272
"arbitrariness of the sign," 257
aromatherapy, 66
arousal, 115
artificial intelligence, 141
ASL (American Sign Language), 96
Aslin, Richard, 264
asociality, 326
ATP (adenosine triphosphate), 88
attachment, 244–45, 328
attentional specificity, 102–3, 108
attention and consciousness, 110–46, 326
  arousal, orientation, detection, execution and, 114–20
  attention deficit hyperactivity disorder and, 125–29
  brain states and, 133–36
  connection between, 111
  easy and hard problems of, 144–46
  emotional tags and, 121–22
  fringe aspects of, 136–39
  genes and, 123–25
  memory and, 130–33
  movement and, 171–74
  neurotransmitters and, 122–23
  process of, 111–14, *119*
  shared attention, 301, 307
  as theater of exploration, 337, 341, 343–44
  see also consciousness
attention deficit disorder (ADD), 27, 80, 125–26, 243, 255, 321, 350–51, 363

attention deficit hyperactivity disorder (ADHD), 66, *67*, 117, *119*, 120, 123, 124, 125–29, 197–98, 238, 254, 256, 309–10, 327, 339
causes of, 126
medications for, 128
as reward deficiency syndrome, 127
attraction, 244–45, 327–28
auditory cortex, 91, 93, 94, 95
auditory nerve, 93
auditory system, *92*
see also hearing; sound
autism, 6, 49, 78–84, 116, 301, 339
areas of the brain affected by, 78, 82–83, 325–26
behaviors caused by, 79–81, 325
features of, 78–79, 82–84, 325
genetics and, 12
of Grandin, 14–16, 82, 83, 91, 298, 301
hearing and, 91
neuron migration and, 24
social brain development and, 306–7, 325–26
autonomic nervous system, 171–72, 229, 232, 312
Axelrod, Robert, 304
axons, 9, 19–20, 24, 25, 134

babies, 336
attention problems and, 116
individuation and, 301
innate temperament of, 230
language exposure and, 264–65, 277, 287
ties between mothers and, 300
touch and, 76–78
Baddely, Allan, *197*
Balint's syndrome, 116
Bard, Phillip, 224, 225
basal forebrain, 227
basal ganglia, 116, 118–20, 143, 157, 158, 160, 161, 162, 175, 177, 179, 200, 244, 248
bedwetting, 6, 32
beer, taste and temperature and, 70–71
Begin, Menachem, 16
"benign masochism," 72
Benson, Herbert, 374
beta amyloids, 367
beta-blockers, 173, 222–23, 367
Bickerton, Derek, 260, 285
biofeedback machines, 6
blackjack, 231, 293
Blakeslee, Sandra, 317
blind people, 59–60, 99
blindsight, 102, 137
blind spot, 57
blood pressure, 74, 84
Blum, Kenneth, 123, 243
body language, 228
bombykol, 65
bonobos monkeys, 259
Boor, Mary Esther, 43
Borges, Jorge Luis, 196
Boston Globe, 46
Braille, 59
brain:
attention and consciousness and, 110–46
care and feeding of, 356–77
composition of, 5, 9–11, *10*, 19–21, 22–23, 24
damage and repair and, 166–71
development of, 14–47
emotion and, 222–51
environmental influences on fetal development of, 26–30
evolution of, 9, *10*, 21–22, 64, 280

brain (*continued*):
feedback between layers or levels of, 164, 180
fetal growth of, 23–24, 40
four theaters of, 336–55
hemispheres of, *see* left hemisphere; right hemisphere
hybrid developmental view of, 17
individual uniqueness of, 4, 7, 106
interaction between front and back of, 156, *159*
language and, 252–89
mapping of, 143–44, 161–65, *163*
memory and, 182–225
movement and, 147–81
nature-nurture question and, 31–34
neurological self-awareness and, 12–13
neuroscientific mapping of, 4
perception and, 48–109
plasticity of, 11–12, 17, 20, 21, 31, 35, 38, 39–43, 56, 85–86, 106, 165, 219, 277–79, 365
as prediction machine, 112–13
as "reflexive" organ, 112
regeneration of, 44–47
regional sharing of functions in, 148
rewiring ability of, 38–39
as selection system, 141
social nature of, 23, 26, 290–335
systems-based view of, 4
three basic components of, 22
brain disease, 42–43
brain function, as theater of exploration, 337, 341, 344–45, 352
brain states, 133–36
brainstem, 115, 158, 174, 229, 233, 327
brain tumors, 74
Brandeis University, 95
Breuer, John, 19, 288
"Bridge Too Far, The" (Breuer), 288
Broca's area, 96–97, 207, 210, 267, *268*, 271, 274, 286
Brothers, Leslie, 316
Brynner, Yul, 374
Bucy, Paul, 225
Bush, George, 45, 167
butterfly effect, 357

Calvin, William, 280
Camp David peace talks, 16
canalesthesia, 326
Cannon, Walter, 224, 225, 247
Cannon-Bard theory, 224–25
Capgras's syndrome, 227, 317
capsaicin, 72
captopril, 74
Caramazza, Alfonso, 267
carbohydrates, 369
Carlson, Mary, 77
Carter, Jimmy, 16
cataracts, 35
caudate nuclei, 118–20, 319
cerebellum, *10*, 143, 161, 200, 345
autism and, 78, 325, 326
functions of, 122, 151, 157, 158, 160, 162–63, 204–5
Latin derivation of term, 163
social brain coordinated by, 305–9, *315*
social processing and, 295–96
Cheek, D. B., 132
Chicago, University of, 271
child abuse, 184, 212–15, 278
child development, music and, 18–19
child prodigies, 33
children, 336
with amnesia, 203
benefits of motor activity for, 180–81

blood nutrient glucose and, 35
cognitive cross-training and, 365
and language ability and acquisition, 39–42, 262
learning-disabled, 28, 254
OCD and, 153–55
self-talk and, 254–57
social brain development and, 299, 300, 301–2
see also autism; dyslexia

chili peppers, 72
chimpanzees, 257–58
chocolate, 245, 327
cholesterol, 369
cholinergic neurons, 313
Chomsky, Noam, 261–62, 263
Churchill, Sir Winston, 36
Churchland, Paul, 130, 134, 144
cilia, 64, 91
City Project, 235
Clinton, Bill, 45, 167
Clinton, Hillary, 18, 19
cloning, 45
"closet kids," 40–41, 278–79
cocaine, 28, 117, 123, 126, 244, 245, 344
Coccaro, Emil, 333
cochlea, 91
"cocktail party conversationalists," 286
cognition:
language and, 285–89
as theater of exploration, 337, 341, 343–44, 350
see also thinking

cognitive acts, 148–51
Collins, James, 58
color blindness, 99, 102
color vision, 101–2, 104, 106
competitive inhibition, 87
complexity theory, 357
computers, 371–72
brain compared with, 5, 17, 131, 140–41, 194

cones, 57, 101–2
confabulation, 208–9
Congress, U.S., 166
"connectionists," 17
consciousness, *135*, 185
development of, 139–44
symphony analogy for, 129–30
see also attention and consciousness

contextual conditioning, 234–35
convergence zones, 187–88
Coors Brewing Company, 359
Corballis, Michael, 259
Cordo, Paul, 58
corpus callosum, 137, 139, 274, 284
cortex, *10*, 71, 194, 280
auditory, 91, 93, 94, 95
emotions and, 225, 234–35
evolution of, 22
40–cycle oscillation in, 133–34, 135, 136, 143
frontal, 99, 107, 115, *119*, 147, 149, 150, 172, 186, 201, 208, 227, 231–32, 233, 234, 237, 239, 254, 290, 312
insular, 76
in learning process, 20–21, 37–38, 149, 157–60
motor, 97, 149, 156–57, 158, 161, 164–65, 175, 230
music and, 35, 60, 205–6
occipital, 147
olfactory, 63, 64
orbitofrontal, 75, 151, *315*
parietal, 107, 147
prefrontal, 107, 130–31, 149, 150, 179, 193, 198, 230, 237–38, 242, 254
premotor, 160
role of, 23

cortex (*continued*):
sensory, 156–57, 312
six basic layers of, 283
somatosensory, 85–86
temporal, 107, 300, 324
visual, 59, 99, *100*, 102, 107
cortisol, 211, 362
Cosmides, Leda, 304
Courchesne, Eric, 116, 306
CREB (protein), 195
CRF (hormone), 172
Crick, Francis, 139–40
criminal behavior, 28
cross-modal influences, 37
Curtis, Martha, 189–90
Cylert, 117, 128

Damasio, Antonio, 187, 188, 210, 211, 230–31, 290–92, 311, 319
Damasio, Hanna, 187, 231, 290
Darwin, Sir Charles, 228, 302
Davidson, Richard, 229
Dawson, Geraldine, 35
Dbh gene, 334
deafness, *see* sign language
decision-making areas, 232, 290–91, 310–11
D8/17 genetic marker, 153
dementia, 43, 46, 166, 217
dendrites, 9, 19–20, 25, 43
Dennis, Maureen, 276
dentate nucleus, 151
Denver, University of, 45, 167–68
depression, 28, 49, 80, 126, 229, 239, 241–42, 249, 339, 359
mother-infant influence and, 35, 42
smells and, 67–68
symptoms of, 241
see also antidepressants
Descartes, René, 158–59
Descartes's Error (Damasio), 187
detachment, 328–29
developmental disorders, 24–25
developmental optometry, 49
DeVries, Geert, 334
Dexedrine, 117, 128
diabetes, 33
Diener, Edward and Carol, 226
diet soda, 74
Dilaudid, 117
dipyridamole, 74
direction, sense of, 108–9
disgust, taste and, 75–76
disinhibition, 128
dissociation, 213–17
divorce, 329, 331
Donald, Merlin, 258
Donne, John, 297
dopamine, 27, 44–45, 116, 117, 122–24, *125*, 127, 168, 245, 250, 360
conflicting duties of, 122–23
effects of blocking of, 118, 128
hunger and, 71–72
limbic tagging and, 121–22
memory and, 218, 219–20
pleasure and, 243–44
dopamine beta-hydroxylase, 334
Dow, Robert, 151
Down syndrome, 180
Draine, S. C., 220
dream (REM) sleep, 134, 189, 192–93, 195
Dronkers, Nina, 272
drug therapy, 338–39, 353–54
drug use, 25, 27–28, 30, 123, 126–27, 238, 243, 327, 344–45
see also addiction
D2R2 receptor gene, 123, 124
dysgeusia, 74
dyslexia, 24, 37, 52, 282–85, 351
hearing and, 90–91, 93, 95
Irlen lenses and, 98

language and, 270, 271, 275
theories on, 282–83
variations of, 282
visual processing and, 103–6

East Boston, Mass., gerontological study in, 43
ECT (electroconvulsive therapy), 49, 242
ectopias, 91, 283, 284
Edelman, Gerald, 11, 30, 140, 141, 143, 144, 145
Eden, Guinevere F., 103, 104
Edison, Thomas, 105
EEG studies, 59, 201, 269, 324
Einstein, Albert, 36, 105, 206
electroconvulsive therapy (ECT), 49, 242
"emotion," Latin derivation of term, 149, 227
emotional intelligence, 250–51
Emotional Intelligence (Goleman), 251
emotional prosody, 273, 322
emotional state, 336–37
emotions, 63, 71, 76, 120, 136, 148, 222–51, 257, 290, 302
attention and consciousness and, 121–22
brain locations of, 229–32, 275
and brain vs. body, 223, 224–27
expression of, 174–75
four basic, 226, 232–47
language, movement and, 270–73
motivation and, 247–50
movement and, 171–74, 227–29
set point for, 226–27
smell and, 63, 65, 66
social brain and, 311–14
empathy, 296, 309
endocrine system, 229, 312
endorphins, 66, 72, 117, 127, 243, 245, 328, 360
engrams, 313
"environmental dependency" syndrome, 120
environmental influences, on brain, 26–30, 327
alcohol, 27–28
cocaine, 28
malnutrition, 29
smoking, 27
toxins, 29–30
epilepsy, 24, 68, 137, 138, 162, 246
epinephrine, 223, 362–63
episodic memory, 201–3, 214
essential tremor, 372–73
estradiol, 329, 331
estrogen, 246
eugenics, 31
euthenists, 31
Evans, Denis, 43
executive organization, 115, 118–20
exercise, 35, 178, 222, 223, 359–61
explicit memory, 199–201
extended selective attention, 326
eye contact, 301–2

face recognition, 314–18, 320
facial agnosia (prosopagnosia), 316–17
facial expressions, 228, 275, 300, *315*
facial memory, 198
false memories, 182–85, 199, 213, 215–17
False Memory Syndrome Foundation, 216
Falwell, Jerry, 375
fatty acids, 29
fear, 66, 226, 232–36
as basic emotion, 226, 232
contextual conditioning and, 234–35
neural pathways for, 233–34
Ferris, C. F., 333

fetal alcohol syndrome (FAS), fetal alcohol effects (FAE), 28
fetal transplants, 45–46, 167–68
fiber-pattern theory, of taste, 69
Field, Tiffany, 77, 363
fight-or-flight response, 66, 88, 114, 161–62, 171, 228, 229, 232
"filling-in" phenomenon, 85
fish, evolution of brain in, 21–22
Fisher, Helen, 244–45
Flach, Frederic, 48
flashbulb memory, 192
"flooding" technique, 235
fluvoxamine, 30
folic acid, 25, 29
forebrain, 22
fovea, 56–57
Frederick II, Holy Roman Emperor, 77
Freed, Curt, 45, 168
Freeman, Walter, 64–65
free will, 11, 17, 318–19
Freud, Sigmund, 6, 210
Freyd, Jennifer, 216
fringe awareness, 136–39
Frith, Uta, 301
frontal cortex, 99, 107, 115, *119*, 147, 149, 150, 172, 186, 201, 208, 227, 233, 234, 237, 239, 290, 312
  language and, 254
frontal lobes, 118–20, 121, 126, 130–31, 174, 185, 206, 231–32, 234, 246, 267, 272, 291
  decision-making and, 310–11
  function and importance of, 309–10
  supplementary motor area of, 271
  ventromedial area of, 290
frontal parietal attention circuits, 116
Fuller, Buckminster, 55
"Funes the Memorious" (Borges), 196
Gage, Fred, 169
Gage, Phineas, 231–32, 290, 292, 294, 311, 323
Galaburda, Al, 103, 285, 290
Gall, Franz Joseph, 185
gambling, 123, 231, 293
Gazzaniga, Michael, 19, 288
GDNF (growth factor), 169
Gehrig, Lou, 34
generativity, 259
genes:
  attention and consciousness and, 123–25
  developmental influence of, 31–34
geniculostriate pathway, 102–3, 104
Genie ("closet child"), 41, 278–79
Genome Project, 32
George, Mark, 241
Georgia, 18
German measles, 25
ginkgo biloba, 367–68
glial cells, 24, 47
globus pallidus, 177
glucagon, 77
glucose, 35, 368
Goldman-Rakic, Patricia, 188
Goleman, Daniel, 251
Good Morning America, 19
Goodwyn, Susan, 180
"gourmand syndrome," 74–75
Grabowski, Thomas, 231, 290
Grandin, Temple, 17, 20
  autism of, 14–16, 82, 83, 91, 298, 301
  human squeeze machine built by, 15
  supermarket door self-training of, 14, 16, 21, 298
Gray, Gregory, 367
Gray, Jeffrey, 170
Graziano, Michael, 113

Greenwald, A. G., 220–21
Guthrie, Woody, 45–46

Haldol, 118
haloperidol, 243
Hamilton, William, 304
handedness, 275
happiness, 229
Hart, Betty, 287
Harvard Medical School, 46, 103, 168
hatha-yoga, 361–62
Haxby, James, 198
Head Start, 32
hearing, 88–97, *92*
  autism and, 91
  blindness and, 59
  dyslexia and, 90–91, 93, 95
  tinnitus and, 89–90
  vision and, 95–96
  see also sound
heart attacks, 237
heart disease, 32
heart failure, 74
Hebb, Donald, 218–19
heroin, 117
"hierarchy of needs," 247
Hildegard, Ernest, 213
hindbrain, 22
hippocampus, 68, 78, 115, 131, 143, 210, 211, 221, 227, 248, 316, 324, 325, 326
  memory and, 185, 188, 192, 194, *197*, 202–3, 235
Hobson, Allan, 192
"Home Brain Gym," 373–74
homeostasis, 54
hormones, 55, 74, 77, 172, 211, 223, 232, 245, 312, 327, 329, 331
Huntington's disease (HD), 45–46, 168–69, 176, 200
Huttenlocher, Janellen, 41
hydrocephalus, 286
hyperactivity, 28, 128
hypothalamus, 66, 88, 171, 172, 225, 229, 312, *315*, 331
  hunger regulated by, 71–72
  pleasure and, 243
hypothyroidism, 74

identity, 336
  as theater of exploration, 337, 341, 345–46, 350
ideomotor apraxia (IMA), 166
imipramine, 241
immune system, 42, 64
implicit memory, 199–201
infections, 25
Insel, Thomas, 329
insula, 227, 272
insular cortex, 76
insulin, 77
intralaminar nuclei, 134–35, 136
ionizing radiation, 29
Irlen lenses, 98
iron, 29
Isabelle ("closet child"), 41

James, William, 56, 136, 224
James-Lange theory, 224
Janet, Pierre, 209–10, 211–12
Japanese language, written forms of, 281
Jenkins, William, 170
Johns Hopkins University, 45, 178, 179, 230, 264, 370
Johnson, Tony, 46
joy, 242–47
  as basic emotion, 226, 232, 242–43
  laughter and, 245–47
  love and, 244–45

Kalil, Ronald, 370–71
Kandel, Eric, 191

Kanner, Leo, 325
Kant, Immanuel, 254
Kaplan, Melvin, 49–50, 52
Karni, Avi, 21, 188–89, 192
Khantizan, Edward, 126
Kihlstrom, John, 212, 213
Kimura, Doreen, 272
Kirkpatrick, Robert, 336*n*
Kirschvink, Joseph, 109
Kluver, Heinrich, 225
Kluver-Bucy syndrome, 121
Kodaly training, 272
Kohn, Bruno, 276
Kuhl, Patricia, 40

Lahey Hitchcock Medical Center, 46
Landis, Theodor, 75
Langton, Chris, 5
language, 94, 137, 148, 230, 252–89, 302, 341
  brain's processing of, 96, 253–54, 266–70, *268*, 273–77
  cognition and, 285–89
  development of, 257–61
  dyslexia and, 282–85
  importance of, 253, 260
  movement, emotion and, 270–73
  nature-nurture question and, 261–66
  plasticity and, 35–42, 277–79
  reading and writing, 279–82
  self-talk and, 254–57
  social brain and, 322–23
  thought, action and, 252–53
  see also phonemes
language-based learning impairments (LLIs), 96
language memory, 203, 207–9
laughter, 245–47
L-dopa, 46, 250
lead, 29
learning disabilities, 28, 254
learning process, 63, 148, 185, 244
  brain affected by, 34–39
  hard-wiring in, 20–21, 37–38, 149, 157–60, 178–79
  language and, 39–42
  memory and, 190–91
  movement, memory and, 177–81
  physical description of, 20–21
  "windows of opportunity" in, 40
LeDoux, Joseph, 226
left anterior prefrontal cortices, 232
left frontal region, role of, 35
left hemisphere, 155, 175, 205, 240–41, 316, 375
  conflicts between right hemisphere and, 138–39, 208–9
  emotions and, 229–30
  language and, 96, 137, 230, 273–77, 286, 322–23
  laughter and, 246
  music and, 97, 206
Leiner, Henrietta and Alan, 151, 157
Lewis, Alan, 44
Lewis, Jim, 33–34
Life, 18, 43
limbic system, *10*, 63, 66, 71, 76, 114, 116, *119*, 121, 248, 311, *315*, 318, 328
  autism and, 78
  composition of, 227
  emotions and, 223, 225, 230, 234, 241, 290
  evolution of, 64
  and novelty detection and reward, 115
  pleasure centers located in, 65
Lindvall, Olle, 45, 168
lithium, 223, 338
Livingstone, Margaret, 103, 104, 284
Llinas, Rodolfo, 112, 133–34
LLIs (language-based learning impairments), 96

Loftus, Elizabeth, 183–84, 213, 215, 221
long-term depression, 220
long-term memory, 122–23, 130–31, 194–95
long-term potentiation (LPT), 190–93, 194, 220
lordosis response, 331
Lorenz, Edward, 357
love, 334–35
  biochemistry of, 327–31
  categories of, 244–45
"love drug," 245
Luders, Hans, 189
Lund, University of, 45, 167
lust, 244–45
Lynch, Gary, 368

macaque monkeys, 316
McDonnell Foundation, 19
McGuire, P. K., 254
MacLean, Paul, *10*, 225
magnetic resonance imaging (MRI), 21, 35, 96, 103, 154, 187, 189, 192, 267, 271, 286
magnocellular neurons, 91, 95, 104, 284
magno deficit, 105–6
malnutrition, 29
mammalian brains, 22
Managua, study of deaf children in, 279
Mandal, M. K., 230
Mankato, Minn., nuns of, 42–43, 363–64
mapping techniques, 21
maps:
  brain composed of, 5
  communication between, 143–44
Marcel, Antony, 166
marijuana, 27, 123, 126, 344
Mark, Victor, 138–39
Maslow, Abraham, 247
massage, 77, 362–63
Massaquoi, Steve, 141
Mateer, Catherine, 272
mating patterns, 329, 331–33
Mayberry, Rachel, 262
Mayburg, Helen, 241
Mayer, John, 251
Meaningful Differences in the Everyday Experience of Young American Children (Hart and Ridley), 287
medial amygdala, 331
medial preoptic area, 330
medial thalamus, 227
medications, 29
meditation, 376
medulla, 71, 82, 85, 94, 331
melatonin, 6
memory, 63, 71, 115, 120, 126, 139–40, 148, 182–225, 235, 248, 312, 341
  attention and consciousness and, 130–33
  dopamine and, 122–23
  episodic vs. semantic, 201–3, 214
  explicit vs. implicit, 199–201, 220
  false, 182–85, 199, 213, 215–17
  of the future, 177–81, 196, 198
  language, 203, 207–9, 261
  long-term, 122–23, 130–31, 194–95, 261
  long-term potentiation and, 190–93
  motor, 203, 204–6
  movement and, 175–77
  old age and, 217–20
  process of, 185–90, *197*
  sensory, 203–4
  short-term, 123, 178, 194–95, 261, 338, 344
  smell and, 62, 64, 68, 83

memory (*continued*):
subjective, 198–99
subliminal messages and, 220–21
trauma and, 209–11
traumatic amnesia controversy and, 211–17
tribal elders and, 201
visuospatial, 203, 206–7
working, 123, 130–31, 139–40, 195–98, 208, 309–10
meningioma, 291
menstrual cycle, 68
synchrony and, 65–66, 330
"mentalese," 285
mental health, 336–77
experiential approach to, 7–8, 336–55
mental exercise and, 363–68
movement and, 361–63
nutrition and, 368–71
passion and, 376–77
personal responsibility and, 356–57
physical exercise and, 359–61
psychodynamic vs. biological approaches to, 6–7
reasons for new perspective on, 337–39
spirituality, meditation and, 374–76
technology and, 371–74
tipping points and, 357–58
mental retardation, 27, 286
Merzenich, Michael, 20–21, 37, 85, 157, 170, 191, 283–84, 372
mesolimbic pathway, 116, 117
"metadata" software, 260
methylphenidate, 251
Miami, University of, 76
mice, 44
midbrain, 22, 94
Miller, Zell, 18
Milner, Peter, 243
Mind's Past, The (Gazzaniga), 288
Minnesota, University of, 271
Minnesota Center for Twin and Adoption Research, 34
Minnesota Multiphase Personality Inventory, 291
"mirror neurons," 267
Mohs, R. C., 220
monkey studies, 20–21, 22, 37, 85, 113, 117–18, 157, 170, 191–92, 259, 311, 312, 316, 319
monogamy, 329, 331–33
montane voles, 332, 333
Moral Animal, The (Wright), 303
Morley, John, 367
morphine, 117, 244, 328
mothers, 300–301, 302, 334
infants affected by depression of, 35, 42
motivation, emotions and, 247–50
motor cortex, 97, 149, 160, 161, *163*, 164–65, 175, 230, *268*
motor functions, 41, 148, 312
see also movement
motor homunculus, 162, *163*
motor memory, 203, 204–6
motor orientation, 115–16
motor system, 123, 229
movement, 147–81, 244
attention, emotion and, 171–74
brain-mapping and, 161–65, *163*
cognitive acts and, 148–51
disorders and repair, 166–71
expressing emotions and, 174–75, 227–29
hierarchical organization and, 160–61
language, emotion and, 270–73
and learning and memory of the future, 177–81

and memory, thinking, and learning, 175–77
mental health and, 361–63
planning, motivation and, 165–66
process of, 155–60, *159*
pure "mental" functions and, 152–55
"Mozart effect," 37, 42
MRI (magnetic resonance imaging), 21, 35, 96, 103, 154, 187, 189, 192, 267, 271, 286
Murphy, Dennis, 236
muscle spindles, 223
music:
brain's processing of, 97, 265
child development and, 18–19
cortex development and, 35, 60, 205–6
memory and, 189–90
"Mozart effect" and, 37, 42
myelin, 24
Myth of Repressed Memory, The (Loftus), 184

Nagler, Stephen, 89
naming, 280
narcotics, 118
National Institute of Mental Health, 103, 153, 225, 329, 373
nature-nurture question, 31–34
language and, 261–66
Naval Academy, 56
NBC, 89
neomammalian brain, *10*
nervous system, autonomic, 171–72, 229, 232, 312
"neural Darwinism," 11, 30–31, 141–42
neural stem cells (neuroblasts), 47
neurogenesis, 70
"neuroid network," 141
neuroleptics, 338
neuronal "burnout," 239
neuronal groups, 142–43
neurons, 5, 19–21, 22, 54, 65, 69, 338
cholinergic, 313
connections between, 9, 19–20, 23, 25, 26, 27, 35, 43
continuous random firing of, 57–58, 60–61, 112, 133
dopamine and, 122
environmental effects on fetal development of, 26–30
learning process and, 20–21, 34–39
magnocellular, 91, 95, 104, 284
migration of, 23–25, 27, 28
"mirror," 267
number of, 9, 19, 26
parvocellular, 94, 104
pruning of, 24, 25–26, 27, 30–31, 82
Purkinje, 307
"neurons that fire together wire together" principle, 31, 55
neuropeptides, 66
neuroplasticity, 167
neuroscientific terminology, 8–9, 12
neuroscientists, elite role of, 8
Neurospheres, 47, 169
neurotransmitters, 22, 25, 55, 60, 117, 127, 239, 243, 245, 312, 338, 360, 369
age and, 218
attention and consciousness and, 121–22
function of, 11
number of, 11
Newsweek, 17
New York Times, 16, 251
Nicaragua, 279
nicotine, 25, 27, 30, 118, 123, 126, 245, 327, 344
see also smoking
Nimoy, Leonard, 90

nociceptor, 87
noise:
  background, 57, 88, 89
  internal, 57–58, 60–62, 79, 83–84, 133
nonlinear thought, 5
nonverbal clues, 320–22
nonverbal learning disorder, 321
norepinephrine, 122, 245, 334, 360
novelty detection, 115, 116–18
NSAIDs, 366–67
nucleus accumbens, 117, 118, 227, 243–44
nucleus solitarius, 71
nutrients, nutrition, 25, 29, 101
  mental health and, 368–71

obsessive-compulsive disorder (OCD), 36, 75, 118, 152, 153–55, 319, 339, 373
occipital cortex, 147
occipital lobes, 161, 193
"off-line thinking," 260
Olds, James, 243
olfactory cortex, 63, 64
olfactory epithelium and receptors, 64, 65
olfactory system, *see* smell
Ontogeny, 47, 169
"On Watching Myself Act Old" (Hebb), 218–19
opiates, 88
optic nerve, 102
orbitofrontal cortex, 75, 151, *315*
Origins of the Modern Mind (Donald), 258
Ornstein, Robert, 218
orphans, 77–78
ossicles, 91
oxytocin, 245, 326, 328, 329–30, 332, 334

pain, 76, 87–88, 319
paleomammalian brain, *10*
pallidotomy, 177
Palmites, Richard, 334
PANDAS, 153, 154
panic disorder, 66, 222, 223, 232, 235
Papez, James, 225
papillae, 68–69
parallel processing, 195
parasympathetic nervous system, 172
parental behavior, 332, 333–34
parietal cortex, 107, 147
parietal lobes, 82, 267, *315*, 320, 326
Parkinson's disease, 44–45, 46, 162, 168, 176–77, 179, 200, 250, 361, 372–73
paroxatine, 30
pars triangularis, 286
parvocellular neurons, 94, 104
Pascual-Leone, Alvaro, 21
passion, mental health and, 376–77
patellar tendon, 156
Penfield, Wilder, 162
perception, 48–109
  autism and, 78–84
  brain affected by, 53–60
  and change and adaptation, 72–73
  individual psychology and, 53, 55, 81–82
  plasticity, phantom limbs, pain and, 84–88
  Rickie's problems with, 48–53
  sensory integration and, 82
  and signal and noise, 57–58, 60–62, 79, 83–84
  sixth and seventh senses in, 108–9
  smell and, 62–68
  sound and, 88–97
  taste and, 68–76

as theater of exploration, 336–37, 341, 343, 347–49
touch and, 76–78
vision and, 97–108
Percodan, 117
peripheral vision, 56–57
Perl, Edward, 76
pesticides, 29
Petersen, Steven, 116
PET scans, 59, 77, 97, 107, 164, 185, 192, 198, 201, 217, 230, 241, 254, 267, 276
phantom limb syndrome, 58–59, 86–87
Phantoms in the Brain (Ramachandran and Blakeslee), 317
phenylethylamine (PEA), 245, 328
pheromones, 64, 65–66, 109, 330, 331
phonemes, 93, 94, 95, 253, 257, 265, 271, 275, 277, 278
phonics, 281–82
photoreceptors, 56–57, 101–2
phrenology, 185
pigs, as brain-cell transplant donors, 46, 168–69
pimozide, 243
Pinker, Steven, 285
pinna (external ear), 94
pituitary gland, 172, 329
planum temporale, 286
Platel, Henri, 97
pleasure, 63, 65, 243–44, 245
polygamy, 329, 331–33
Pons, Timothy, 85
Posner, Michael, 116
posttraumatic stress disorder (PTSD), 66, 212, 216, 233
prairie voles, 331–32, 333
prefrontal cortex, 107, 130–31, 149, 150, 179, 193, 198, 230, 237–38, 242, 254
pregnancy, 74, 283, 330, 365–66
environmental effects on brain during, 26–30
premature births, 27
premotor cortex, 160
Prenatal University, 365–66
Pribam, Karl, 156
"prisoner's dilemma," 304
progesterone, 329, 331
propanolol, 173
prosopagnosia (facial agnosia), 316–17
Provine, Robert, 246
Prozac, 6, 36, 223, 238, 327, 338, 353, 360
birth anomalies and, 30
psychoanalysis, 6, 241–42, 339–40, 346, 354
psychopharmacological revolution, 223, 353
PTSD (posttraumatic stress disorder), 66, 212, 216, 233
Purkinje neurons, 307

qualia, 145

radiation, 29
Raine, Adriane, 324
Ramachandran, Vilayanur, 85, 86, 208–9, 317
rape, PTSD and, 233
rat studies, 22, 42, 44, 71, 243–44, 360
reading, 272–73, 279–82
"ready-to-access" module view of brain, 17
"ready-to-respond-to-environment" machine, brain viewed as, 17
recognition threshold, taste and, 73–74
recovered memories, 216

reductive materialism, 144–45
reentrant signaling, 143
Reeve, Christopher, 166, 227
Regard, Marianne, 75
Reier, Paul, 170
relaxation techniques, 222, 223, 374–76
REM (dream) sleep, 134, 189, 192–93, 195
"remembering, the" (of Seneca Indians), 201
repression, 213–17
reptilian brain, 9, *10*
retina, 101
reward deficiency syndrome, 123, 126, 128, 243
reward system, 65, 115, 117, 122–23, 126, 191–92
Reynolds, Brent, 47, 169
rhesus monkeys, 312
rheumatic fever, 151
rhodopsin, 101
Rickie (Flach), 48–53
right hemisphere, 137, 155, 205, 286, *315*, 316, 375
  conflicts between left hemisphere and, 138–39, 208–9
  emotions and, 229–30, 275
  hearing and, 96
  language and, 265, 273–77, 322–23
  music and, 97, 265
  nonverbal clues and, 320–22
right hemisphere deficit syndrome (RHDS), 321
right temporal lobe, 189–90
Ritalin, 117, 128, 250
Rodier, Patricia, 326
rods, 101–2
Romania, orphans in, 77–78
rooting reflex, 77
Rosen, Glen, 284
Rozin, Paul, 72
Rumbaugh, Duane, 259

Sadat, Anwar, 16
sadness, 229, 239–42
  as basic emotion, 226, 232, 239
  brain activity and, 239
Saffran, Jenny, 263
Sagi, Dov, 188–89, 192
Saint Vitus' dance, 151
salience, 101
Salk Institute for Biological Studies, 276
Salovey, Peter, 251
salt, 69
Sapolsky, Robert, 232
Saussure, Ferdinand de, 257
Savage-Rumbaugh, Sue, 259
Schacter, Daniel, 199, 202, 212, 213–15, 217
Schacter, Stanley, 225
Schacter-Singer theory, 225
schizophrenia, 24, 49, 67, 338
School Sisters of Notre Dame nunnery, 42–43
Schwartz, Jeffrey, 36
Schwartz, Mark, 363
Scoville units, 72
Searching for Memory (Schacter), 214–15
sea slugs, 191, 195
seasonal depression, 22
sea squirts, 156
selective listening, 90
selective serotonin reuptake inhibitors (SSRIs), 238, 327
self-actualization, 247
"selfish gene," 261
"self-medication hypothesis," 126
self-talk, 254–57, 260
semantic memory, 201–3, 214
sensory cortex, 156–57, 312

sensory integration, 82, 94
sensory lobes, 174
sensory memory, 203–4
sensory neglect, 319
septum, 243
sequencing, 177
serotonin, 42, 117, 122, 127, 236, 245, 311, 314, 333, 338, 360, 369
sertraline, 30
set point, 226–27
sex, 65, 66, 233, 330–31, 363
  sense of, 108–9
sexual abuse, 215
shadow syndromes, 294
shared attention, 301, 307
Sharper Image, The, 373
Shatner, William, 89, 90
Shaywitz, Sally, 271
Shepard-Kegl, Judy, 279
Shereshevski, S. V., 196, 203–4
shock treatment, 49, 242
short-term memory, 123, 178, 194–95, 261, 338, 344
Sieburg, Hans, 235
signal and noise, 57–58, 60–62, 79, 83–84, 89–90
Signal Pharmaceuticals, 44
sign language, 96, 254, 262, 266, 271, 276, 279
Singer, Jerome, 225
skin, as sensory organ, 84, 85
SLI (specific language impairment), 286
smell, 62–68, 74, 331
  brain functions and, 66–68
  emotions and, 63, 65, 66
  memory and, 62, 64, 68, 83
  process of, 64–65, *67*
  taste and, 62, 63–64, 70
  training and, 62–63, 64
"smell brain," 64
Smith, Katy, 369
smoking, 25, 27, 30, 123, 127
  see also nicotine
Snowdon, David, 43
Snyder, Evan, 44
social brain, 23, 26, 290–335, *315*
  aimlessness and, 292–95
  antisocial personality disorder and, 323–24
  autism and, 306–7, 325–26
  biochemistry of love and intimacy in, 327–31
  cerebellum as coordinator of, 305–9
  development of, 300–302
  emotion, the amygdala and, 311–14
  and evolution of social behavior, 302–5
  face recognition and, 314–18
  feelings, frontal lobes and, 310–11
  free will and, 318–19
  and hope for the socially awkward, 296–99
  insight and social behavior in, 309–10
  language and, 322–23
  mating patterns and, 329, 331–33
  nonverbal clues and, 320–22
  recent thoughts on, 295–96
  social success and, 334–35
social dyslexia, 320, 345
somatosensory cortex, 85–86, *268*
somatosensory system, *67*, 312
"sonic hedgehog," 47, 169
sound, 88–97, *92*
  background noise and, 89–90
  oversensitivity to, 81, 83
Southeastern Comprehensive Tinnitus Clinic, 89
spatial reasoning tests, 37, 42
specificity theory, of taste, 69

specific language impairment (SLI), 286
Speigel, David, 213
sperm, life span of, 30
spicy foods, 72, 74
spina bifida, 29
spinal cord, 22, 23, 85, 166, 171, 331
spiny anteater, 193
spirituality, 374–76
Springer, Jim, 33–34
Squire, Larry, 201
squirrel monkeys, 20–21
SSRIs (selective serotonin reuptake inhibitors), 238, 327
startle response, 232–33
Star Trek, 90
Stern, Dan, 42
stochastic resonance, 58–59, 83–84
stress, 12, 42, 53, 117, 226, 364–65
stress hormones, 172, 211, 232, 362
striatum, 45, 168, 244
stroke, 38, 43, 44, 74, 175, 185, 274, 277, 286, 316, 364
Sturge-Weber syndrome, 276
subliminal messages, 220–21
substantia nigra, 44–45, *125*, 168, 179
sudden infant death syndrome, 27
superior colliculus, 94
superior olivary nuclei, 94
supplemental motor cortex, 246
"survival of the fittest," 302
Sutton, Jeffrey, 336*n*
Swedo, Susan, 153
Sydenham's chorea, 153
Sylvian fissure, 266–67, 270, 283
sympathetic nervous system, 171–72, 173
synapses, 25–26, 39, 60, 142, 194
  function of, 9–11
synaptic cleft, 25
synaptic connections, weakening of, 219–20
synesthesia, 203–4
syntax, 266, 279, 322

tactile perception, *see* touch
tactile receptors, 84–85
tai chi, 361
Taiwan, 271
talk therapy, 241–42, 354
Tallal, Paula, 96, 283–84, 372
Tamagotchi, 372
taste, 68–76
  absolute and recognition thresholds of, 73–74
  adaptation and, 72–73, 74
  categories of, 63, 69
  deficiencies and disorders in, 74–76
  number of receptors for, 68
  process of, *67*, 69, 71–72
  regeneration cycle of receptors for, 70
  smell and, 62, 63–64, 70
  spice and, 72, 74
  temperature and, 70–71
technology, mental health and, 371–74
tectopulvinar pathway, 102
teenagers, emotions and, 234
temporal cortex, 107, 300, 324
temporal lobes, 118, 161, 189–90, 266–67, 272, 317, 326
Terr, Lenore, 213
testosterone, 238, 329
thalamus, 59, 63, *67*, 71, 82, 85, 86, 99, *100*, 103, 104, 107, 134, 174, 248, 283, 312
  emotions and, 224–25, 232, 233, 234
  hearing and, 91, *92*, 94–95

thalidomide, 326–27
theaters of the brain, 336–55
  description of, 337, 341–46
  interdependence of, 347–52
Therapeutic Learning Program, 372
thinking:
  brain affected by patterns of, 36
  movement, memory, and learning in, 175–77
  see also cognition
Thinking in Pictures (Grandin), 82
Thomas, Steven, 334
Thompson, Richard, 218
Thorazine, 118, 223
Time, 17
Timeless Healing (Benson), 374–75
"time out," to defuse anger, 257
tinnitus, 89–90
tipping points, 357–58
Tit for Tat computer program, 304
TMS (transcranial magnetic stimulation), 21, 242, 373
Tobias, B. A., 212
Today, 89
toe-walking, 151, 155
Tooby, John, 304
Toronto, University of, 29
touch, 58, 76–78, 108
  autism and, 15, 78, 80–81, 82, 83
  babies and, 76–78
  culture and, 78
  importance of, 76, 78
  process of, 84–85
  three sensory levels of, 84
Touch Research Institute (TRI), 76
Tourette syndrome, 33, 127, 151, 153, 319
toxins, 25, 29–30
transcranial magnetic stimulation (TMS), 21, 242, 373
trauma, memory and, 209–11
traumatic amnesia, 211–17
Traumatic Stress (van der Kolk), 209
Trehaub, Sandra, 18–19
trigeminal nerves, 72
triune brain, *10*
trophic factors, 25
trust, 5
tryptophan, 369
Tucson Consciousness Conference (1994), 145
Tulving, Endel, 185, 202, 341
Turin, Luca, 64–65
Tuszynski, Mark, 169
twins, 33–34, 325
Type-A personalities, 237

ulcers, 74
Ulrich, Dale and Beverly, 180
umami, 69
unconsciousness, 137, 139
"universal grammar," 262
"use it or lose it" principle, 6, 24, 31, 47, 54, 59, 60, 364

vagus nerve, 77
Valium, 223
Van de Carr, René, 365–66
van der Kolk, Bessel, 209, 210, 212
Vargha-Khadem, Faraneh, 203
vasopressin, 328, 329–30, 332, 333
Velvet Harpoon, 249
ventral tegmental area, *125*
ventromedial cortex, 310–11, *315*
ventromedial prefrontal cortex, 232
vervet monkeys, 311
Vicary, Thomas, 221
virtual-reality treatment, 235
vision, 97–108, *100*, 161, 347–49
  brain's ability to learn, 35–36, 41, 56

vision (*continued*):
hearing and, 95–96
overload and, 55, 83
peripheral, 56–57
Rickie's problems with, 48–53
see also dyslexia
visual cortex, 59, 99, *100*, 102, 107
"visual dropouts," 306
visuospatial memory, 203, 206–7
vitamins, 29, 101, 366, 369–70
vomeronasal organ (VNO), 109

Wada test, 274
Waterhouse, Lynn, 326
Weiss, Sam, 47, 169
Wellbutrin, 128
Werner, Emmy, 365
Wernicke's area, 96, 207, 267, *268*, 274, 286
White House child development conference (1997), 18, 19
whole-language reading techniques, 281–82
Williams, Donna, 82
Williams, Linda Meyer, 214
William's syndrome, 286
"windows of opportunity," of brain development, 40
Wisconsin, University of, 45
Wisconsin Card Sorting Task, 291
women:
language and verbal IQs of, 274–75
memories of sexual abuse of, 215
menstrual synchrony and, 65–66, 330
rape and, 233
superior sense of smell of, 68
see also pregnancy
working memory, 123, 130–31, 139–40, 195–98, 208, 309–10
worry, 226, 230, 236
Wright, Robert, 303
writing, 279–82

X-rays, 29

Yale University, 244
Young, Andrew, 225
Yurgelun-Todd, Deborah, 234

Zachman, Marcella, 43

THE MORAL ANIMAL

Evolutionary Psychology and Everyday Life

Robert Wright

'This is not a book of breathless reportage from some frontier of science; it is an eye-opening, thought-provoking, spine-tingling, mind-boggling, wish-I-had-thought-of-that sort of science book'
Times Literary Supplement

A lucid and provocative study of the ground-breaking new science of evolutionary psychology, *The Moral Animal* examines the significance of this extraordinary shift in our perception of morality and what it means to be human. Taking the life and work of Charles Darwin as his context, Robert Wright brilliantly demonstrates how Darwin's ideas have stood the test of time, drawing startling conclusions about the structure of some of our most basic preoccupations. Why do we commit adultery, express suicidal tendencies and have the capacity for self-deception? Wright not only provides the answers to such fundamental moral questions from the perspective of evolutionary psychology, but challenges us to see ourselves anew through the clarifying lens of this fledgling and exciting science.

'An engrossing guide, written with wit and an eye to inducting the ignorant into evolutionary psychology'
Guardian

'This is sociobiology as you've rarely seen it – slowly matured and richly textured, yet with all its abrasiveness intact'
New Scientist

'*The Moral Animal* overturns old ways of thinking'
Independent on Sunday

Abacus
978-0-349-10704-2

FASTER

The Acceleration of Just About Everything

James Gleick

Do you hit the 'door close' button because the lift doors are taking too long to shut? Did you know we work longer, commute longer, shop longer and sleep 20% less than we did a century ago? We are obsessed with making more time, yet a microwave oven saves just four minutes a day – the average amount of time we spend having sex or filling in government forms.

With acute insight and mordant wit, James Gleick dissects our unceasing struggle to squeeze as much as we can into the 1,440 minutes of each day. From one-minute bedtime stories to Federal Express, from multi-tasking to the double-edged benefits of e-mail, *Faster* is the perfect self-help book for the twenty-first century, a description of our lives and an answer to our fundamental complaint that there is never enough time.

'*Faster* will make you think . . . if you want to understand why your To Do list might be brimming over, you'll add *Faster* to it'
Financial Times

'A fascinating meditation about the modern definition of time that helps to define the age we live in'
Daily Telegraph

'There is no way a reviewer can do justice to the density of anecdote and fact here . . . a delightful read'
Times Higher Education Supplement

'James Gleick, a science writer with a healthy pop sensibility, has written a highly readable dissection of our speed-obsessed age'
The Face

Abacus
978-0-349-11292-3

Now you can order superbs titles directly from Abacus

	Title	Author	Price
☐	Paradigms Regained	John L. Casti	£9.99
☐	The Quark and the Jaguar	Murray Gell-Mann	£10.99
☐	Faster	James Gleick	£8.99
☐	Nonzero	Robert Wright	£12.99
☐	The Moral Animal	Robert Wright	£12.99

The prices shown above are correct at time of going to press. However, the publishers reserve the right to increase prices on covers from those previously advertised, without further notice.

ABACUS

Please allow for postage and packing: **Free UK delivery.**
Europe; add 25% of retail price; Rest of World; 45% of retail price.

To order any of the above or any other Abacus titles, please call our credit card orderline or fill in this coupon and send/fax it to:

Abacus, P.O. Box 121, Kettering, Northants NN14 4ZQ
Fax: 01832 733076 Tel: 01832 737526
Email: aspenhouse@FSBDial.co.uk

☐ I enclose a UK bank cheque made payable to Abacus for £ . . . .
☐ Please charge £ . . . . . . to my Visa, Delta, Maestro.

Expiry Date Maestro Issue No.

NAME (BLOCK LETTERS please) . . . .

ADDRESS . . . .

. . . .

. . . .

Postcode . . . . Telephone . . . .

Signature . . . .

Please allow 28 days for delivery within the UK. Offer subject to price and availability.